Engaging the Cosmos

Astronomy, Philosophy, and Faith

"Neville Brown brings a distinctive voice and a lifetime's reflection to the discussion of both science and spirituality. He provides an engaging compendium of information on human evolution, emphasising the critical role astronomy has played in redefining human self-perception. Eastern as well as Western cultures fall within his purview and he provides the grounds for welcoming the spiritual experimentation taking place within and across different religious traditions. This is a book for all who share his concern that in a shrinking world the choice between fanaticism and conciliation is becoming ever starker." *John Hedley Brooke, Andreas Idreos Professor of Science & Religion, University of Oxford*

"Professor Neville Brown has written a magisterial book on current relationships and debates involving astronomy, philosophy, and theology. His work on this broad and daunting topic is a highly creative and accessible analysis by a writer with unusual interdisciplinary strengths. He explores the interactions among the main religious faiths of the world, and between them and astrophysics and astrobiology. He examines the implications of this exploration for such questions as the concept of a 'just war', the non-weaponization of space, and effective measures of arms control.

Professor Brown argues that the preservation of Nature could be one valuable consequence of a heightened dialogue and discussion among the world's great faiths. And he has produced a work that has sweeping implications for social and political policy-making. Political, religious, and civic leaders – world-wide – would do well to heed this book's message." *Milton C. Cummings, Jr., Professor Emeritus of Political Science, Johns Hopkins University*

"One can always be certain that any book by Neville Brown will exhibit breadth and depth of experience and learning. *Engaging the Cosmos* does not disappoint. This book is absorbing. The narrative detail serves to place all the major historical information in a fascinating context.

Neville Brown makes it natural that astronomy, philosophy and faith in their widest interpretation should be discussed together. Each reinforces the other in the development of the book. It is rare to find an author who is equally comfortable with modern astrophysics as with the motivations of the ancient world. If you want to know who is or was who in these fields you will find a discussion of them in this book. The issues are not left to ideals. We are brought to face issues of modern conscience and responsibility. Well worth reading." *Professor Yvonne Elsworth, HiROS, School of Physics and Astronomy, The University of Birmingham*

Engaging the Cosmos

Astronomy, Philosophy, and Faith

NEVILLE BROWN

sussex
ACADEMIC
PRESS

Brighton • Portland

2 4 6 8 10 9 7 5 3 1

First published 2006 in Great Britain by
SUSSEX ACADEMIC PRESS
Box 2950
Brighton BN2 5SP

and in the United States of America by
SUSSEX ACADEMIC PRESS
920 NE 58th Ave Suite 300
Portland, Oregon 97213-3786

British Library Cataloguing in Publication Data
A CIP catalogue record for this book is available from the British Library.

Library of Congress Cataloging-in-Publication Data
Brown, Neville.
Engaging the cosmos : astronomy, philosophy, and faith / Neville Brown.
p. cm.
Includes bibliographical references and index.
ISBN 1-903900-66-2 (hc : alk. paper) — ISBN 1-903900-67-0 (pbk. : alk. paper)
1. Religion and faith. 2. Cosmology. 3. Astrobiology. 4. Astrophysics. I. Title.
BL240.3.B77 2006
201′.65—dc22

2005031337

Typeset & Designed by G&G Editorial, Brighton & Eastbourne
Printed by TJ International, Padstow, Cornwall
This book is printed on acid-free paper.

CONTENTS

PREFACE

This text arises from a burgeoning conviction that, in order to survive in tolerable shape, our planetary civilization urgently needs to consolidate the forces of reason within it but also to be infused with the virtues we see as aspects of spirituality. This means more positive interaction between the major belief systems as well as between them and natural science, above all cosmology. At the heart of this enquiry will be the relationship across the Cosmos between Life, Consciousness and Material Being.

A good deal of accommodation as between all perspectives will ultimately be necessary. Customary religious beliefs, in particular, must be subject to searching scrutiny. None the less, much understanding may come from reviewing the historical roots and subsequent development of the great religious traditions. Beforehand, the history of astronomy merits attention for similar reasons. Many of our forefathers had valuable insights. They deserve to be honoured in our generation, not least to temper our own hubris.

There is little need to amplify here what I have said elsewhere – particularly in *Global Instability and Strategic Crisis* – about where one comes in from oneself: academically, professionally and philosophically. Regarding religion, one can perhaps say this. My grammar school at Thame in Oxfordshire subtly but insistently encouraged a broad "via media" in such matters. John Hampden had been an old pupil as perhaps had – so we were advised – the deistic radical, John Wilkes. The William of Wykeham tradition also weighed in appreciably through a Board of Governors link with New College, Oxford.

Meanwhile, my small home town of Watlington on the Chiltern spring line was still a community within which the local priests – Anglican, Methodist and (with the Stonor connection) Catholic – were *ipso facto* people of note. Having been christened an Anglican, I was confirmed in due course because my parents felt I should be in contact to that extent. But there was never any pressure towards regular church attendance. Nor any such impulse from within.

My father had been reared in a strongly Baptist and Anglican environment. Also he had volunteered for military service in April 1916 and been disabled in the Arras Offensive just twelve months later. He always spoke well of the British Army and especially of the hussar regiment into which he was initially inducted. Some years later, however, he reacted strongly against the way the pulpit at his local Anglican church had been used by the incumbent priest (himself no military veteran) for "conscription sermons", averring the imperative need for young English Christians to go out and kill young German ones. Dad duly evolved his own brand of Protestant hymnal agnosticism, this considerably informed by the literature – some of it very jejeune – of the Rationalist Press Association.

This study, like predecessors, reaches across disciplinary divides. My hope is that it has been effected in such a way as to leave those who peruse the text free to make up their own minds about the interpretation here proposed or about whatever alternatives may appeal. The key terms "cosmos" and "universe" are here treated as interchangeable. The same goes for "astronomy" and "cosmology". As explained in the text, the term "symmetry" is used in its customary and classical sense, not in the more esoteric manner physicists tend to favour. Chinese words are spelt in *pin yin* except when alternative renderings are very familiar in the West. A light year (the distance electromagnetic radiation travels in a year) is in excess of five million million miles.

A convention we should now address is that concerning numeration. The phrase "an order of magnitude" is accorded its formal mathematical meaning: that is to say, a close to ten-fold change or difference. Also regular use is made of the index number above a 10 to indicate how many orders of magnitude. Thus 10^6 is one with six zeros to the decimal point – i.e. 1,000,000 or a million, six orders of magnitude. A billion is taken throughout to be a thousand million or 10^9. The minus sign is introduced to the index to indicate a reciprocal. Thus 10^{-6} means one part in a million.

As regards institutional back-up, the Japanese religious community has again been well to the fore in support of my wife, Yu-Ying, and myself. This time round, one recalls with warm appreciation dialogues with Professor Katsuhiko Yoshizawa and Mr Thomas Kirchner of the International Research Institute for Zen Buddhism at the University of Hanazono. So does one the extended discussion on Esoteric Buddhism with the Reverend Chisho Namai, Vice Chancellor of the University of Koyasan. Also Yu-Ying and I were deeply honoured to be invited by the Director, Mr Michio Umemoto, to the relaunch in summer 2005 of Oshajo Art Museum in Hiroshima. All else apart, this relationship has brought home to me like nothing else the dreadful dilemma of how to end dependence on nuclear weapons without opening the door to biological ones. A sense of common cultural heritage may be an overriding precondition of success for this and other vexatious issues.

In Britain, valuable support has been forthcoming from the Oxford University School of Theology and, indeed, the Bishopric of Oxford. One has especially in mind Charles Brock, John Brooke, Richard Harries, Philip Kennedy and John Muddiman. Valuable insights have also been afforded by Professor Chandra Wickramasinghe, Director of the Centre for Astrobiology at the University of Cardiff; and Nicholas Campion, Director of the Sophia Centre at Bath Spa University College. Libraries in Oxford that have been helpful as ever include those of Mansfield College and the History Faculty plus, of course, Radcliffe Science. Those in London of which the same can be said include the British Library, the London Library, the Royal Astronomical Society (RAS), and the Royal Society. Special mention should be made of Mary Chibnall, Deputy Librarian at the RAS, for assiduous application well beyond the normal call of duty. May one add that individuals who have contributed singularly through informal conversation include Jocelyn Bell, Ivette Fuentes-Curidi, Mitsutoshi Horii, Frederick Schuller and Michael Sherwood.

Either one or two individual chapters have respectively been read, regularly to good effect, by the following: John Adair, Colin Baker, Julia Bishop, Stephen Blundell, Yu-Ying Brown, Margaret Christie-Brown, Michael Freeden, Jerry Losty and Ron Nettler. My cousin Michael Brown has again read a number, continuing a scholarly dialogue between us going back to our mutual grammar school experience. Again, too, Mrs Jill Wells, my literary secretary of thirty-five years' standing, has been a pillar of strength translating my palaeographic long-hand into elegant typescript and in countless other ways. She has, as ever, enjoyed the solid backing of husband Trevor and of their extended family ranging from bouncy grandchildren to prehistoric sheep. Lest this all sounds just too miraculous, I should perhaps explain that it is the sub-species which is prehistoric, not the individuals.

London, December 2005

THE AUTHOR

Since 1994 Professor Neville Brown has been a senior member of Mansfield College, Oxford. Thanks to several strokes of extreme good fortune, his career has come to be based on an unusual interaction between the humanities and physics, especially the sky sciences. After grammar school majors in the physical sciences, he read economics with geography at University College London (UCL) followed by modern history at New College, Oxford. For about half of the time he then spent as a forecasting officer in the meteorological branch of the Fleet Air Arm (1957–60), he specialized in regional upper air analysis. But other assignments included extensive experience on a front-line coastal air station plus some with the Mediterranean Fleet in a gunnery trials cruiser. He was a field meteorologist on two British Schools expeditions to sub-polar regions.

In 1980, he was elected to a chair in International Security Affairs at the University of Birmingham. He has held Visiting Fellowships, or the equivalent, at the UK National Defence College, then at Latimer; the School of Physics and Astrophysics at the University of Leicester; the International Institute for Strategic Studies in London; the Stockholm International Peace Research Institute; the Australian National University, Canberra; and (2001–4) at the Defence Engineering Group, UCL. From 1965 to 1972, he worked part-time though quite proactively as a defence correspondent in the Middle East and South-East Asia, successively accredited to several leading journals.

From 1981 to 1986, Professor Brown was the first Chairman of the Council for Arms Control, a British all-party body drawn from parliament, the churches and other professions and committed to a serious but multilateral approach to arms control. He thus became involved in the multinational debate about Ballistic Missile Defence (BMD). In 1985 and again in 1987, he paid extended visits to the Strategic Defense Initiative Organization (SDIO) in the Pentagon. The first was at the invitation of Allan Mense, then Acting Chief Scientist; and the second as the guest of O'Dean

Judd as Chief Scientist. From April 1994 to the summer of 1997, he was attached half-time to the Directorate of Sensors and Electronic Systems (within the Procurement Executive, UK Ministry of Defense) as the Academic Consultant to the official Pre-Feasibility Study on BMD policy. A declassified version of an 87,000-word *Fundamental Issues Study* he wrote in this connection was published by Mansfield College in 1998. Throughout, the vexed question of Space-based missile defence has been on the agenda.

He has authored twelve books or major reports. Among them has been *The Future of Air Power* (Beckenham, Croom Helm, 1986). But with the award-winning *Future Global Challenge* (New York, Crane Russak, 1977), he began to give economic, cultural and ecological factors some salience in the quest for a peaceable world. This thrust has continued with *New Strategy Through Space* (Leicester, Leicester University Press, 1990) through to *Global Instability and Strategic Crisis* (London, Taylor and Francis, 2004). His chief contribution to date to the scholarly study of sky science has been *History and Climate Change, a Eurocentric Perspective* (London, Taylor and Francis, 2001). It reviews the last two millennia.

In 1990, Professor Brown was elected a Fellow of the Royal Astronomical Society, partly in recognition of his work at Leicester and on *New Strategy Through Space.* In 1995, the University of Birmingham conferred on him an official Doctorate of Science in Applied Geophysics.

Engaging the Cosmos

Astronomy, Philosophy, and Faith

Dedication

David Chandler, 1934–2004

To my mind,

the finest military historian of my generation.

David faced a long and uncomfortable

decline with all the fortitude he

had so well observed in others.

PART ONE

The Heavens in History

OVERVIEW

The Heavens in History

This overview is focussed very particularly on the early Near East and on Western Europe, the zone from which modern sky science essentially derives. It indicates the psychological drives that give rise to cosmic curiosity and religious belief. These drives considerably arose out of the grim struggles our forebears had to engage in to survive ecological stress by undertaking territorial expansion. What strikes one from the earliest civilizations onwards is the frequency with which philosophic issues come up which we still regard as urgent but unresolved: Creation, Idealism versus Materialism; Life elsewhere; other universes; reductionist logic versus holistic insight . . .

Special emphasis is placed on how the late seventeenth century represents the follow through of the Renaissance quest for greater knowledge, not least of the Heavens. Then as one considers twentieth-century cosmology, one is strongly reminded of a comment by the pioneer geneticist and philosopher of science, J. B. S. Haldane (1892–1964). It was to the effect that the universe is not merely stranger than we realize, it is stranger than we can realize.

1

THE AGE OF STONE

DRAMATIC ADVANCES MADE THESE LAST SEVERAL YEARS in the use of DNA to delineate our human and humanoid ancestry are sure to stimulate further a mounting interest in our early history. All bar several per cent of our human history to date was spent within the Old Stone Age (the Palaeolithic); and everyone these days appreciates how large a part this *longue durée* must have played in shaping our basic psychology. Furthermore, our Palaeolithic ancestors could be said to have shared with ourselves a pattern of frustration not experienced in quite the same way by the several hundred generations in between.

The allusion here is to a peculiar difficulty in forging distinctive local cultures able to consolidate social bonding within clannish groups a thousand or so strong. Such cohesion was operationally vital to our Stone Age forebears; and is inwardly needful to us, their descendants. Early humans found local identities hard to delineate because they were short on skills and resources for buildings, textiles, pottery, metalwork, painting, music making, cuisine . . . Obversely for modern mankind, cultural distinctiveness is being eroded by pell-mell globalization. Individuals may buck the trend, partially a while. But it is getting harder and harder for a tribe, workplace or extended family to respond thus.

Closely related to this identity crisis are profound doubts, personal and metaphysical, about ultimate aim and object. As for our early ancestors, they might often have wondered whether a struggle as unremittingly harsh as was theirs could, in any sense, be worthwhile. Ritualized burial is well recognized to relate to a quest for wider purpose.

Twin-tracks

From the Stone Ages through to the present, religious answers have usually been sought via a twin-track approach. The one has been to gaze upwards to the overarching heavens. The other has focused on local spirits especially as harboured within natural features. The former have afforded a glimpse of the infinite and eternal. The latter have aided societal definition, not least when communities have dispersed for hunting and collecting. In pre-Christian mythologies, Teutonic and Celtic, particular deities (especially male ones) could be confined very much to one locality and tribe. The Celtic cult of Lug, extending from Iberia to Ireland,[1] would have overlorded a few, as would Stonehenge earlier.

In a famous letter to Abbot Mellitus in 601, Pope Gregory the Great cautioned missionaries in England against forbidding recourse to ancient sanctuaries. Better to rededicate them to saints and martyrs. Often the early Christians would build churches on pre-existing holy mounds or even within stone circles; and not a few heathen springs were renamed as well. Admittedly, in 963 the Canons of Edgar banned further worship of trees, stones and fountains. But what was more usual was that "ancient heathen beliefs remained visible through the white surplice of Christianity".[2]

Meanwhile, recourse to the heavens above came naturally to small communities innocent of building cities too light-polluted to access the heavens. The late Sir Frederick Hoyle would tell how he came to astronomy through being gripped one fine night in his Cumbrian boyhood by a sky "powdered with stars". In my own early childhood, spent bestride the South Chiltern escarpment along the Oxfordshire–Buckinghamshire border, there were compensations for having those precious years constricted by war. Prominent among them were powdered skies pristinely sustained by the nightly 'black out'.

Upper and Lower

To comprehend how our palaeolithic ancestors viewed the skies, a distinction to draw is that between Upper Sky and Lower. Obviously, the former encompassed "the firmament/the vault of heaven": a crystalline sphere on the undersurface of which the permutations appear far more regular and constant than any terrestrial fluctuations.

Conversely, the 'sublunar' Lower Sky was the weather zone. Its scenic changes appear much more chaotic and transient than most on the Earth's surface. For the West's literati from Aristotle onwards, clouds have epitomized the insubstantial and elusive, even dissolution in death.[3] Correspondingly, to landscape artists they have been background features it has never been important to depict exactly. Do not bother, for example, to give convective cumulus the flat base it reveals when naturally viewed obliquely. Even Constable lapses thus. The philosopher Karl Popper contrasted clouds with clocks, the former betokening limits to scientific prediction.

Dualism

From Late Antiquity through the High Middle Ages, the disposition across Christendom was to cast in very absolute terms the contrast just drawn. Still, this was but an extension of how our basic psychology reduces much of our understanding to stark alternatives: life and death; Heaven and Hell; Heaven and Earth; male and female; land and water; frost and thaw; light and dark; alive or dead; friend or foe; yin and yang; silver and gold . . . Binary structures of this kind were always going to figure prominently in language as it shaped up. No doubt, too, they have enduring uses, not least in conflict prosecution. On the other hand, they can engender negation, not least apropos conflict avoidance or resolution.

Lunar vs. Solar

Binary polarization will have further been encouraged by what to us is a cute coincidence but will have struck with awe our forebears. The Sun is 397 times as wide as the Moon – if you like, the Full Moon – but is averagely 389 times as far away. Therefore, these two spherical bodies appear to be effectively the same size in the sky.

What St Francis of Assisi was to call "my brother, the Sun" has successively invited worship in a variety of cultures from Pharonic Egypt to Inca Peru. The appeal of the Moon has been more subtle but hardly less considerable. Witness its timekeeping function. What may be rudimentary lunar calendars have been found from the Upper Palaeolithic – maybe from as far back as 20,000 BP.

Instructive in this regard are the pre-European cultures in and around Central America. By such yardsticks of progress as writing or wheeled transport or applied metallurgy, each and every one had blind spots to the last. But watching the night sky and, above all, the Moon, they between them "encompassed a vast range of skills and abilities, from the informal lunar calendars of many American hunting tribes to the startling precision of Mayan bark books".[4] Unfortunately, only three or four of the latter now survive. In the sixteenth century, many were burned (to the deep distress of the Mayans) by Spanish friars resolved to extirpate the "superstitions and falsehoods of the Devil". But even the few survivals carry data sufficient to indicate that the keeping of lunar records was a major preoccupation. Accompanying tables for the prediction of lunar eclipses would have hit the right date about half the time.

All of which well relates to the widespread evidence that the prime concerns of early scholars devising calendars were sacerdotal. Seek to keep on good terms with heavenly authority by timing religious festivals appropriately and by detecting forthwith any signs of heavenly displeasure. Navigation across deserts and the regulation of farming were important yet ancillary requirements.

This was just as well for the lunar specialists. For in these latter realms, the value of the Moon as a chronometer was and is vitiated by two plain facts. The lunar month is not a whole number multiple of the Earth day. Nor is it a whole number divisor of the Earth year. The former truism means the Moon rises too variably to be satisfactory as a navigation beacon. The latter limits its value as a farm regulator. To an extent, however, religious festivals gain mystique from being movable. To this day, the Moon determines the respective dates for Easter, Passover, Ramadan and most Buddhist festivals.

Lunar eclipses are easier to predict than solar ones, as well as more frequent. But these differences never led to their being considered more commonplace. Almost the reverse obtained. In early cultures from East Africa to China to North America, utensils or implements were rattled or banged to check the perceived depredation. Fire also figured extensively in strategies of eclipse curtailment.[5]

One factor in the foreboding was the presumption that, seeing how the Moon governs the tides, it must similarly exercise a commanding influence over the fluxi of heat and moisture that determine our weather. Unfortunately, nearly all the correlations adduced across the centuries are quintessential moonshine, much more consistently so than is the case with weather lore as a whole. There are, however, exceptions. In Western Europe folk culture, rainy weather is held to be in prospect if the circular outline of the whole Moon can be seen within the cradle of the luminous crescent on nights close to New Moon. As the point was put by the Romantic poet Samuel Coleridge in his *Dejection Ode*: "For lo! The New Moon winter bright and overspread with phantom light . . . I see the old Moon in her lap, foretelling the coming-on of rain and stormy blast." He sees the storm in prospect as a turning point, the time when his dejection may turn to inspiration.

The optical effect in question is real. It is caused by refraction mainly due to humidity fluxes within the atmosphere. It can be a pointer to rain near term, albeit not a very categoric one. A rider to add, obvious but necessary, is that the cradling crescent may as well represent the final phase of an Old Moon as the first phase of a New.

One lunar cycle receives curiously little overt mention in folklore, whether in its own right or as a harbinger of weather change. It is the 18.6 year one in the declination of the Moon's path across the sky. An apparent exception is that traditional Somali weathermen identified 28 stations (*anwa*) through which the Moon might transit. When it was well northwards, dire droughts were apprehended.[6]

Demonstrably, this cyclicity does have tangible influence, at least on the oceans. In 1965, two Soviet scientists found that, in the North Atlantic area, sea level at 75°N was 13cm higher (relative to 45°N) when declination attained its peak (28° 40′) as compared with its lowest reading (18° 20′). They surmised that this peaking would cause a subtle retardation in the progress northwards of the North Atlantic drift.[7]

Quite possibly astrologers of old discouraged lay discussion of this cycle because it affected the prediction of lunar eclipses and was therefore best retained as part of their professional "mystery". One could also say that the theme has been subsumed by the considerable interest later shown, both popularly and in academe, in perceived 20-year cycles in the economic realm. This time span was the mean of the alternation Simon Kuznets proposed in 1930 in regard to American economic history.[8] If his thesis does have general application, this could be due to endogenous tendencies – i.e. ones inherent in the economic system. Additionally or alternatively, it could relate to the secular lunar cycle here being considered and/or the 22-year double sunspot cycle, another phenomenon well authenticated.[9]

At all events, a lack of popular recognition of the declination swings contrasts with the veritably universal and intense interest long taken in the progress of the Moon through its close to 29-day cycles, a subject already noted in detail on bone-markings dated at 25,000 BP.[10] The waxings and wanings suggest birth, maturation, decline, demise and generational rebirth: an association of ideas strengthened by the close correspondence between the lunar cycle and the median length of the human menstrual one. Moreover, the mental mapping readily extends to encompass as well fertility, infertility, destiny, intuition, madness, death, eternity . . . All in all, the Moon is a veritable guardian of life.

The many ramifications are definitively explored in Jules Cashford's beautiful evocation of the part the Moon has played in our consciousness across the millennia. Take the Moon goddess of Classical Greece, Selene. In the New Testament, Jesus is described in Greek by St Matthew as healing the *seleniazomenoi.* This term was translated as "lunaticks" in the King James' Authorized Version of the Bible (1611) but as "epileptics" in the 1881 Revised Version. Epilepsy was thought to be caused by water on the brain; and liable to turn worse at Full Moon.[11]

In the light of what has been said, one might conclude (as the Austrian psychologist Carl Jung apparently did) that the Moon is archetypically feminine in the human imagination, the Sun duly being male. Likely, too, the Moon was very generally "the Lady who walked in the woods" of prehistoric Europe. Yet overall the gender ascription within pre-modern societies is mixed. Take indexed references in the 1959 *New Larousse.* There are 37 to Sun gods but ten to Sun goddesses. There are five to Moon goddesses; and 25 to Moon gods.[12] Similarly, an anthropological survey of the Americas by Claude Lévi-Strauss found the Sun to be feminine and the Moon masculine in eight or nine tribal groups while the "inverse relation prevails" in a further nine.[13] Then again, there is "the Man in the Moon". Likewise a hare is perceived on the Moon in far-flung cultures; and might tend to be taken as male except perhaps in China and Japan where it is said to pound rice eternally.

It does seem that our ancestors found the Moon more alluring than the Sun. If so, this may have owed something to a vague sense that the Moon had assumed a salient role in the extension of life from the seas to the land as, indeed, it likely did tidally. There is, besides, a sense of the Moon's imaging a wider range of human sensations, negative as well as positive. "Ill-met by moonlight", Shakespeare had "proud Titania" advised. "Ill-met by sunlight" would not have resonated the same way.

Climatic Disturbance

Our enduring disposition to speculate about fundamentals, concerning this world and the next, must owe something to the trials our human or prehuman ancestors endured, not least those due to climate vagaries. In regard to which, may one first say that, irregular though climate has been throughout on decadal to millennial timescales, the overriding trend since the Miocene epoch (which ended ten million years ago) has been towards coolness.

Mean air temperature is a basic indicator overall. Too rapid or extreme an alteration over time bodes ill for any ecosystem, regardless of the direction of change. Within that context, the specifics of a local situation will be determining, changing rainfall patterns usually being decisive. Secular trends remain hard to elucidate fully. But over deca-millennial timescales, rhythmic alterations in the lie of the Earth's axis can be very important. Long solar cycles of maybe 800 years have also been mooted.[14] Then again, the amount of volcanic dust varies across centuries and millennia as well as more abruptly. Exceptional vulcanism helps explain the marked coolness of the last two million years.[15]

Entering with emphasis an "other things being equal" caveat, one can note how the latitudinal climate belts tend to shift polewards when the Earth's getting warmer but equatorwards when it is cooling. Notably reactive are the subtropical desert zones, the perennial existence of which is confirmed by modeling and, across hundreds of millions of years, by geologic observation. Their migration during the Wisconsin–Würm glacial maximum (*c.* 18,000 BP) was the backcloth to an altitude decrease of the Amazonian rain forest widely in excess of 700 metres. Overall there will have been, too, substantial area shrinkage and parcellation.[16]

The Hominid Sequence

Our hominid (i.e. near human) antecedents evolved in tropical Africa as the balmy, lush conditions of the Miocene progressively gave way to cooler and more arid conditions. In other words, their ancient homeland – the rain forest – was being progressively undermined. Therefore adaptation to life on the surrounding grasslands became ever more imperative. One response strategy within our evolutionary progression was walking upright on what previously had been hind legs. Several rationales have been adduced, among them a good field of vision and keeping a largish brain cool. To me it does seem that leaving the forelimbs free to acquire more manipulative abilities must have been a prime driver. The earliest hard evidence of a walking ape is afforded by skeletons of *Australopithecus anamensis* some four million years old found in Kenya in 1995. Tendencies in that direction probably began several million years before.

In 1924, Raymond Dart of Witwatersrand University made the keynote discovery of *Australopithecus africanus*, a species within this genus some three million years old. Subsequent investigations showed australopithecans to possess crude stone and bone implements. Also, a considerable proportion of the identified skeletons had fractured skulls. The inference may be that certain blunt implements figured in intra-genus or intra-specific warfare about access to limited food and water.

Well by then, however, an evolutionary divide had opened up within the hominid family. One branch, *Paranthropus* (of which the australopithecans were initially a part), evolved to the monkeys and apes, a group still essentially arboreal and herbivorous. The other, *homo* progressed to our omnivorous and ubiquitous selves.[17] Little growth in representative brain size was to take place en route among the *Paranthropus* plant eaters. But by 1,000,000 BP, the representative *homo* brain size had risen from 400 to 1000 cubic centimeters, the latter being only a fifth less than with *Homo sapiens* today.[18] A sizeable fraction of this increase was directed to front limb management. Brain size is by no means the sole predictor of mental virtuosity. Still it has to be quite indicative in this sectoral context.

Just over two million years ago, *Homo ergaster* became the first member of the *Homo genus* to advance beyond Africa, evolving into *Homo erectus* as it did so. The Neanderthals (who, in Europe, survived to 30,000 BP) were a strong offshoot of this very dominant species.

Despite the corruption of evidence over time, DNA analysis is already of real value in definitive succession mapping. Quite its most decisive contribution to date has been the report from Berkeley in 1987 concerning mitochondrial DNA gathered from five disparate human populations. It showed all to have derived from the one female who lived 200,000 years ago, very likely in Africa.[19] The clear inference was that the *Homo sapiens* species as a whole is descended from this African "Eve", not from the Neanderthals or their close cousins spread across Eurasia. Subsequent DNA and conjoint materials have well confirmed this "out of Africa" thesis, not least by failing to yield evidence of fruitful miscegenation between humans and Neanderthals, despite our having advanced from Africa into Eurasia sometime around 85,000 BP.

So why did Neanderthals dwindle to extinction as *Homo sapiens* settled in close proximity to them, a juxtaposition at perhaps its most evident in the last enclave held by the former, South-West Iberia? After all, what we used to speak of as Neanderthal Man was well endowed in various ways. Average brain size was 1450 cc; and the species/subspecies was robust in the face of weather adversities. They used stone tools for hunting and butchering; and would often live in caves. A stone-and-bone face mask (from 32,000 BP) has been attributed to them.[20] They buried their dead; and maybe in emulation of their human neighbours, sometimes placed flowers on graves. My own suspicion is that our strategically mobile species proved all too efficacious at spreading diseases it had gained a goodly measure of immunity from itself. Obviously an effect originally registered accidentally might sometimes be contrived later on.

Planetary Colonization

The territorial and social progress of *Homo sapiens* was conditioned throughout by erratic climate swings. Indicative of this fluctuating background has been a seabed core extracted from mid-Atlantic at 53°N. It reveals a summer sea-surface peaking at 16°C in *c.* 124,000 BP. Then irregular progression to 75,000 BP culminated in a plunge to 6° around 72,000. This dramatic cooling was a prime consequence of a supervolcanic explosion at Toba in Sumatra, easily the worst eruption any time in the Quaternary – i.e. the last two million years. The biological impact worldwide was great. DNA analysis indicates that the human population was reduced to several thousand. In particular, it may have put paid to early human endeavours to colonize the Levant via the Sinai.

Many details of the sequence still await elucidation. But as regards those Atlantic core findings, the post-Toba recovery gave way to another peak-to-trough fall (from 12 to 7°C) between 54,000 and 46,000 BP. During this cooling phase the Cro-Magnon people entered Europe via what then was the Bosporus land bridge.

With this advance came an explosion of creativity, involving both standardization and diversification. Witness fabrication skills ranging from rudimentary hearths to sewn clothes. Witness, too, artistic expression. The cave paintings of Chauvet and Lascaux and finely-carved Venus figures found throughout Europe date back to this Aurignacian culture. Undoubtedly language syntax and vocabulary were advancing faster, too, aided by a subtle jaw mutation.

However, one is talking about the accelerated development of human advantages the roots of which went way back. Processed pigments were used in Africa well before 100,000 BP. By then, also, *Homo sapiens* was probably establishing a solid lead over fellow primates linguistically.

Homo Sapiens Ascendant

Cultural elevation will have lent impetus to, then gained impetus from, the annihilation of the Neanderthals. A 1995 Stanford study dates the final decline to extinction of them as being from *c.* 50,000 BP.[21] Likewise human predation interacted with climate vicissitudes to effect the extensive extinction of large mammals – i.e. those well over human body weight. However, one region where the intervention of our ancestors was decidedly secondary is Siberia. A high proportion of what will have been a huge mammoth population simply died of cold and starvation while precluded by the terrain from southerly migration.

Given what has been said above, one may allow that ecological disturbance figured considerably in the tropics as well. But human involvement was instrumental. In Africa, the island of Malagasy conspicuously excepted, the extinctions came early (mainly between 110 and 70 millennia BP) though only two-fifths of the megafaunal species went into oblivion.

North America is something of a special case. What appears to have been the greatest extension of northern hemisphere ice in the past 200,000 years climaxed in 18,000 BP. An associated drop in sea levels left a Behring land bridge available for the decisive human entry of the Western Hemisphere, starting two or three thousand years before. It was largely in the twelfth and eleventh millennia BP that over two-thirds of the megafaunal species were destroyed.

DNA evidence from the Baja peninsula in Lower California suggests a previous Asiatic littoral intrusion a good 40,000 years ago. So do fossil footprints in Mexico, reported in 2005. In due course, this led to assimilation, not continental dominance. In which connection, it is mildly interesting that, by medieval times and maybe long before, the planet Venus connoted among both the Japanese and the Aztecs a belligerent god, not the lovely goddess known elsewhere.

Seeds of Faith

A sense of human exceptionalism may have been ingrained in our forebears by how they prevailed not only over their near relations, the Neanderthal but also over the larger mammals and the broader forces of Nature. Moreover, this sense may have been reinforced by a burgeoning ability to provide themselves with life's comforts as exemplified in Europe by the Aurignacian culture. Such a sense may in its turn have engendered a desire to place *Homo sapiens* in a wider context midst the eternal verities.

A pertinent question to ponder is how far rock paintings were an expression of this yearning. *Homo sapiens* entered the Australian continent via a then narrower Timor Sea around a chilly 70,000 BP. Within twelve millennia (perhaps much less), rock paintings had been executed in Western Australia. But, in 1996, a remarkable research result was reported by a triumvirate from the Australian Museum in Sydney, the Australian National University (ANU), and Cambridge. The gist was that *c.* 6,000 BP, scores of rock paintings were crafted in northerly Arnhem Land which depicted a "Rainbow Serpent" evidently based on *Haliichthys taeniophora*, a member of the seahorse family. Apparently, this fish had been unfamiliar to the aboriginal Australians before the recent post-glacial rise in sea levels brought it into the comparatively shallow waters extending across the continental shelf. Now it had become a sacred symbol of unified Creation though also of general destruction.[22]

Mainsprings of Civilization

Around 10,000 BP, this "interglacial" (i.e. prolonged interlude of relative warmth) had peaked. For two millennia or more, what we know as the Saharan and Arabian deserts would be considerably grassy. Accordingly, South-West Asia was as well placed as anywhere to usher in the Neolithic or New Stone Age. This involved *inter alia* plant cultivation; animal domestication; settled habitation; trade; and, as it phased into the Bronze Age, monumental architecture plus mini-cities. By 7000 BP, Neolithic culture, broadly defined, had spread extensively across temperate Eurasia. By 5500 BP, if not before, the Near East had entered the Bronze Age, this soon to be succeeded by the Iron: these terms referring to the said metals being respectively cast or smelted, mainly for use in personal artifacts. Stone remained of primary importance in construction, save in Mesopotamia where bricks had assumed pride of place from 7000 BP.

The Neolithic threshold was often quite abrupt, not least in regard to matters sacred. Yet well ahead of it, there are some indications of graves being aligned (by Neanderthal as well as *Homo sapiens*) towards the Sun, either rising or setting. However, heavenly bodies as such seem not to figure in early cave paintings.[23]

Intercontinentally prevalent in Upper Palaeolithic cave art, intermingled with animal portrayals, are simple geometric patterns – grids, lines of dots, curves, chevrons, rectangles, triangles . . . These have been variously interpreted as symbolizing (a) hunting paraphernalia and stratagems, (b) sexuality, male and female or (c) shamanistic hallucinations.[24] Come the Neolithic, this abstract tendency waxes stronger, finding expression in elaborate and imaginative designs on urns and vases. Now shamanism seems less pertinent since those concerned clearly "rejoiced in the revelation of congruence, symmetry and similarity".[25]

As the Neolithic era progresses, many monuments appear which consist of carefully positioned and erected big stones – "megaliths". The oldest known, at Nabta Playa in southern Egypt, will have been in place by 4800 BP. Their construction does not correlate closely with local development in other respects; and is widely spread both in time and place. Other early locations in the eastern hemisphere are in Western Europe, Algeria, Palestine, the Caucasus, Persia, Baluchistan, Kashmir, the Deccan and Sumatra. In Japan, megalithic tombs were built from the second century BC to the seventh AD when Emperor Kotoku banned them as a waste of labour.[26] Similar structures occur in the Pacific though the stone heads of Easter Island are usually deemed another genre, maybe one relatively recent.

On the High Plains of North America, about 50 "medicine wheel" arrays have been identified. Comprising local rocks, they measure up to several hundred feet across. Their most characteristic layout is circularity accented by a hub from which spokes radiate. This hub may be a stone assembly weighing 100 tons. The best-known individual example is the Big Horn Medicine Wheel found at an altitude of nearly 10,000 feet in Wyoming. It was in use, well into the second half of the millennium just passed, to register and celebrate the summer solstice.[27] Suffice to remark that a striking aspect of medicine wheel proliferation is its taking place among thinly spread peoples to whom agriculture was still decidedly secondary to hunting.

However, the most remarkable megaliths are in Western Europe. The Grand Array at Carnac in southern Brittany comprises three separate "thoroughfares" of "menhirs" – i.e. standing stones. One of these, the Kermario, consists of ten near-parallel lines of menhirs extending axially over 1,000 metres; and one fallen menhir has a mass estimated at 340 tons. The Grand Array might have been applied to lunar eclipse prediction. Then again, a structure at Newgrange in Ireland (dated by radiocarbon calibration at 5150 BP) apparently served to register the winter solstice.

Still, quite the most remarked exemplar is Britain's Stonehenge, a circular structure basically built a few centuries after Newgrange but successively enhanced to 3,500 BP. The computerized decoding of its layout began in the early sixties but was to be studded by tart exchanges about exactitude and inference. Registration of the winter and summer solstices does seem to have been an initial aim and object early on. Just how reliably those concerned could thus "signal the danger periods for an eclipse of Sun or Moon"[28] remains a moot point. As elsewhere, much would hinge on how firm a grasp the astronomers in question had on that 18.6-year lunar oscillation. On which score, a sanguine interpretation came from a father-and-son partnership who were admirable trailblazers in this field. No doubt sustained by a relevant Stonehenge alignment, they found it "obvious that the megalith builders knew of the . . . cycle and measured it . . . Their astronomical potential . . . was great enough to permit prediction of eclipses".[29] There are also indications of wide regional correspondences in design philosophy – e.g. across Britain and France and well beyond.[30] By 100 BC, indeed, the styling of tableware could sometimes be thoroughly intercontinental. For instance, itinerant Thracian artisans appear to have crafted an ornamental silver bowl rich in motifs ranging from the Baltic to India.[31]

Also to ponder is what purpose the specific insights gained were sought for prehistorically and what upshot there may have been. Did those involved seek to demystify eclipses by subjecting them to measured assessment? Or was their purpose rather to supply priesthoods and governments with esoteric knowledge enabling them to appear all the more omniscient? It is tempting to presume the latter motive was heavily dominant. Could this explain why the said cycle has not survived in the folklore of Britain?

To be remarked, too, is an emergent concern with geometric symmetry. Stonehenge and Big Horn are conspicuous manifestations of circular symmetry. Meanwhile, surviving evidence about Vedic fire altars has square symmetry featuring as well. Whereas altars for the Earth were usually drawn as circular, those for the sky were square.[32] A similar distinction was more broadly drawn in East Asian design.[33] More intercontinentally later on, the "four corners of the Earth" would be widely taken as square, the sky as circular.

In the Near East, the pyramids and ziggurats embodied square regularity on the grander scale. Their apices, very high and central, represented a perhaps too obsessive quest for freedom, in whatever sense, from the constraints of terrestrial existence. The same has been said of modern high-rise.[34]

Acts of God

Still, monumental buildings harmoniously designed can betoken a one universal God or, at the least, a paramount deity presiding over a pantheon. Ra, the Sun God of Ancient Egypt, was a case in point. Assyria's Asshur had similar connotations. Very likely, too, Stonehenge will have subordinated individual tribal gods to a quest for political unity across the regional down-lands.

In such paramountcy may one discern a precursor of Judaism, Christianity and Islam – the Abrahamic monotheisms originating on the north-west fringes of the Arabian desert. This setting is considered in the geography of Holy Land history George Adam Smith wrote in 1894. He was intrigued by how the starkness of steppic and desertic landscapes, viewed under scintillating skies, encouraged single tribal gods which at a given stage might translate into an encompassing monotheism. Smith had gleaned this notion, a shade equivocally, from Ernest Renan (1823–1892), the celebrated French historian.

The Indian subcontinent invites comparison. Agreed, a Hindu pantheon could hardly emerge against a background too austere. Nor might a paramount god such as Vishnu with his four arms and multiple incarnations. Nor, indeed, would a Buddha seeking benign accommodation with a regenerative Nature. However, the presumptive corollary that monotheism abhors the rain forest holds less true. A good third of the people on the subcontinent today are Muslim, at least half of them in places subject to monsoonal downpours. None the less, the Renan thesis has marched on. T. E. Lawrence was among those subscribing to it.[35]

A further opinion aired by George Smith was that a desertic milieu breeds "seers, martyrs and fanatics".[36] In this dimension, however, cause and effect are contingent. During the Islamic expansion, its horsemen regularly fought with fervour against the infidels or, were none to hand, one another. Yet in the wake of conquest, they as regularly turned to co-existence.

Nevertheless, it was not for nothing that the faiths here being considered are sometimes dubbed "the three warrior religions". Moreover, their combative inclinations may have been strengthened by vague ancestral memories of two regional "acts of God" on a satanic scale. The first had been the surge of water across the Turkish Straits *c.* 5,550 BC to create the Black Sea (see Chapter 11). The second was the cataclysmic eruption of the volcano Thira on the Aegean island of Santorini (now dated in 1628 BC) which was to cause *tsunamis* within the eastern Mediterranean. It may well have been the inspiration for Plato's Atlantis, the insular utopia (ultimately destroyed in an earthquake) that Plato concocted to develop his critique of the wilting city-state system of Ancient Greece.

Tectonic disturbance (i.e. quakes or eruptions) has little affected in historic times the Levant–Hejaz sector save that (since John Gastang's expedition, 1935–6) it has come generally to be accepted that the collapse of Jericho's walls before Joshua's troops (*c.* 1250 BC) was caused by an earthquake in the Jordan rift valley.[37] Lethal earthquakes are infrequent but by no means unknown in or near Palestine – e.g. 1837, 1906 and 1927, not to mention Sodom and Gomorrah or other Biblical events.[38] Besides which, on twentieth-century evidence, the most active earthquake zone in the world is nearby, the Pacific ring of fire excepted. Its axis sinuously runs from the Adriatic and Aegean through Turkey and the Fertile Crescent, then across Persia.

Seismic instability may have been the background to a striking feature of both the New Testament and the Qur'an: namely, an acute dichotomy between Heaven above and Hell beneath. However, Judaism came to this portrayal less readily and more equivocally. Classically the Greeks had thought of the underworld as a very gloomy, lonely and sterile ambience, a perception undoubtedly encouraged by the character of their many limestone caves. Eroded by water working down cleavages, not a few of these features afforded near surface sites habitable by Palaeolithic Man. But their backs often led into minor labyrinths of narrow channels, precipitate drops and treacherous waters – all desperately cold and dark.

Occasionally cave mouths could be where oracles resided. The most fabled one, at Delphi, was customarily linked to Apollo. Some brilliant fieldwork has lately corroborated classic statements about how its authority was confirmed by effusions of psychogenic gas. To be specific, intersecting seismic faults periodically effect the release over extended intervals from bituminous limestone below of a cocktail of gases including ethylene.[39] The Romans were influenced by the Greek view of things. But so were they, too, by fiery Etruscan imaginings, perhaps derived from what may have been their ancient homeland of Anatolia, a territory set well within a zone of earthquakes and volcanoes.

By comparison, the skies above the clouds will have looked considerably benign and predictable. But they were nowhere entirely so. Whether considered to be above or below the Moon, the occurrence of meteor showers or comets was bound to be disconcerting. The more so were any meteorites observed. Highly suggestive of divine displeasure, too, will have been the dark and/or lurid skies liable to be generated worldwide by heavy volcanic explosions throwing dust into the stratosphere. It does seem that the vivid red sunset in Edvard Munch's painting "The Scream" was inspired by one such event as the Krakatoa dust veil extended over Christiania (now Oslo) in the winter of 1883–4.[40] One can only imagine what the post-Toba firmament looked like, never mind all the associated weather.

2

HAMMURABI TO PTOLEMY

NEXT WE SHOULD LOOK FOR DEEPER UNDERSTANDINGS of the relationship between Heaven and Earth. This involves focusing on how cosmology developed, irregularly yet insistently, in the Near East and Europe. The story can best commence with the first Babylonian dynasty, the family tree started by the visionary Hammurabi (1792–1750 BC). It takes us through the Bronze Age well into the Iron.

An aspect to consider is why Babylon not Egypt assumed the lead. Nodality in part explains this. Mesopotamia was focal to the exchange of ideas between India and points West, the initial transfer being very much to them. Take the synodic periodicities (i.e. cycles relative to the Sun) of the planets. The Vedic Indians apparently had several good values by 1900 BC.[1] The Babylonians had obtained one for Venus, "the bright lady of Heaven", by 1575 BC.[2]

Their astronomy drew further encouragement from progress with meteorology, rainfall prediction being at a premium. In Egypt, regime stability could readily be compromised by variations in the Nile's flood. But the causes thereof were too distant and mysterious to admit of serious prognosis. Across the Fertile Crescent from Palestine through Mesopotamia, however, significant winter rains are and will have been borne by wintertime depressions coming off the Mediterranean. Typically in this last century, Mosul's annual rainfall was 42 centimetres and Baghdad's 18. Cairo's was barely three. A Babylonian text from *c.* 1700 BC advises us that "if the North Wind blows across the face of the sky before the New Moon, the corn will grow abundantly".[3] What the visibility of a thin lunar crescent could betoken were the clearing night skies to be expected, along with the northerly winds, on the rear side of a depression. Moreover, if one depression is passing through, a family of several likely will. The Chaldeans *c.* 1000 BC shrewdly took account of haloes, too.[4]

Recourse to the Moon naturally fed into its salient role in astrology. By 1600 BC, eclipses were being recorded; and consideration given to which days in the lunar month might be portentous. By 500 BC, the prediction of lunar eclipses (basically by arithmetic extrapolation) was accurate, to within a few hours, in a good half of recorded cases. Reference to the 18.6-year lunar cycle was routinely made to this end.[5] Likewise, a cuneiform tablet from *c.* 100 BC registers quite precisely the angular velocity of the Moon across the sky.

The Assyrian kings who ruled over Babylon from the ninth to seventh centuries BC have been described as "potty" about astrology. Should the signs turn negative, a monarch might retreat into rural seclusion for some weeks. On his resuming the throne, his stand-in would be at risk of precautionary execution.

From the sixth to the third centuries BC, much translation of Babylonian astronomy took place in India. It thenceforward "formed the basis of the Indian astrological tradition".[6] From the sixth century, too, transference to Egypt stimulated the generation of indigenous astrological theory. A pyramid was intended to secure, for a pharaoh and his entourage, good placements in the afterlife. A primary way to lock on was to have a narrow shaft lead from a key tomb to the skies above in such a direction as to capture the Sun or Sirius or some other star at a critical juncture.

Enter the Greeks

Alexander the Great's occupation of Egypt in 332 BC had two very visible consequences. One of his generals, Ptolemy Soter, would soon establish himself as king, thereby launching a new dynasty. Meanwhile, the city port of Alexandria was founded. Its free population by 75 BC was 300,000; and it was a center of Hellenistic and, indeed, Judaic learning. Its libraries carried 700,000 scrolls. Among the great scientists or mathematicians who had been or would be working there were Aristarchus of Samothrace, Euclid and Claudius Ptolemy (see below). Later it hosted radical religious tendencies, Christian and otherwise.

By AD 118, Rome had taken over Mesopotamia; and well by then, too, controlled the entire Mediterranean coastline. These new realities marginalized Mesopotamian culture while allowing Greek to come into its own at last. Quintus Horatius Horace (65–08 BC), lyric poet and unabashed philhellene, bespoke a Roman readiness to accommodate this.

Turning to Greece, one addresses a culture about which Western civilization has evinced successive waves of filial piety interposed with rejection. Lately it has been going through such a cycle in very short order. Come 1965, the Graeco-Roman classics were in steep decline, in high school and above, in countries like France, the United States and Britain. A decade or so later, a nadir was reached. Two decades after that a strong revival of interest in the Classics, at least as read in translation, was under way. This has owed something to how two more universal themes out of fashion in the sixties have attracted attention again. These are (a) the role of chance in History and (b) what it means to be a hero.

Aristotle, the Exceptional

Undeniably Greece produced, through Classical Antiquity, its fair share of intellectual heroes. *Primus inter pares* was Aristotle (384–322 BC) who to me remains peerless in respect of how he melds comprehension, observation, reason and originality. Admittedly he cannot compare with Albert Einstein in terms of how insightful their cosmologies respectively appear at this juncture. Nevertheless, the former outstrips the latter when it comes to considering political and social change, not to mention biological science.

Furthermore, he brought even to everyday aspects of natural science a voracious spirit of enquiry tempered by tight control of data collection and deduction. Thus in *Meteorologica*, a book normally ascribed to him, Aristotle observed how, since the Trojan War some valleys in Greece had become too moist for agriculture while others (e.g. Mycenae) had become too dry.[7] Then again, there is his explanation of dew and hoar-frost formation on a clear and still night: "Some of the vapour that is formed by day does not rise high because the ratio of the fire that is raising it to the water that is being raised is small. When this cools and descends at night it is called dew and hoar frost".[8] As a succinct exposition, this remains hard to fault. About *Meteorologica* in the round, Joseph Needham conceded, in his own magisterial study of early Chinese science, that there was no Chinese counterpart "similar in scope".[9] It is salutary to be shown by the likes of Aristotle as by the Chinese (see Chapter 14) how far one may progress in the specifics of sky science without benefit of instrumentation or data networking.

Even so, his impact in this regard palls into insignificance compared with his dominating influence the next two millennia over logic and metaphysics. While charging Aristotle with manifold shortcomings in his exposition of the syllogism, Bertrand Russell urged we try "to remember how great an advance he made on all his predecessors (including Plato) or how admirable his logical work would still seem if it had been a stage in a continual progress" instead of being so long left a dead end.[10] Arguably, however, Aristotle fell short of modern philosophy in a more crucial and perhaps less excusable respect. He may have taken too much for granted straight concordances between mind, language and external reality. That tends these days to be seen as risky question begging.

SYMMETRY AND ECONOMY

One respect in which Aristotle made, to my mind, an enduring contribution lay in how he underlined the interpretative importance the Greeks attached to "symmetry". In modern physics, this word is often accorded esoteric usage. It is so in order to define criteria for elucidating the behaviour of elementary particles (or anti-particles) within fields subject to relativity but also to quantum mechanics. What Aristotle meant, however, was symmetry in the more everyday sense in which the word is still employed by biologists. They say a form has bilateral symmetry if it can be bisected into two halves which mirror-image one another. It has "radial symmetry" if its layout exhibits an all-round uniformity akin to that of a starfish or a buttercup. Many Europeans from the time of Dante and again that of Newton would have well understood. In this text, too, is on the customary usage, carefully though it needs be applied.

Another Aristotelian precept, closely related to the accent placed on symmetry, is that "Nature does nothing in vain". Not far distant conceptually is the stress another notable Greek philosopher, Heraclitus (born *c.* 544 BC), had placed on "cosmic justice": in effect, equipoise between contrary tendencies. One likewise recalls the principle of "efficiency" that the great classical scholar Sir Alfred Zimmern saw as important to the Classical Greeks in the ordering of their daily lives. By it was meant leading one's own life to the full by achieving due balance in a range of activities.

Harmony and proportion are notions which inform each and every one of the several precepts just cited. One could therefore say that all Aristotle was doing was putting his distinctive stamp on a tradition of enlightened moderation already deeply ingrained in Classical Greek intellectualism. But within that context had been generated a number of specific proposals about cosmology and metaphysics, some of which may strike us now as remarkably modern.

A Greek Renaissance

When Lord Russell further opined that "in all history, nothing is so surprising as the sudden rise of civilization in Greece",[11] he will have expressed the feelings of many of his contemporaries in the mid-twentieth century. It must be said, however, that any surprise expressed stemmed in part from a prior disposition to see the Greek experience as even more singular than it was. Thus Athenian democracy tended to be treated as at once more thoroughgoing and more typical of the Greek statelets than the record really showed. There was similarly a slowness to acknowledge how much Classical Greek thought, not least about cosmology, owed to the Near East and India.

And yet there was also a tendency, until the sixties at least, to overlook a dramatically traumatic entrée to this Classical era: namely, quite a precipitate collapse *c.* 1200 BC of the Mycenean civilization which had flourished in the Aegean–Peloponnese region the previous eight centuries. It was part of an acute crisis around the Eastern Mediterranean. Witness how Egypt sank into hapless apathy for a full 400 years, following the nine plagues and the Jewish Exodus. A fair measure of support has been engendered for the thesis that this regional Dark Age, lasting several centuries, was mainly due to the rather abrupt onset of recurrent droughtiness.[12] Seismic disturbance also played a part.

The Dark Age into which Mycenae therefore sank looks grim indeed. Though wine-making was upheld resolutely, much else was abandoned, a syllabic script included. Then in the eighth century BC began – once more, somewhat suddenly – a pronounced renaissance complete with an alphabetic script. In due course, the Greeks acquired a sense of a short if colourful history, mistily preceded by a pre-Homer golden age.

The said renaissance strongly revived a pre-existing taste for geometric visual art as expressed in temples and other important buildings as well as in pottery and other artifacts. It also gave rise early on to Homer then soon to Hesiod plus their rival schools of poetry. Homer mentions various stars and constellations by name. Hesiod relates stellar movements to agrarian obligations. Thus when the Pleiades first rise above the horizon pre-dawn (about 10 May), the corn should be cut; and once they are setting then (around 12 November), ploughing should commence.

Ionia

Systematic astronomy first flourished in Ionia on the western littoral of Asia Minor – a Graecian outpost oriented towards the Near East and not overly awed by the emergent Olympian pantheon nor, of course, traditional local deities. Miletus, the capital and main port was decidedly cosmopolitan. What we know as "Science" could therefore take root, further encouraged by the local need for nautical skills – not least folk meteorology.[13]

Two Ionian scientists-cum-philosophers merit special attention. Thales (624–547 BC) saw water as the first principle of all things. Therefore he had the land float upon the deep. His near-contemporary, Anaximander (611–547 BC), thought more adventurously. He saw our universe as but one of a limitless number, all of them transient creations comprised of a ubiquitous material basic to water, air, fire and everything else substantial. The Earth was at the centre of our cosmos, kept there by, essentially speaking, the dictates of symmetry. The heavenly bodies are mobile vents of fire:[14] a proposition which, alas, raised more problems than it solved.

Across the water from Ionia was its commercial rival, the island of Samos. There was born Pythagoras (*c.* 572–497 BC), someone acclaimed by Arthur Koestler as "the founder of Science as the word is understood today",[15] yet somebody whose involuted persona is shrouded in mystery. Said to be the son of a silversmith-cum-gem engraver, he was likely a pupil of Anaximander. Around 530 BC, he abandoned his establishment position within the Samian enlightened despotism and settled in Kroton, a fractious Greek colony in southern Italy. His repute and charisma enabled him to found there a Brotherhood which soon ruled the city. But a backlash late in his own life led to meeting houses being sacked and members slain or exiled. Pythagoras himself was banished awhile.

Nor is there any denying that this Pythagorean "rule of the saints" had its morbidly crazy aspects. Its best known precepts were the transmigration of souls and the inadmissibility of eating beans. Among many other taboos were sitting on a quart measure; having swallows in the roof; touching a white cockerel; and stirring the fire with an iron piece. Breaking bread was forbidden. Yet so was eating from a whole loaf. Modern dictators have similarly appreciated that the exercise of power can be all the more intimidatory when it is arbitrary and otiose.

Koestler interpreted Pythagorean deviance altogether too generously, commending its apparent linkages with the Orphic Mysteries in vogue among the populace at large within the Greek homeland. He suggested that under Pythagoras "religious intuition and rational science were brought together in a synthesis of breathtaking originality",[16] the stated aim being release through catharsis from various forms of mental enslavement. Yet how far Pythagoras' own thinking had taken him down this edifying path is not clear. What is clear, however, is that Orphic mythology (about Dionysian good and Titanic evil struggling for control of the universe and human souls) was altogether too tortuous and tortured to allow of a positive intuition/emotion/reason synthesis; and that, in any case, the Brotherhood had imposed its involuted self much too precipitately.

The achievements of Pythagoras in Mathematics and Science might be easier to gauge had he himself committed them to writing. Nevertheless, they were undoubtedly considerable though sometimes gratuitously flawed. Remarkably, he discovered that the pitch of a musical note depends on the length of the taut string producing it; and that the harmony, *armonia*, between the notes on a scale is attained through simple numerical intervals. But he and certain disciples then illogically proposed that the perceived distances between heavenly bodies could yield a musical "harmony of the spheres". This expression has marched on down the centuries, for instance in the works of John Dryden and John Milton. It is beautiful allusion but also illusion. Still, Pythagoras did elevate "proportion" and "harmony" to prominence within an invisible verbal pantheon which also included symmetry, justice, balance and efficiency: all words which connoted keeping everything in due perspective, be this in pursuit of operational aims or causal explanation.

He thereby gave mathematics and, in particular, geometry a sharper definition. Even so, it is doubtful whether he effected or even contributed to the original formulation of the theorem that famously bears his name. However, either he himself late in life or some immediate disciples ascertained there to be certain values (e.g. *pi* or the square root of 2) which must be judged "irrational" or "incommensurable" because they seem not to lend themselves to final resolution. Since when, *pi* or Π, the standard ratio between the circumference and diameter of a circle, has been calculated to millions of decimal digits yet remains an approximation. Against the background of the political pressures they themselves were then subject to, irrational numbers seem to have caused the Pythagoreans considerable anxiety. Still, the advance thus made did in a sense serve to underline their cardinal principle that numeration was the arbiter of all things.

The master himself was satisfied that our Earth, like other celestial bodies, is a sphere. One presumes he was thus persuaded by (a) considerations of symmetry, (b) the "below the horizon" effect, and (c) eclipse shadows. Meantime, he may also have been the first to divide our spherical abode into polar, temperate and torrid zones. More specifically, he may have achieved this ahead of Parmenides of Elea (*c.* 504–450 BC), the close associate of the Pythagoreans often credited with this analytic advance.

Pythagoras himself always remained persuaded that our cosmos is geocentric. But before the Pythagorean school finally faded (around the time of Plato) a broad consensus had emerged within it to the effect that the Earth was revolving, albeit around a central hearth, *hestia*, which is other than the Sun. A rider added awhile included a revolving "counter-Earth" intended to explain eclipses.

Other contributors could be considered. Two must be. Democritus (*c.* 460–*c.* 370 BC) was the leading proponent of the Atomist theory, first promulgated (says Aristotle) by a near contemporary – Leucippus of Miletus. Democritus surmised that all things everywhere are comprised of atoms – minuscule particles, imperceptible to the senses. Though these do vary in size and shape, they are always underived, indivisible and indestructible. When he said "everywhere", he meant across an infinity of worlds, each and every one brought into being through the constant atomic motions – worlds that were endlessly different in size, character and distribution. It is a mechanistic account in which deep simplicity engenders infinite variety. That rings remarkably modern.

Democritus can also be said to have followed on from Anaxagoras of Ionia (*c.* 500–428 BC), a scholar who spent thirty years in Athens, having been invited – it seems – by Pericles. When the latter was waning, Anaxagoras was run out of the city under a new law proscribing theorizing about "the things on high". But he had stayed in Athens sufficiently long to have consolidated the city's position as second to none among the founts of Greek cosmology and philosophy. At the level of what we can term "astrophysics", he made a signal contribution. He was the first firmly to aver that moonlight was reflected sunlight. He did further deduce definitively that lunar eclipses were shadows, usually or always cast by the Earth.

Anaxagoras perceived Mind as a prime mover in two separate respects. Though a distinct entity, it is spread throughout the animal kingdom with striking uniformity, the exceptional intelligence human beings apparently possess being considerably a function of their manual dexterity. Then again, Mind exercises an influence over all atomic motions, tending to organize them into a centrifugal rotation that dictates cosmic evolution. First Socrates then Aristotle were to complain that Anaxagoras never developed properly his thinking on this score.

Plato versus Aristotle

The thought experiments of the Classical Greek thinkers were remarkable, above all their appetite for speculation far beyond what was testable then or maybe now. It all owed something to how the Greek system of small city states functioned. Friction between them usually ensured that a free thinker in difficulties in one could, however grudgingly, be granted asylum in another, an uncertain prospect but one sweetened somewhat by shared language and culture. However, the Peloponnesian War (431–404 BC) between Sparta and Athens highlighted and accentuated the shortcomings of this common weal. A basic rethink was unavoidable.

Since the European Renaissance, successive generations of students of politics and philosophy have been encouraged to encapsulate this rethink as a dialectic between Plato (*c.* 427–347 BC) and his erstwhile pupil, Aristotle (384–322 BC). Plato was horrified by the war and by its ending in the defeat of Athens, his main academic base. He shared a widespread concern about the general demoralization of this Greek world; and was personally traumatized by the judicial execution of his own mentor, Socrates (469–399 BC). Yet he felt close enough to the pre-war world to be able to advocate reform and revival. In fact, he enunciated in *The Republic* proposals for the social engineering of a new-style city state. By definition, this utopian exercise elevated the part ideas play in human affairs.

However, Plato went a deal further. He so developed and generalized this notion as to arrive at the proposition that, throughout creation, Ideas or Form (especially as mathematically expressed) are prior to substance. Therefore he is not merely to be seen as the progenitor of utopian political thought. He can also be dubbed the founding father of the long and surviving tradition of philosophic Idealism. It may have more to contribute in the twenty-first century than most of us have yet appreciated, always provided one does not simply assume an automatic and exclusive connection between the cosmic mind and the human one, as made manifest in some disembodied or transcendental mode. That Plato did, in fact, make this very assumption is evident in his treatment of Immortality: "having got rid of the foolishness of the body we shall be pure and have converse with the pure, and know of ourselves the clear light everywhere, which is no other than the light of truth".[17] It is part and parcel of his very Socratic commitment to taking care of the enduring *psyche* (if you will, "soul"), this in the context of the city but ultimately of the cosmos.[18]

Avowedly, his aim was to effect a monumental grand design. But how far did he succeed? His cosmogony is characterized by "the greater or less admixture of myth, romance and poetry";[19] and draws heavily on Pythagoras. Where his Idealism may have led him to part company with all predecessors is in depicting the firmament we actually see not as heavenly perfection but merely a blurred and imperfect expression of the pure mathematical aesthetics underlying it: "Yonder broideries in the heavens . . . are properly considered to be more beautiful and perfect than anything else that is visible; yet they are far inferior to those which are true."[20] How did he know? And in these circumstances, how might perfection be gauged?

Odder still is the proposition that the possession by stars of intelligence is demonstrated by the fact that "they always do the same things because they have long been doing things which have been deliberated upon for a prodigious length of time".[21] Nor can one simply rest content with the averration that, into the universe, the Creator "put some soul and spread it throughout the whole, and also wrapped the body with the same soul round about on the outside; and he made it a revolving sphere, a universe one and alone".[22]

The difficulties thus encountered bear directly on Plato's endeavours to delineate the *Republic*, his vision of utopia. Granted, one has always to ask with him how far affectation and hyperbole shape the narrative. Then, as with any utopian portrayal, one must ask if it is but a mirror-image critique of a particular status quo. Actually, *The Republic* does seem genuinely intended to serve, however forlornly, as a blueprint for a new society. Moreover, the notion that a philosopher overlord should draw such a proposal up is in line with Classical Greek tradition. Thus Solon, law-giver and poet (*c.* 639–559 BC), had achieved the very comprehensive reform of the legal codes of Athens.

Still, Plato's blueprint has to be dubbed conservative totalitarian, more or less as much so as the Spartan regime had in practice been. A supreme authority would have to inculcate, across a generation or two, the "noble lie" that mankind could be divided, neatly and enduringly, into three classes – gold, silver and baser metals. The Gold were the guardian class, liable to be very largely self-perpetuating. Silver comprised the warriors. The third class were, of course, the commoners. The education system was designed to ensure acceptance of this framework and the appropriate performance of individuals within it was rigidly repressive. The aim was to perpetuate indefinitely a very static interpretation of the "good" as incorporated into a perfectly hemispherical heaven in which all motions were circular. One cannot progress beyond perfection.

Plato and Aristotle are *alter egos* to an extent. Both were, outwardly for sure, imbued with the standard presumption that Greeks were entirely superior to barbarians, likewise men to women. With each, too, authentic writings survive in considerable bulk, the first philosophers of Classical antiquity of whom this can be said. But acknowledging this raises the question of how impact is influenced by successive translations. Thus one is advised that "the precision of Latin cannot do proper justice to the flexibility and the nuances of the Greek".[23]

More distractive, however, is transposition into a profoundly different social order. The Greek word "polis" has given rise to a tranche of English-language political terms, applied wherever. Yet it pristinely possessed a non-reproducible flavour as applied to, say, the 40,000 or so full citizens of the Athenian city state. Meanwhile, the meaning of *demos* evolved even within Greek Antiquity from country folk to all freemen acting politically.

Polymathic Aristotle

None the less, it was not good enough for Arthur Koestler to lump Plato and Aristotle together as concerned to "build a walled-in universe, protected against the Barbarian incursions of Change" and with its Earthly "centre of infection safely isolated in the sub-lunary quarantine".[24] By the same token, it was conceding a point too grudgingly to remark that Aristotle's attitude to change was "not quite as defeatist as Plato's".[25] In fact, the two were qualitatively different. In Plato's fantastical utopia, "change" was synonymous with "decay". For Aristotle, on the other hand, "change" or "motion" was the one great constant in human life just as in Nature.

Instinctually the latter tended to see change as teleological (i.e. goal oriented) much as it usually was in the biological studies he so enjoyed. Nevertheless, his depiction of political change was what we would call "Popperian" – i.e. continual ebb and flow. He develops considerably further than Plato the notion of different kinds of government and the decadent forms they lapse into: "Tyranny is the perversion of Kingship; Oligarchy of Aristocracy; and Democracy of Polity."[26] He then crafted a brilliant typology of revolutionary threats and responses to them.[27] Among the options he found a tyranny turns to is a secret police.[28]

Undoubtedly his understanding of government was honed by his having been personal tutor for three years to the talented though obsessive teenager whom History would come to know as Alexander the Great. Lately our understanding of their relationship has become firmer than it once was. During the tutelage, the two were drawn together by a mutual interest in knowledge acquisition. But afterwards Alexander waxed impatient with Aristotle's disposition to admonish from afar; and to inveigh a bit too pointedly against recklessness, drunkenness and angry self-concern on the part of people in authority.[29] Theirs was clearly a tough-minded interaction within a fractious milieu. It hardly gels with the Koestler perception of Aristotelian defeatism.

Nor do Aristotle's cosmological formulations. Usually he did most thorough reviews of evidence and arguments, albeit within a high profile rendering of Greek Classical culture. The results were written up in what we still term true academic style. He satisfies himself the Earth is spherical, mainly by how the starfield coverage varies with latitude. He cites received opinion as reckoning its circumference at 200,000 stades (*c.* 22,200 miles), a calculation within 15 per cent of what we now know it to be.

He further concludes the Earth is motionless at the center of everything. Its centrality seems confirmed by how (a) heavy things fall towards it and (b) the Moon (like, presumptively, all heavenly bodies) always shows it the same face. Its being without lateral motion (otherwise an Aristotelian attribute) confirms this nodality. Every year a given star sets at the same point on the horizon on a given day.

Meanwhile, the rest of the Cosmos is seen as exhibiting perfect radial symmetry. Beyond it, one can but speak of God, "the unmoved mover". Otherwise there is not even empty Space. After all, a "space or void is only that in which a body is or can be".[30] Nor can other universes exist. They would distract the unmoved mover and compromise the symmetry principle. Efficient motion for all things heavenly must be on circular courses. At which point, a distinction is drawn between general regularity above the Moon and randomized motion below. Meteors, Aristotle realized well enough, were liable to be sublunar.

One explanation for the sublunar world being so inchoate would be that it was understood to contain four primary elements – earth, water, air and fire. Above the Moon there was just the one ubiquitous substance, *quinta essentia.* How ubiquity related to the forms of the heavenly bodies was not made clear. Nevertheless, Aristotle came up with the best hypothesis until then for explaining heat and/or light from heavenly sources. He put it down to friction between their moving surfaces and the quintessential atmosphere. Best yet desperately inadequate, even on a contemporary showing. Why, for instance, was the Moon so cool compared with the Sun?

Nor was this the only problem insufficiently resolved. During the fifth century BC, Greek scientific opinion had come to accept that the Earth was considerably smaller than the Sun. Now, Aristotle suggested it was outdone thus by the stars above. This was hardly what you would expect of a centerpiece. Nor could one feel entirely reassured by a geocentric view of how the *planetes,* "the wanderers", fitted into the divine scheme. Their motions relative to the Earth were intricate and individualized. Yet within the sequence of 50 or so concentric crystalline spheres which Aristotle proposed rotated round the Earth, theirs lay intermediately between ones bearing the stars with their uniformly regular motion and those respectively bearing the Sun and the Moon, each locked forever into intricate though comprehensible movement. Interposed thus, the planets did not accord well with Aristotelian notions about gradual alteration.

Besides, Aristotle offers everything and nothing on the subject of divine purpose. His disposition was to see "theology" and "metaphysics" as synonymous but often to treat this whole subject light of hand. However, he does see love of God as the ultimate driver of all living creatures. He proposes the human soul has two parts, the irrational and the rational. The former comprises (a) a vegetative section similarly found in all living things and (b) an appetitive one present in all animals. Obversely, the rational part is strictly a human attribute. Blessed with it, some may through sublime contemplation prepare themselves to receive what can only be a strictly impersonal share in God's immortality. Meantime, the material cosmos could be trending towards greater articulation of form, a kind of cosmic teleology. But it is more different than Aristotle acknowledged from the generative emergence he knew as a biologist. In any case, he never gave a divine teleology the impress some Christian theologians have read into him.[31]

His enduring strength lies in enunciating general principles rather than elaborating particular themes. The loose ends left by the latter, others could gather up. His enunciations were driven by boundless curiosity tempered with stringent objectivity; and the whole enriched with an awareness of Symmetry and Economy as principles of Nature.

What one cannot admit is the Koestler contention that he was but preparing the way for a neo-platonic "Mandarin Universe",[32] the backdrop to something between an oligarchy/aristocracy and an emperorship/tyranny. All else apart, the proposition is ludicrous in relation to how trenchantly Aristotle wrote about political forms.

Through Late Antiquity

A contemporary of Aristotle's, Heraclides of Pontus (388–315 BC), "saved the phenomena", as they were wont to put it, by averring that the Earth rotated. He further proposed that Venus and Mercury orbited circularly round the Sun. He thereby paved the way for a full commitment to heliocentricity by Aristarchus of Samos (*c.* 310–230 BC), a doyen of science and mathematics. It was an advance Copernicus himself was to acknowledge a debt to. Yet in those times, the only astronomer known to have endorsed it was Seleucus from Mesopotamia: a territory which still lay well outside Rome's imperium.

Reasons can be conjured for this near to total contemporary rejection. Samian associations may have counted against Aristarchus in minds partial to prejudice. Perhaps his own personality did, too. Nor will the astrologers have been much enamoured of his radical astrophysics. Above all, however, his thesis eventually faced the blocking opposition of Hipparchus of Rhodes (*c.* 162–126 BC), the cerebral giant who comes down to us (especially via Claudius Ptolemy) as the father of systematic astronomy. Ptolemy credits him *inter alia* with discovering equinoctial precession and the eccentricity of the Sun's apparent orbit. During a solar eclipse in 129 BC, he used observations in the Hellespont and Alexandria to estimate lunar parallax. He duly deduced that the distance of the Moon from the Earth was medianally 68 Earth radii. Actually, it is just over 63. Also he generated a catalogue of nearly 850 stars, these very accurately located in azimuth and bearing. To a revised version, Ptolemy is thought to have added another 170.

To explain the anomalous motions of the planets, Hipparchus developed further the concept of epicycles (in effect, wheels within wheels) current by the third century. Already epicycle doctrine was too elaborated. Even so, it was bound to win out against heliocentric thinking in a situation in which all concerned were assuming heavenly orbits must be circular. Admittedly, it did not well account for the fluctuating brightness of the planets, notably Venus. But this was then seen as secondary.

However, instrumental refinement was liable to put epicycle doctrine under ever tighter scrutiny. Take an artefact (dated at 80–50 BC) recovered from a Mediterranean shipwreck at this last turn of the century. With its 30 gear-wheels, it was apparently intended for festival and astrological calculation, not primary data collection. Nevertheless, students of these matters see it as a remarkable manifestation of how equipment quality had improved.[33] In these circumstances, geocentricity came under renewed pressure. Arthur Koestler was able to cite pertinent comments successively made by Cicero (106–43 BC), Pliny (*c.* AD 23–79) and Plutarch (*c.* AD 46–*c.* 120) which do seem to dethrone the Earth. Nevertheless, under the leadership of Claudius Ptolemy at Alexandria, the geocentric principle was to be reaffirmed.

In his magisterial *Almagest* text, Ptolemy (an outstanding mathematician by background) sought to consolidate existing astronomical knowledge. In a parallel work, *Tetrabiblos*, he did as much or more for astrology. Babylonian astrologers had understood the Gods to communicate their concerns and intentions via the stars. But Ptolemy had the stars directly influence, at least contingently, situations on Earth. The Moon and the Sun clearly did. So why not all heavenly bodies?

In fact, by AD 200, what one might dub main frame Greek science had pretty much run its course in astronomy and geometry just as it had in acoustics, optics and mechanics. By responding with consolidation Ptolemy – along with Galen (*c.* AD 130–200) in medicine – gave the whole area of Science more cohesion and long-term perspective, *scientia aeterna.* The problem was that, while so doing, he compromised unduly on data discrepancies. Most especially this applied *vis-à-vis* Venus, a planet which Copernicus was likewise to find nigh unmanageable.[34]

In any case, the political and social climate of the Graeco-Roman world was turning less conducive to scientific leaps forward. Ptolemy himself was at work during the great Antonine age, the Roman Empire's golden afternoon. But already such Oriental cults as those of Mithras (the pugnacious Indo-Persian sun god) and of Jesus the Messiah were challenging imperial charisma. The third century witnessed a prolonged urban crisis more or less throughout the Empire. None the less, the fourth was to be seen by contemporaries as an "age of restoration", *reparatio seculi.* Nowhere was this more so than in the cities of the Greek East.

But the terrible defeat of the legionaries by the Goths at Adrianople in 378 heralded irreversible decline against a background of philosophic confusion, epitomized come the sixth century by a wave of irrational neo-Platonism which placed much stress on astrology and alchemy. However, since Emperor Constantine experienced Christian revelation in the heat of battle in AD 312, Christianity had effectively been the established religion, a state of affairs which survived the formal division of the Empire between Rome and Constantinople in 395.

3

LATE ANTIQUITY TO COPERNICUS

THE ROMAN EMPIRE CAN BE SAID TO HAVE PEAKED early in the third century. A campaign to conquer Scotland, launched by Emperor Severus, was halted on his death in 211. Rome's age of expansion thus came to a close. Next year, Emperor Caracalla extended Roman citizenship to all freemen within the borders. This marked Rome's climax as a political experiment.

The religious movements, Christian and otherwise, that waxed and waned tended to be eclectic. Nevertheless, several themes recurred. A good-versus-evil polarity was usually basic. So were puritan values, sexual restraint above all. Blissful immortality was in prospect for righteous believers. Astrology and cosmogony figured, but not that prominently. Charismatic leadership and chiliastic prophecy featured. But a disposition, through the third century, to identify canonical scriptures gave Christianity added solidity and cohesion. It was becoming a religion of the history book in the best Judaic tradition.

After Emperor Constantine experienced Christian revelation in 312, he made his new-found faith a state religion in all but name, leaving it free to launch wave after wave of coercive conversion. This elevation did nothing for the Church's pristine concern for the underdog, gender equality, the end of war and like causes. Furthermore it involved, into the sixth century, the strong discouragement of all sky science lest it allowed of subversion via astrology. By the late fourth century, farmers and fishermen were said to be afraid to study the stars lest this be seen as pagan.[1] However, this new conformity did make of Christianity a powerful agent for social cohesion especially in the Greek-speaking East. There a more even spread of wealth and power further favoured solidarity.[2]

Earlier streams of influence from the East to the Latin West had included the Stoics, a movement founded *c.* 300 BC and named after its original meeting place in Athens. Socrates was their role model. The poet Seneca (*c.* 3 BC–AD 65) and the Antonine Emperor, Marcus Aurelius, were to be their most renowned Roman adherents. Along with their concept of an all-pervasive God went their belief in "universal love", rather austerely defined. Eventually, so they advised, a cosmic conflagration will consume all. Yet they mostly believed this would but complete a cycle destined to be endlessly repeated: "Everything that happens has happened before; and will happen again . . . countless times."

St Augustine (354–430)

The title of Augustine's most famous work, *City of God,* he took from Marcus Aurelius. Each of them endorsed the early Stoic view of the cosmos as one big city shared with God. Likewise did St Paul and, before Christ, Posidonius (*c.* 135–*c.* 51 BC) – a Ptolemaic astronomer and Platonist philosopher with Stoic affiliations. Whereas most Stoics had believed the soul perishes with the body, Posidonius proposed it lived on in the air, usually till the next great conflagration.

The Roman Africa into which Augustine himself was born was but a lingering remnant of Pax Romana, as yet intact but beset by looming menace. Hard upon a tormented reaction to his own hedonistic youth, he was baptized in 387. He was bishop of Hippo from 395 till his death there during the Vandal onslaught. He was never at Alexandria; and his bishopric was not a senior one. He gained his repute through intra-faith didactics. By 400, he was condemning contemporary millenarianism (see below) for its vulgar understanding of paradisal reward.[3] Of more import, however, was the lead he assumed in combating the dissentient liberality of Pelagius (*c.* 355–425) and his followers. This included rejection of Original Sin, a more sympathetic approach to sexuality in general, and insistence that even non-believers can freely seek Grace.

City of God was written (from 413 to 426) in response to the relatively restrained sacking of Rome by Alaric, an Arian Christian Visigoth, in 410. Since the second century, the Christian mainstream had been moving away from the "great whore" image of Rome portrayed in the Book of Revelation. Accordingly many Christians (e.g. Pelagius and Jerome) found the sacking horrific. For his part, Augustine regarded a *Pax Romana* as important for the progress of Christianity, but saw all political structures as transient in a Popperian sense – i.e. products of random contingencies not historical cycles nor progressive secular trends. Technological advance he saw as ambiguous: "for the injury of man, how many kinds of poisons, how many weapons and machines of destruction have been invented".[4] More specifically, he was keen to stress that the origins of the 410 disaster lay in a moral decline which had long preceded the rise of Christianity.[5]

All the great and good of the Middle Ages, it seems, were to read *City of God*. It permeates the theological discourse of those times. It is a medley of chapters (some of just one paragraph) addressing didactically, though not tritely, virtually every theme in the current polemics. Certain loose ends lie unresolved. Thus Plato is portrayed as the philosopher who "comes nearest" to Christianity (Book VII, Chapter IX). What we are not told is whether he could therefore receive Grace.

In various works, Augustine emerges as a resolute Christian critic of the astrologers while admitting that, in his misguided youth, he "did not cease openly to consult these impostors".[6] Some of his arguments seem to us pretty standard. Twins may have identical prognoses but contrasting outcomes. How can the stars in heaven engender evil as well as good? To respond that they proffered signs not causes was to run counter to certain classic statements.

As Book XI of *Confessions* highlights, however, Creation and Time were what exercised Augustine above all else. How could Genesis cosmogony be reconciled with a Classical world disposed in the main to believe the cosmos had been and would be existing forever? In particular, what was God about before Creation? In essence, his response is to aver that Creation was a totally comprehensive occurrence, marked by a huge flash of light. Therefore the formlessness that preceded it subsumed the absence of the Time dimension. Accordingly, the question put is meaningless. Correspondingly, he reflects that "I only wish that other useful matters . . . I could know with an assurance equal to that with which I know that no created being was made before every creature came into being" (Book XI, Chapter XI). Yet he goes on eloquently to express his mystification over the basic nature of Time: "who will tell me there are not three times – past, present and future – as we learnt when children and as we have taught children but only the present because the other two have no existence?" (Book XI, Chapter XVII).

In the next book within *Confessions*, he ponders further the pre-Creation formlessness. He puts to the Almighty Himself that, in effect, His mind had imposed a reality on nothingness: "No doubt the Heaven of Heaven which you made in the beginning was a kind of Creation in the realm of the intellect" (Book XII, Chapter IX). This was about as Platonist as they come.

Medieval Christendom

In the eighteenth and nineteenth centuries, Protestant and secularist historians viewed the millenary interval, AD 500 to 1500, as simply the Dark Ages, a long cheerless night ahead of a sunlit Renaissance.[7] Accordingly, the field was left largely to national historians to trace unilaterally the evolution of their respective countries. Jules Michelet in France and Edward Freeman in England are the best known.

By the time my generation of aspirant historians were on campus, the term Dark Ages had virtually been confined to the sixth to tenth centuries and was coming to mean "obscure" not "atrocious". Nevertheless, Arthur Koestler still felt free to dub the whole thousand years in question a "dark interlude". Not that his own text well supported this.[8]

During the five centuries just cited (AD 500–999), the accent was on avoiding Christianity's extinction and then making it territorially secure. This consummation was effected soonest in the Eastern Orthodox realm of Constantinople. It was the more enduringly in the Catholic realm of Rome. It involved containing but then accommodating successive external threats, Islamic, Magyar and Viking. In the Viking case, historians have been particularly struck by how readily aggressive warfare gave way to convergence with indigenous peoples – to intermarriage, the adoption of Christianity, and bilateral trade.

A thesis to conjure with apropos Islam is one enunciated just before his death by Henri Pirenne (1862–1935), a ranking Belgian medievalist. Its gist was that, with the Mediterranean transformed by 850 into "a Moslem lake", the center of gravity of Catholic Europe was displaced northwards. Most visibly was it with the creation of a Frankish Empire, centred in Aachen. On Christmas Day 800, the Pope crowned its ruler Charlemagne as "Holy Roman Emperor". Then the imperial confines extended from the Pyrenees to the Elbe.

The Pirenne verdict was that "without Islam, the Frankish Empire would never have existed; and Charlemagne without Mohammed would have been inconceivable".[9] Nor is there much doubt that, during the eighth century, Mediterranean maritime commerce disappeared "almost altogether" awhile.[10] In the eastern basin, indeed, forceful encounters between Byzantines and Arabs were to be almost continual to 925.

Yet come the ninth century, several Christian ports in Italy were trading extensively with Islam, Jewish communities helping to bridge divides. The city state of Amalfi assumed prominence through dealings with all comers, its population mushrooming to 70,000 by 1020 or so. Harūn ar-Rashīd of

Baghdad made his pen-friend Charlemagne honorary protector of Jerusalem, a compliment symbolic of the times.

Still lacking, however, was scientific dialogue between the great obediences. In Spain at least, there will have been contacts about farm practice. But hardly any literature on that subject would enter circulation before 1500, in western Europe at least.[11] However, the medical school established in Salerno in the ninth century would, in due course, be translating renderings in Arabic of Classic texts as well as the works of Avicenna (see Chapter 13). Otherwise the European resource base in medicine was less firm than in Late Antiquity.

Nor was there interfaith exchange on matters astronomical. Though Islamic scholars observed the skies competently, few were much inclined to widen their enquiries. Meanwhile, apprehension persisted throughout Christendom of recording aberrations on the undersurface of heavenly perfection. St Augustine had warned against it for reasons partly pro-Plato and partly anti-astrology. Much remarked of late has been a disinclination across Christian Europe to note the advent in 1054 of the supernova (i.e. explosive death of a large star) we know as the Crab Nebula. The only surviving records come, in fact, from a Nestorian Christian in Constantinople and a chronicler in Bologna.[12] Yet to the Chinese, this "guest star" in constellation Taurus was visible, even by day, for many weeks. It was recorded, too, in Japan and, somewhat allusively, in *dār al-Islām*. Over twenty pictures have been identified in Amerindian rock art, in locations from Texas to California.

Millenarianism can also be considered within this context, seeing how it sometimes relates to the magic of signal numbers within the popular consciousness. Considered especially characteristic of Judaeo-Christianity and Islam, it has often found expression in times of stress through religious or political channels. An enveloping vision of the future is identified with a great leader, lately arrived or expected soon. He tells of the imminence of History's last great struggle, a chasm of crisis which must and will be traversed to attain the celestial city – a millennium of peace. Archetypical was the "New Earth" suffix appended to Ragnarok, the Norse legend of the "Twilight of the Gods", as Scandinavia came under Christian influence.

Romanticist historians in the nineteenth century (notably Michelet) saw the year 1000 as having been a singular opportunity to air such radicalism. To this inference there has been a reaction, partly on the grounds that millenarian unrest may have been more in the early decades of the eleventh century than at its turn.[13] But nobody can be at all sure of the actual spread, given the concern of the then powers-that-be to play the whole tendency down.

What can be said *a priori* is that structural factors always weighed in. Millenarian outbursts drew support from an "amorphous mass of people who were not simply poor but who could find no assured and recognized place in society".[14] Not too extensive in the Europe of 1000, this stratum became more so with structural change in the ensuing centuries. Millenarianism bore upon the launching of the Crusades, this none too helpfully.

The Crusades

A Christian movement for social peace (to be secured through more equality) was promoted by the great Benedictine abbey founded at Cluny in Burgundy in 910. Initially it drew core support from inner Provence, the most Latinized province of former Gaul. Later its focus shifted more towards central and northern France.

There, the year 1034 came within a cluster of famine years that even gave rise to the "pagan" practice of cannibalism. It was also the millennial anniversary of the presumed date of Christ's Ascension. The social peace or "Truce of God" movement took off as never before. But now it was being seen as a precursor to a crusade against non-believers. Already iconic for the ramifying Cluniac fraternity was the shrine to St James the Apostle at Compostela in Galicia, an inspiration for the *reconquista* of Moorish Spain.

Things progressed but slowly for some time. Then a meteor shower over France in April 1095 was extensive enough to be remarked throughout the land. It seemed heavenly patience with continued inaction was wearing thin. Meanwhile famine had returned and martial belligerency was endemic. Meanwhile, too, Emperor Alexius of Constantinople made what was, in fact, a second plea for papal assistance against the Seljuk Turks.

Pope Urban II, himself a French aristocrat, held council that November at Claremont. There he effectively launched the First Crusade, reportedly arguing *inter alia* that France needed a population outlet. In the ultimate, however, his main aim was to unite the Greek and Latin churches "under the headship of the Bishop of Rome".[15] He campaigned for a Crusade, throughout France and Italy, well into 1096. Enthusiasm for it and the "Truce of God" was engendered far afield. Signs of a bounteous French harvest pending also augured well.

However, Urban looked towards a well-founded expedition assembled in good order at Constantinople. Instead the first big departure (from Cologne that April) comprised tens of thousands of *tafurs* (i.e. vagabonds or *sans culottes*) galvanized by the charismatic fanaticism of Peter the

Hermit or, to quote Anna Comnena (the tough-minded daughter of Emperor Alexius), "Peter the Cuckoo". Their first warlike activity was a pogrom in several Rhenish towns. Then most members of this "People's Crusade" disappeared, one way or another, before reaching Asia Minor. There and in the Levant, Peter regularly proved adept at detaching himself from any serious fray.

The 30,000 cavalry and infanteers of the First Crusade proper showed amazing resolve, forcing their way across Anatolia and through Antioch to capture Jerusalem in July 1099. But the next day or so, alas, all the city's Jews and virtually all its Moslems were massacred. More generally, Sir Steven Runciman, a doyen of medieval Near Eastern history, saw this First Crusade as gratuitously disrupting benign cross-faith tendencies: "In the middle of the eleventh century, the tranquility of the east Mediterranean seemed assured for many years to come. Its two great powers, Fatimid Egypt and Byzantium, were on good terms with each other. Neither was aggressive and both wished to keep in check the Moslem states further to the East."[16]

Subsequent Crusades were even less edifying. In particular, the Fourth Crusade allowed itself to be diverted (despite the vehement opposition of the Papacy) in furtherance of Venetian grand designs, the upshot being the dreadful sacking of Constantinople (with its priceless cultural heritage) in 1204.

These "divine campaigns" to free the Holy Land effectively came to their inglorious end with the Moslem recapture of Acre in 1291. Apologists have long claimed Christendom thereby learned much from the strength and virtuosity of Islamic culture. Also, they strengthened the effective authority of emergent national monarchies as against the Papacy. All the same, there could have been better ways.

A Germinal Century

Ironically, the First Crusade can be counted among the first fruits of what the distinguished Cambridge medievalist G. G. Coulton endorsed as "a very real revival, comparable to the later revival we call the Renaissance".[17] A resurgence of deforestation and of population were among its more primal manifestations. Romanesque churches and castles were among its most expressive. In many spheres, though, progress was evident. While climate improvement contributed, the inherent forces of social change were the primary drivers.

A broad advance visibly continued late into the thirteenth century. One qualitative aspect was a keener awareness of Nature artistically. Take floral motifs in ecclesiastical architecture. Between 1140 and 1230 these progressed from abstract formalism to many species being recognizable.[18] Then again, two sunspots (albeit stylized) featured in a solar image inscribed in a chronicle by John of Worcester in 1128.

The often illiterate master masons who determined the graceful forms of churches and cathedrals lacked mathematical refinements. So a few of the strivings to Heaven ended in tragedy. But thousands survive well today.

Advances in Awareness

In 1216 and 1223 respectively, two monastic orders were founded, the Dominicans and Franciscans, which broadly forsook the principle of "enclosure" (i.e. monastic living). These "friars" were soon contributing signally to new thought and knowledge.

St Thomas Aquinas (*c.* 1225–1274) eventually became, and has since remained, the most influential Dominican thinker. Encouraged by Averröes' commentaries, he endorsed Aristotle rather than Plato. While allowing that these two lead thinkers were sometimes addressed syncretically, one can say Plato had figured more prominently in Western consciousness since St Augustine. Now a reappraisal had begun, stimulated by the reintroduction via Islam of the Classic texts.

The reaffirmation of Aristotle by Aquinas and others could be too uncritical. But to an extent, it facilitated the emergence of scientific method. For it encouraged study of the material world. It also favoured a clear distinction being drawn between revelation by faith and determination by reason. It led Aquinas himself to enunciate, in *Summa Theologica*, five logical demonstrations of the existence of God. Three of these recapitulate or closely correspond to Aristotle's delineation of the "unmoved mover". And whatever else this may or may not have done, it underlined the principle of cosmic coherence which is fundamental to scientific enquiry.

The Franciscans drew their inspiration from the poetic mystic extolled by atheist Bertrand Russell as one of the most likeable men known to History. Nowadays, casting St Francis of Assisi (*c.* 1182–1226) in that mould is routinely part of any endeavour to demonstrate that true saintliness can be defined and encountered. But can it, absolutely? A needle question apropos St Francis is one raised by Julian Green in his empathetic but probing biography. Green asked why there was no record of Francis having opposed that singularly grotesque phenomenon, the Children's Crusade. In 1212, some 50,000 youngsters (from eight upwards) marched joyously down the Rhone to Marseilles or over the Alps to Genoa, destined to board ships supposedly bound for the Holy Land. A few turned streetwise in time to escape. But the great majority died on passage or were sold into slavery. Green reminds us that literary evidence about this ugly episode could have been destroyed, out of shame.[19] But it is still terribly difficult to avoid the conclusion that Francis protested little or not at all, being concerned to launch securely his order of friars.

On a more positive note, Francis evinced a cosmic spirit well ahead of his time. One could say no less of somebody who, as death ineluctably closed in, celebrated "blessed Brother Sun"; "Sister Moon and the stars"; and, indeed, "our Sister, bodily death". Such celebration will have helped nurture among his followers a reverential curiosity about Creation.

An infant Oxford University nurtured three Franciscan school men who did much to pave the way for modern Science. Robert Grosseteste (*c.* 1175–1253) was an Aristotelian quasi-polymath who prepared the ground for "scholasticism" as interpreted by the likes of Aquinas. His interests included astronomy and optics. Roger Bacon (*c.* 1214–94) extolled comprehensive enquiry guided by mathematics, observation and experiment.

However, it is William of Occam (*c.* 1285–1349) who has most visibly influenced attitudes since. He has, above all, through his Aristotelian insistence that "it is vain to do with more what can be done with fewer". He was dogged by factious strife which started with his being hounded by an ex-Chancellor of Oxford. It led on to his being summoned to Avignon to face ramifying charges of theological infidelity. It culminated with his aligning with the Holy Roman Emperor against the Pope.

William applauded Aristotle for his empirical flair: "With the eyes of a lynx, as it were, he explored the deep secrets of Nature and revealed to posterity the hidden truths of natural philosophy."[20] He himself was especially committed to the application of refined logic in the philosophy–theology borderland. Take his demonstration that there can only be one God. If one concedes there might be two, no grounds remain for excluding three, four . . . up to infinity. Obviously, however, there cannot be an infinite number, not for our cosmos.

Not for the last time, metaphysical theorizing was running ahead of data collection. In 1090, a French monk, Walter of Malverne, had calmly described a lunar eclipse. But records of the next conspicuous supernova, in 1181, survive only from East Asia. The earliest known weather diary in Europe was compiled in Oxford and Lincolnshire (1337–43) by Walter Merle, Fellow of Merton. However, by the fifteenth century, the inflow of Greek texts was encouraging proactive sky science. In Vienna, Georg Purbach (1423–61) sought to refine Ptolemy's epicycles.

The Venerable Bede of Jarrow (673–735) had averred (having read Pliny) that the Earth was spherical. But doubt persisted in Catholic Europe as to whether this could be, given the implications. Now Grosseteste aligned with those who reaffirmed the proposition. For one thing, it aided understanding of climate contrasts. The earliest example known in the West of a world map climate-zoned is a work of 1110 by Petrus Alfonsus, a Spanish Jewish convert drawing on Arab sources.[21] However, Dante (1265–1321), who was exercised by cosmo-theology, eventually concluded that the cosmos was an inverted hemisphere with God at its summit and a flattish Earth directly and immutably below.[22]

In 1277, the academics dominant at the University of Paris had formally challenged Aristotle's cosmogony, in particular his view that symmetry considerations precluded life existing beyond the Earth. Seminally, too, Nicole Oresme (1320–82) headed there a group which criticized Aristotle's theory of motion, a theory which showed no inkling of inertia. Meanwhile, independent colleges of higher education were founded across Europe, above all in wartorn France. In an incisive study, 25 new colleges are listed for the thirteenth century of which 20 were in France; 87 are listed for the fourteenth, 54 of them in France.[23] What is more, the striking claim is made for the fourteenth century that "no power – king, pope, bishop, ecclesiastical or lay authority – ever attempted to press its own candidates for college fellowships".[24] Clearly, a strong springboard was being laid for the accelerated cultural development (from *c.* 1450) that we still recognize as being *the* High Renaissance.

The High Renaissance

A visit by Halley's comet in 1456 induced Pope Calixtus III to order noonday prayers that "God save us from the Devil, the Turk and the Comet". But by then moves were afoot to treat such a body as a scientific phenomenon, notably by computing its distance through parallax measurement. By then, too, Italy was assuming a lead in artistic virtuosity nourished by classical revivalism. In 1452, Leon Alberti (1404–1472) had started work on his treatise on classical architecture.

Unfortunately, History proffers many materials showing how prone are times of cultural flux to intolerance and strife. A welter of violence emanated from the Renaissance and Reformation between states and within them. Perhaps the ugliest manifestation was witch-burning. It besmirched Christendom for a good two centuries from the death in 1431 of Joan of Arc, both Catholics and Protestants being heavily culpable. As was stressed by the late Lord Dacre (alias Hugh Trevor-Roper), this craze cannot be seen (*pace* nineteenth-century liberal historians) as mere "delusion detached from the social and intellectual structure of the time".[25]

Astrology, too, featured. A revival thereof had been under way since *c.* 1150, stimulated by relevant Arabic texts appearing in translation. By the fourteenth century, the subject was entrenched in official circles and universities across Europe, despite the endeavours of Oresme and others to curb its pretensions. In 1348, the Paris medical faculty averred the Black Death had been triggered by Mars interacting with Jupiter after their joint conjunction with Saturn in March 1345.

Come the Renaissance, the astrologers positively flourished, sustained by the anti-clerical side of the humanist ethic as well as by generalized uncertainty. A revival of Plato may have assisted, too. After all, in *Timaeus* he explores the rather Persian theme that the "cosmos" (originally a Platonist term) was so interlinked as to embody a single living spirit. But, as always, there were ups-and-downs. One William Parron was *de facto* astrologer to Henry VII of England as the sixteenth century dawned. He vanished after the Queen died in 1503, five months after he had expressed confidence she would live several decades more. In the same tract, he predicted the young Prince Hal (Henry VIII to be) would enjoy marital bliss and sire many sons. He would remain an outstanding servant of the Church of Rome.[26]

The deeper contradictions these aberrations betokened have been discussed by modern writers. In 1992, the late Ted Hughes – then Britain's Poet Laureate – interpreted the narrative poems and tragedies Shakespeare wrote after 1590 as reflecting anxiety lest a macho-reductionism had detached humankind from its emotional roots.[27] Earlier, a similar theme had been identified by Eric Fromm (b. 1900), the celebrated Frankfurt School psychoanalyst who took refuge from Nazism in the United States. Warning how fascism may appeal to both the submissive and the aggressive sides of our human psyche, he saw its being nurtured by the isolation, insecurity and sense of futility all too endemic in modern humankind. This syndrome he traced back to the archetypal Renaissance virtuoso with his ill-judged sanguineness, unsettling learning and impulsive enterprise. His genre formed a new *haute bourgeoisie* of domineering individualists. Distracted by them, the masses may have felt less calm inwardly than ever before. Yet through their self-engendered isolation, the virtuosos themselves may have languished.[28] Similarly, Simone Weil (1909–43), the French cultural patriot and Christian Jewish quasi-mystic, attributed modern "uprootedness" largely to how the "Renaissance everywhere brought about a break between people of culture and the mass of the population".[29] In other words, it had induced a crisis of fulfilment.

Globalization

Still, quite the most dramatic manifestation of the Renaissance *leit motif* was the worldwide projection of European influence. Iberian initiatives had salience. From 1415, "Henry the Navigator" (1394–1460) masterminded a Portuguese thrust down the African coast and related islands. In 1492, Columbus was dispatched by Queen Isabella on his first epic voyage, this to celebrate completion of the *reconquista* of Moorish Spain. Both men were somewhat bestirred by Renaissance learning. Both were committed to extending Christian dominion while forging good sea routes to the spices of the East.

A Papal Bull of 1493 formally divided the extra-European world down a longitude 100 leagues west of the Cape Verde islands. Everywhere to eastward would be Portuguese, and to westward, Spanish. The Pope thus recognized, yet at the same time belittled, the great events in train. Once Magellan's surviving ship had completed its circumnavigation of the world in 1522, Europe could claim global reach, a decisive break-out which stands in surreal contrast with the stance Imperial China had adopted a century before. In 1405, Admiral Zheng had led out of the Yangtse a fleet of 300 ships with a complement of 28,000. He thus began a saga whereby a succession of expeditions bespoke a modulated Chinese dominance within the Indian Ocean and beyond.

As to how far beyond, a challenging hypothesis has lately been argued by Gavin Menzies, an ex-Royal Navy submariner with a strong Oriental background. He says that, between 1421 and 1423, four Chinese task forces traversed between them the High Seas, including the Arctic and the fringes of Antarctica. Among the manifold skills applied was the use of lunar eclipses in longitude determination.[30] Yet on their return from wherever, these bold navigators found the imperial authorities turned off the whole idea. In fact, a voyage to Arabia in 1435 ended such forays. How categoric this rejection became is shown by "the destruction of many of the records of the fleet and its ships' designs".[31] The reasons still await full elucidation. But resentment at Zheng's virtuosity was probably compounded by his being a Moslem eunuch. Also, a serious fire in the Forbidden City in 1421 (apparently due to ball lightning) will likely have been read as divine admonition. Unrest induced by famine was again a problem on the northern border. Besides, the 1421–3 four-pronged foray had incurred heavy losses.

Yet behind all this, one can also discern mandarin anxiety about cultural pollution. Perspectives could be changed too fast within China and, indeed, elsewhere. Perhaps the Chinese were wise in their generation and the West

Europeans foolish. Thanks to the latter, a wind of change ran suddenly across the world and then continued strong. Today we reap the whirlwind. Therefore we cannot allow ourselves to assess the High Renaissance (Alberti to Shakespeare) just in terms of its matchless individualism in the arts. There was a downside holistically.

Strictures on Astrology

Contemporary unease was, of course, played to by the astrologers, flaunting planetary conjunctions as their most esoteric theme. One famous practitioner, Johannes Lichtenberger, told how the influence of a conjunction in 1484 of the "weighty planets", Jupiter and Saturn, could extend over 20 years. Acutely apprehended, too, was a 1524 conjunction event. Hundreds of tracts warned of inundations around the Earth. Not a few arks were laid down, if nothing more. In the event, there was the outbreak of the Peasants' War in Germany. That was bad enough (see below).

The ill-fated Dominican rebel friar Girolamo Savonarola of Florence (1452–98) had been fiercely opposed to astrology as epitomizing the corruptive pagan elitism of the times. A consuming hatred of both Plato and Aristotle informed, in his case, a more general rejectionism. In due course, too, the leadership of the Protestant Reformation – Martin Luther, John Calvin and Ulrich Zwingli – would likewise oppose the astrologers, a notable exception being the suave Philip Melanchthon who remained firmly supportive. As always, however, the radical reformers were considerably creatures of their time. Prefacing a book by Lichtenberger, Luther wrote: "The signs in Heaven and on Earth are surely not lacking they are God's and the angels' work to warn and threaten the godless lords and countries and have significance."

Through medieval times, Catholic opinion had tended to accept that the planets and stars affected human physical and mental health and maybe events more generally. But there was chronically concern lest astrology (a) came between the individual and Christian witness, (b) fostered a pagan revival, (c) became too determinist or (d) became too exploitative. As the sixteenth century advanced, these reservations firmed up. The crunch came with the Papal Bull, *Coeli et Terrae* (Sky and Land), issued by the combative reformer Sixtus V (in post, 1585–90). It condemned as invalid and, indeed, devilish horoscopes and all procedures for foretelling human fortunes. Some years later Savanorola suffered a decisive collapse of his Florentine support after one of his disciples had volunteered for ordeal by fire to make manifest his master's holiness. Mercifully the event was rained off.

The Reformation[32]

Neither the Protestant Reformation nor Rome's Counter-Reformation were cosmology-driven in any direct sense. But there was interaction between religious radicalism and sky science as well as between each and the information explosion, a core feature of this High Renaissance era.

The northern Renaissance led in the exploitation of the printing press. After Luther, an Augustinian friar still, had set the Reformation in motion at Wittenberg in 1517, he secured this primacy by tireless recourse to the printed vernacular. Between 1518 and 1525, he published more works in German than the next 17 anti-Rome publicists did between them. In his lifetime, he published five times as many as did all his Catholic protagonists combined.[33] No doubt northern winters encouraged reading and reflection. Maybe, too, sombre skies and landscapes fostered puritan values.

Everywhere, however, the Reformation made inroads into urban locales, especially those oriented more towards long-distance commerce. In general, after all, the "Protestant ethic was more welcoming to the amassing of capital and lending it with interest".[34] Meantime, the Peasants' War had raged across upland South Germany (1524–6). The lead revolutionists were Ulrich Zwingli from Switzerland and Thomas Münzer of Saxony, the latter being close to the Anabaptist millenarians. The whole movement was remorselessly crushed, Münzer being among the many thousands killed. Martin Luther had solidly backed this Imperial counter-offensive.

Luther died in 1546. Ten years before, Calvin had completed the definition of the fatalistic Puritanism he was launching from Geneva. By then, too, the Lutherans had adopted their separatist creed. On the Counter-Reformation side, the Jesuits were founded in 1540–1; and the Roman Inquisition revived in 1542. Acute polarization was thereby reciprocally effected, regardless of what most people at all levels might have preferred.

Nicolas Copernicus (1473–1543)

The individual whose name History has customarily associated with scientific advance in the sixteenth century often gets these days a scornful reception as putatively lacking intellectual refinement and moral strength. As usual, the pendulum may have swung too far. Born by the Lower Vistula, the progeny of German colonists whose forebears had been invited to settle by the Polish king the previous century, Copernicus studied astronomy and astrology at the esteemed University of Krakow. He later spent a total of nine years in Italy, studying medicine and then law combined with the Classics – Plato, Aristotle, Cicero, Ptolemy . . .

His final three-year stint in Italy was at the University of Padua, very much an epicenter of scientific radicalism. The background is that the Latin Averröists so active in Paris in the thirteenth century had considerably translated to Padua in the fourteenth. Their secularist sympathies led them into natural philosophy, critiques of Aristotle included. Come the sixteenth century, they led Europe in this domain.[35] Indeed, at the Fifth Lateran Council (1512–17), the Padua Averröeists were roundly condemned for contending that reason rather than faith led one to philosophy.

Though at Padua Copernicus majored in Canon Law, he kept alive an interest in astronomy. Considering his background overall, one could have expected him to move in either of two directions: be an astrologer-cum-physician or else an aristocratic patron of astronomy. Thanks to avuncular patronage, however, he was from 1512 Canon of Frauenburg on the Baltic shore. Residing in a watch-tower, he overlooked the Frische Half, a fresh-water lagoon. Too timid, it is said, to step much outside these confines, he nevertheless deplored their misty isolation and, in particular, how they inhibited observation of the night skies. Yet he generally managed enough to get by. But his admitted failure to measure Mercury definitively did hamper the proper development of his heliocentric thesis.

Before 1515, he had distributed circumspectly a first rendering of his hypothesis. Its Introduction included a fervent endorsement of the subject area: "For this queen of the sciences which is most worthy of the attention of free men is based on almost all the branches of mathematics . . . And while it is a characteristic of all the sciences that they turn the spirit of man away from vices and direct it to better things, astronomy can do this in a particularly high degree . . . "[36] Then the main text dilates on his essential tenets. The Earth rotates on its axis diurnally. The middle of the Earth is not the center of the world but only of gravity as we experience it and of the lunar orbit. Every planet and, indeed, every star revolves round the Sun.

The argument was fully developed in *De Revolutionibus Orbium Coelestium.* For basic materials, Copernicus drew heavily on Ptolemy. In significant measure, however, this data base was refined, extended and reorganized. He also challenged Ptolemy's interpretation of fluctuations in planetary angular velocities and in lunar distance. Yet, in the final analysis, he failed to enunciate a compelling alternative to geocentricity. This was essentially because a trite application of the symmetry principle led him to expect celestial orbits always to be circular not elliptical.

Copernicus completed this text in 1532–3. Word of it steadily got around. Hence Martin Luther's after-supper jibe in 1539 at "the new astrologer who wants to prove the Earth goes round". Meanwhile, the author hesitated to go into print for fear, he said, of ridicule. He was eventually persuaded to by Rheticus, a fiery young heliocentrist with Lutheran affiliations. It was produced in Wittenberg in 1543, Rheticus having left the overseeing of everything to a Lutheran colleague. Reportedly, Copernicus at last held a printed copy in his dying moments. It bore no acknowledgement of Rheticus, a slight that young enthusiast found insufferable.

4

A RENAISSANCE CONTINUUM

Eventually, Copernicus recast our cosmic perspectives profoundly. C. S. Lewis (1898–1963), Oxbridge man of letters and conservative theologian, noted how the late medieval *South English Legendary* indicated that the *stellatum* (i.e. high point of the firmament) could be 150 million miles overhead – i.e. about 13 light minutes not billions of light years. However, he refused to be impressed by this contrast, arguing that even ten million miles was a distance humankind could conceive but not imagine. More significant, he suggested, was the fact that the medieval universe was "unambiguously finite" which made "the smallness of the Earth more vividly felt". In a less bounded cosmos, all else seemed tiny too.[1] He took no account of there likely being many other universes.

A simpler comparison he made can be readily appreciated: "to look out at the night sky with modern eyes is like looking out over a sea that fades away into mist . . . To look up at the towering medieval universe is much more like looking up at a great building . . . an object in which the mind can rest, overwhelming in its greatness but satisfying in its harmony".[2]

Others had likewise reacted. Take John Donne (1572–1631), the English "metaphysical" poet who was reared a Catholic but who became, from 1621, Dean of St Paul's Cathedral. His *Anatomy of the World*, published 1611, warns how Copernican "new Philosophy calls all in doubt". His overriding fear is that a cosmology with "all coherence gone" could induce the breakdown of a social order in which everyone knew their place:

> "Prince, Subject, Father, Son are things forgot,
> For every man alone thinks he hath got
> To be a Phoenix, and then can be
> None of that kind, of which he is, but he".

Donne is considered to be preoccupied with death. A related theme is the possibility of the souls effecting a transcendental union with God. His poetry is incisive and insightful; and underwent a revival in the twentieth century.

A Slow Lift Off

Thomas Kuhn saw the Copernican revolution (which he had previously studied) as exemplifying his famous thesis that big scientific advances occur through delayed then sudden shifts in "paradigm".[3] Initially criticized for not making clear what this meant, Kuhn responded by stressing how a professional community may share not just formal rules but "examples of successful practice". As and when these are found wanting, a new corpus of experience may displace them quite abruptly.[4] But in his specialist study he had concluded that "the final victory of *De Revolutionibus* was achieved by infiltration".[5]

"Infiltration" may be near the truth. For one thing, the book was no best-seller. The thousand copies in its first edition never sold out. Moreover, the text had but four more printings the next four centuries. Philip Melanchthon's *Doctrines of Physics*, published in 1553 to oppose Copernicus, was printed ten times before *De Revolutionibus* was twice.[6] But the latter was, through the second half of the sixteenth century, a basic reference for astronomical research. The irony is that many who used it thus either ignored Earthly motion or else treated it as a convenient fiction. In this they were not discouraged by the low-key neo-Classical idiom Copernicus himself employed. As late as 1594, the English astronomer Thomas Blundeville could write that "Copernicus . . . affirmeth that the Earth turneth about and that the Sun standeth still . . . by help of which false supposition he hath made truer demonstrations of the motions and revolutions of the terrestrial spheres than ever were made before." Through 1600 it was Averröist critiques of Ptolemy that were impacting most among educated laymen.[7]

A respect in which Copernicus was unabashedly unconventional was the ordering of academic disciplines. Regarding Astronomy as the "Queen of the Sciences", he gave it analytic priority over Natural Philosophy – in effect, over Physics. The orthodox were unamused. His stress on Geometry likewise seemed subversive.[8]

One would think the astrologers had everything to lose from the new perspectives. Yet even they took Copernicus pretty much in their stride. Indeed, several were among the early converts. Johannes Lichtenberger saw him as "proceeding with the sure and resolute step of a genius, always going in a straight line toward the truth".[9] Persuaded, too, was John Dee (1527–1608), prominent at the Elizabethan court and judged by Peter Whitfield to be "without doubt . . . a serious scientist", astronomically and even at first in regard to the black magic which compromised him eventually. Meanwhile, Francis Bacon's magisterial *The Advancement of Learning* (1605) included a varied agenda for reforming (but not abolishing) astrology.[10] The book as a whole was central to his concern to make natural science the driving force of national development. Take also William Gilbert (1544–1603), the pioneer researcher of "electricity" (a word he coined) and magnetism. He scorned talk of metals being ruled by the planets but never doubted the stars influenced children at birth.[11]

The Religious Factor

Most striking is the alacrity with which the bible-conscious leaderships of the main Protestant sects inveighed against Copernicus, a pitch not all their young activists (starting with George Rheticus) were happy about. In the 1539 jibe already alluded to, Luther recalled how "Joshua commanded the Sun to stand still and not the Earth". A decade later, Philip Melanchthon, his suavely diplomatic lieutenant (his Chou En-lai, if you like), argued that "The eyes are witnesses that the heavens revolve in the space of 24 hours. But certain men, either from the love of novelty or to make a display of ingenuity, have concluded that the Earth moves . . . the example is pernicious". Later on he looked more for compromise.

In the Catholic realm, doubts were expressed less overtly for some decades. That great reforming institution, the Council of Trent (1545–63), never discussed the matter. A critique by Jean Bodin (1530–1596) was published posthumously (1597). A radical Toulouse philosopher, Bodin had formulated *inter alia* the Quantity Theory of monetary inflation. About the new cosmology, his text says no right-thinking person could imagine that the unwieldy Earth "staggers up and down around its own center and that of the Sun; for at the slightest jar of the Earth, we would see . . . towns and mountains thrown down". Besides, things "finding places suitable to their natures remain there, as Aristotle writes".

Otherwise, Bodin looked towards nation states strong enough within and without to allow of religious toleration. Yet during the sixteenth century, the incipient emergence of nationhoods had actually led to the battle lines of prejudice being drawn more sharply. For one thing, the Holy Roman Empire and the Vatican were drawn into close if informal alliance as the other great Hapsburg power, Spain, emerged as the anvil of the Counter-Reformation. At popular level, tension found outlets in quite exceptional intra-European violence in the name of Christ. Witness a resurgence of witch-hunting post-1550; the Wars of Religion in France (1562–98); and politico-religious strife in the Netherlands from *c.* 1567.

Meanwhile, Ottoman Turkey was waxing expansionist. Granted, Christian forces checked its proclivities awhile with a ferocious defence of Malta, 1565, and the sea battle of Lepanto, 1571. However, Spanish naval power (decisive at Lepanto) was terribly discredited by the failures against England from 1587 onwards, papal blessing notwithstanding. France and the Ottomans were on goodish anti-Hapsburg terms.

Nothingness Accepted

One positive aspect of the High Renaissance is how the kind of people who went to playhouses were coming to terms with "zero" and with "nothingness" in general. Shakespeare's plays mention the theme frequently, sometimes as unsubtly symbolic of female sexuality but often more generally.[12]

The Valley of the Shadow

The spirit of the age hardly favoured toleration in whichever State or Church. Soon Rome was feeling constrained to get tough with Copernicanism. Giordano Bruno (1548–1600), a militant Dominican from Naples was burnt at the stake for heresies including the bowdlerized Copernicanism he had defended before a hostile Oxford audience in 1583. Bruno had further divined that the stars must be suns surrounded by planets which could be life-bearing. To cap everything, he saw an Anglo-French geopolitical axis as a natural corollary of his "true philosophy".[13]

In 1613, Galileo Galilei (1564–1642) committed himself to Copernicus more emphatically than ever before or, indeed, after. So it was in 1616 that he was first required to treat Copernicanism as mere hypothesis, not to be affirmed or defended. This same year the new cosmology was endorsed by Thomas Campanella, a Dominican apostate imprisoned and tortured for leading a revolt against the Hapsburgs in South Italy in 1599. Then again, Galileo's trial by the Inquisition for circumventing the 1616 proscription was set in motion in 1633 by Pope Urban VIII, a Renaissance virtuoso who had once penned him complimentary sonnets. Urban's switch followed

Galileo's publication of *Dialogue concerning the Two Chief World Systems - Ptolemaic and Copernican*, a text which *inter alia* confirmed that his own tidal theory required the Earth's rotation. But it also followed two near-disastrous years for the imperial forces leading the Catholic side in the Thirty Years' War in Germany, 1618–48.

Galileo saw himself as a Catholic Platonist drawn to astronomy by its mathematical geometry.[14] However, his signal contribution derived from his skill at lens grinding. His first telescope was made in 1609, just several years after such instruments first appeared. It was a superior design. Soon his observations (at magnification 8 to *c.* 20) called into question or debunked outright pre-Copernican astronomy. The Moon had a rugged surface, not the smoothly crystalline one customarily expected on heavenly bodies. Recording four of the moons of Jupiter in 1610 afforded added proof that heavenly bodies did not *ipso facto* revolve round the Earth. Additionally, the eclipses of these moons were easier to systematize if Jupiter was understood to orbit the Sun.

Two other astronomers figured prominently in the Copernican claim to recognition. One was the eccentrically uncouth Danish nobleman, Tycho Brahe (1546–1601). Himself very loyal to Aristotle and, indeed, astrology, he generated excellent data for the Copernican revisionist, Johannes Kepler (1571–1630) – his successor as court mathematician to the Holy Roman Emperor. Tycho's input was much enhanced by his running (1576–9) a superb observatory on an islet near Copenhagen, an installation for which he received royal funding. He, at least, was undeterred by northern mists.

Fortified by the bequest of all Tycho's records, Kepler enunciated three laws which seemed at the time definitive enough to clinch things. Two were promulgated in 1609. The first said every planet revolved in an elliptical orbit with the Sun at one of its two foci. The second said that the area swept through in relation to the Sun is the same, in a given span of time, in whichever part of a planetary orbit. The third law was promulgated in 1619. It said that every planet shows the same ratio between two quantities. The one is the square of the time taken to complete one planetary revolution. The other is the cube of the semi-major axis of the elliptical orbit.

Through the Nadir

In the meantime, the ambient scene darkened as Protestant–Catholic intolerance polarized further, driven by reciprocal hatreds though also perhaps by a sense that true religion was being compromised all round by Copernicus and other modish tendencies. A signal departure apropos

Galileo and others came during the 1616 showdown with Rome's placing on its Index of proscribed literature of all works promulgating Copernicanism.

The Galileo episode certainly disturbed Europe's intelligentsia. For "ordinary people" in and around Germany, however, the Thirty Years' War waged across that fragmented country was much more traumatic. Big Protestant successes in 1631 and 1632 were largely due to the military involvement by Gustavus Adolphus, the King of Lutheran Sweden. After his being mortally wounded late in 1632, the Imperial forces could rally. Therefore in 1635, Catholic France intervened openly (with Swedish collusion) to preclude Hapsburg hegemony. All the Holy Roman Emperors between 1438 and 1740 were Austrian Hapsburgs.

What with this and all the other strife (including the English Civil War, 1642–8 and the 1649 regicide), it is not surprising to find apocalyptic pessimism prevalent come mid-century. In England, not just the Fifth Monarchy millenarians talked of having been "born in this setting of time". In London, panic by rich and poor about the "Black Monday" solar eclipse of 29 March 1652 was extreme in relation to currently received wisdom on such events. Meantime, some feared a final crisis might be triggered by a cometary impact or planetary collision. How closely such concerns related to revolutionary radicalism awaits elucidation.[15]

Gresham and Oxford

At all events, these years marked a turning point. Soon a new mood was burgeoning across Europe – one of reason and reasonableness. The Thirty Years' War in Germany had ended pretty much on the basis of *cuius regio; eius religio*, "to each prince his own religion" – the principle first adopted for the German lands in the 1555 Peace of Augsburg. Also one could surmise that, across England, Cromwellian autocracy acted as an aversion therapy to extreme Puritanism.

Meantime scientific leadership within Europe was shifting from Italy to England. Through the first half of the seventeenth century, London's big teaching centre for Science and, above all, Mathematics was Gresham College, founded in 1597. It could address the need for a better understanding of military ballistics, military architecture and, above all, navigation. In her remarkable account of how at long last (*c.* 1765) the chronometric determination of longitude was satisfactorily resolved by John Harrison, Dava Sobel quotes the anonymous "Ballad of Gresham College" (*c.* 1660):

"The College will the whole world measure;
Which most impossible conclude,
And Navigation make a pleasure,
By finding out the Longitude."[16]

Besides which, Copernicianism now interacted positively with mainstream Calvinism in that both schools of thought were formally anti-hierarchical. A further consideration was the overriding by Copernicus of the old distinction between the heavens where movements were supposedly circular and the terrestrial ambience where rectilinear motion was considered the norm. Already this advance had been conducive to William Harvey, physician to Charles I, consolidating the theory of blood circulation in the human body.[17]

Oxford had lately been considerably Aristotelian and High Church as well as proactively Royalist. Nevertheless, a certain inclination within him towards a managed liberalism led Oliver Cromwell to encourage the university's modernization via links forged with London. He introduced men from Gresham College on transfer or joint appointment. Most particularly, the able mathematician and puritan divine, John Wilkins, was made Warden of Wadham College. *Inter alia* he pushed Copernican astronomy hard, notably by persuading Seth Ward to accept the Savilian chair in 1649.[18] Seven years later, Wilkins prevailed on Cromwell to help secure Christopher Wren's election to a Professorship at Gresham, it to run alongside his Fellowship at All Souls'. This was despite Wren's father having been Dean of Windsor till 1646. With Wilkins himself becoming Lord Protector Cromwell's brother-in-law in 1656, Wadham was the epicentre of this Oxford renascence.

Within two years of his return from exile in 1660, Charles II had incorporated an "invisible college" of eminent scientists as the London-based Royal Society. Several of its early doyens, notably Isaac Newton and Edmund Halley, were brilliant mathematical astronomers. More generally, this scientific elite was disposed to show how God presided over a law-abiding universe, a model for society on Earth. This, plus something of an anti-Puritan backlash in post-1660 Oxford, evoked hostility from the more millenarian radicalism emanating from the Civil War era: "Copernican astronomy had ended the distinction between heavenly and sublunary: the radicals aimed at completing this by ending the distinction between specialists and laymen. They wanted to drive scholastic theologians out of the universities to end the dominance of Latin, Greek and Hebrew; but they did not want Science to be handed over to a new set of mumbo-jumbo men."[19]

However, the Royal Society was also concerned to promote various practical arts. Isaac Newton was a fine optical engineer while the workaday achievements of Christopher Wren defy summation. Well publicized Society experiments in 1664 failed to establish neat correlations between chord lengths and music intervals. Had the results been positive, they would have been deemed to confirm modish neo-classical notions about the harmony between music, mathematics and, indeed, astronomy – "the music of the spheres".[20] But, overall, the Society was still very much seen as a model for other states to follow the next several decades. In 1666, the Paris Academy of Sciences held its first meeting. In 1700, its Berlin counterpart did under Leibniz (see below). In 1725, Peter the Great established the Moscow Academy of Sciences. A Florentine academy had been founded as early as 1657 but had lasted a mere decade. More generally, this was also an era of quality drawing (often from microscope viewing[21]) and of statistical compilation all round.[22] From 1654 an experimental network of synoptic weather stations was established across the Duchy of Milan. Seven stations were within the duchy and a further four abroad. The network survived until 1670.[23]

The Newtonian Revolutions

In the new ambience, one could expect pure science to flourish as well as applied. In England alone were doyens who might have excelled in any time or place: Robert Hooke and Robert Boyle at Oxford, Isaac Newton at Cambridge, John Flamsteed at Greenwich, Edmund Halley . . . Of these, Newton (1642–1727) most fully encompasses the aspirations of the age.

Varied perspectives on Isaac Newton emerge from the numerous studies. He was the most solitary of figures, persuaded that "Truth is the product of silence" though given to bitter quarrels with colleagues over rival claims to primacy. His philosophic attitudes were inchoate. Shocked by Restoration frivolity, he regarded both Anglicanism and Rome as corrupted. He was fascinated by Stonehenge but still more by what he believed himself to know of Solomon's temple. He saw the latter as symbolizing the cosmos but also as the re-entry point for the Second Coming – maybe in 2060. The Trinity blasphemed, he felt, against the First Commandment. He likened the Earth to "a great animal or rather vegetable". From 1668 to 1693, he covertly indulged in a passion for alchemy. He was also given to numerology, not to mention acute sexual fantasy. Small wonder John Maynard Keynes, having perused his papers at Cambridge, dubbed him "the Last Magician".

Through his early twenties, Isaac Newton showed no exceptional promise.[24] None the less, the two years he spent during the Plague (1664–6) secluded in his native Woolsthorpe laid the foundations for three major breakthroughs, all important to sky science as well as more broadly. He and Gottfried Leibniz, the German philosopher and mathematician, evolved independently the calculus: the mathematics of minuscule changes in series. Leibniz went into print in 1684 and Newton in 1687. Also, Newton developed a corpuscular theory of light, discussing the while how a prism separated "white light" into its component colours. He published *Opticks* in 1704.

Evidently, however, his reputation rests heavily on how he recast and codified contemporary thinking about gravitation. The key proposition was and is that the pull one body exercises on another will be directly proportional to its own mass and inversely proportional to the square of the distance between them. Apparently this inverse square law was lighted on more or less concurrently by Newton, Hooke and Halley. Moreover, in a 1673 presentation, Christiaan Huygens from Holland envisaged a "centrifugal" force which was possessed by a revolving body which offset the gravitational pull from the orbital centre. Speaking instead of a "centripetal" force, always tending to pull an orbiting body tangentially out of its path, Newton likewise argued that orbital equilibrium would be maintained if this tendency were balanced by gravitational pull from the centre. In addition, he demonstrated that Kepler's planetary laws gave really accurate results only when it was allowed that the gravitational pull between each planet and the Sun works in both directions.

Indeed, one had unreservedly to accept that any body anywhere in the universe would ultimately exert a pull on every other one wherever. Newton confessed he was mystified as to how this exertion was achieved. But he wisely insisted that the prior question had to be whether refined observation and calculation confirmed the correlation. They did and have ever since, subject only to Einsteinian qualifications feeding in at extreme values. Newton's three-part *Philosophiae Naturalis Principia Mathematica*, published in 1686–7, was therefore an epic advance. Even so, this new paradigm of universal gravitation was not to gain general acceptance much before 1780, by which time it was considerably beholden to French *philosophes*, glad thereby to relegate God either to the passivity of a constitutional monarch or else to complete oblivion.

To appreciate how obtuse the scientific community could be on this score, one may refer back to Galileo. He had pondered to some effect acceleration due to terrestrial gravity. Yet merely extending the gravitational principle to the Moon he had seen as occult *fancifullezze*. A century or two later, many students of science were still locked in similarly negative mindsets. It is all curiously akin to the general rejection for half a century of the "continental drift" hypothesis first proposed in 1915 by Alfred Wegener

(1880–1930), an estimable German geophysicist and explorer. A rationale usually advanced was his alleged failure to identify the requisite motive power: "Contrary to all the available physical and geological evidence (*sic*), he blandly postulated that the basalt of the ocean floor is so weak that it cannot resist deformation even under the action of infinitesimal forces."[25] Today every geologist believes in continental drift.

One sign of the new times was the sudden and seemingly terminal collapse of astrology after 1650. Until then, it had been bolstered by a pervasive edginess; confusion as to what and was not "occult"; and links with other fringe pursuits – among them, physiognomy.[26] In 1650, most medics still felt astrology retained at least a placebo value *vis-à-vis* choice of days for letting blood or administering potions. However, the professional astronomers were turning less indulgent. Old objections were being accorded more weight. Robert Hooke, once a quasi-believer, now found it "vain". To Seth Ward, it was *inanis et illicita,* a "ridiculous cheat". Isaac Newton could not care less. Pascal remarked that, with the world how it is, any eclipse was bound to precede dire misfortune.

Using similar metaphors, two historians have well captured the character of this paradigm shift. Astrology died "like an animal or plant left stranded by evolution. It was not killed".[27] In the main, "the subject was left to die a natural death. The clergy and the satirists chased it to its grave. The scientists were unrepresented at its funeral".[28]

America turned away from astrology later than Europe but by a few decades at the most. Soon Reason applied to Nature was driving serious discourse both sides of the ocean. However, this did not mean the end of apprehension. Nobody was that sure what Reason and Nature respectively were.[29] Besides, geopolitical revaluation was called for in a world of new asymmetries. In London, the commercial competitiveness globally of the Calvinist Dutch engendered a desire to curb these "new Papists" militarily. Meantime, Paris was an emergent superpower.

Nor was the new-found precept of religious toleration easily applied internally. Things were still too chaotic for that. In England, the 1680s saw a short but sharp upsurge of chiliastic prophecy as well as (1678–85) contrived anti-Catholic frenzy. More maturely in vogue were the Mortalists, people whose Christianity embraced the notion that the soul perished with the body. John Milton and William Harvey were among those so persuaded. So, too, was Richard Overton, a prominent ultra-Left Leveller from Civil War times. One inference he drew from Mortalism was that, theologically, "man hath no pre-eminence above a beast". To him, too, heliocentricity connoted God's residing in the Sun. Some saw this as leading to atheism via paganism. Then from 1685, France was distracted by waves of covert emigration by Huguenots, its large Calvinistic minority. This followed the revocation that year *pour raisons d'état* of the 1598 Edict of Nantes which had accorded Protestants a fair measure of toleration when reasonably interpreted.

Thus were the bounds of the seventeenth century effectively set. Historians talk of its "general crisis". It was characterized by "large-scale agricultural calamities" and revolts "all over the world". China's last imperial dynasty was founded by Manchurian invaders who, exploiting a spate of indigenous insurrections, seized Beijing in 1744. A global secular trend towards lower temperatures (linked, on the whole, to worsening weather overall) was deepening into what we today call the "Little Ice Age".[30]

The Cartesian Age

But whether in spite of this ambience or because of it, the century witnessed a further intellectual/cultural Renaissance in Europe plus one artistically in India and Japan. Europe's advances in cosmology have been remarked. However, her intelligentsia similarly made big strides in metaphysics, the study of the underlying attributes of Creation. Among the luminaries who thus graced these times were Francis Bacon, Descartes, Hobbes, Leibniz, Locke, Pascal and Spinoza. Of these, Hobbes and no doubt Locke are best remembered as political thinkers. The others were philosophers in what could be called a pure sense.

All in all, "we need not scruple to say that, in the realm of knowledge and thought, modern history begins in the 17th century".[31] The source of this opinion, John Bagnell Bury (1861–1927) was a professor at Trinity College Dublin and then Cambridge, specializing in the Classical world and its more evident derivatives. To him this seventeenth-century Renaissance *inter alia* involved the first endeavours to conceptualize the notion of Progress – philosophic as well as material.

The person often seen as having spearheaded the forward movement was René Descartes (1596–1650). A reflection of this is the extent to which Cartesianism has since been prayed in aid by a diversity of interests, some in subject areas he never wrote about. Then again, although Descartes was a committed Catholic, his works were placed on the Index of proscribed books in 1663; and in 1685, Louis XIV banned in France anything considered Cartesian philosophy. Obversely, a few Calvinist theologians cited him.[32]

Well educated, especially in mathematics, in a Jesuit college, Descartes went on to the University of Poitiers. He early appreciated the salience of mathematics in cosmology. Moreover, he sought to apply mechanistic analyses throughout physics, biology and social studies. Next, he hoped military service might proffer the kind of privacy and quietude he sought for hard thinking. Initially he enlisted in the Duchy of Nassau's army.

However, the outbreak of the Thirty Years' War shattered any prospects of soldierly seclusion. He concluded an erratic military career in 1628 to retire to Holland. Then after twenty years, he agreed to join the Swedish court. Taxed there by cold and an onerous timetable, he soon died.

His matchless point of philosophic departure, *cogito ergo sum* ("I think therefore I am") was contrived to serve two ends. It indicated how free from preconceptions he claimed to be. It affirmed that the mental identity of himself (and, by extension, of all other humans should they truly exist) was unique, a mind-and-soul twinning. Nor did Descartes' mechanistic envisionings leave room for some nebulous life force. Nevertheless, he sought to define a place for God. In this he was guided, it seems, by personal faith but also by political circumspection, especially after the censuring of Galileo. His God would be presiding over a universe operating in accordance with unerring mechanical logic.

Yet although he never really develops this theme, he does appear to have been at odds with the seventeenth-century quasi-consensus concerning it. What received wisdom said was that when God decided to create the world, he did so out of nothing, *ex nihilo*. In the nature of things, however, there were constraints on how He could proceed. Contrariwise, Descartes submitted that God shapes the Laws of Nature much as a monarch shapes laws for his people. As a monarch can alter laws, so can God.

His thinking seems to have been that if, for example, God resolved there should be five spatial dimensions instead of the customary three, he could so ordain. In other words, God chooses the "eternal verities". The fact that, in our experience, He alters these little, if at all, shows they suit His purposes, obscure to our mere selves though these may be.[33] Putting this construction on things he felt took cognizance of Divine Will whilst accommodating a burgeoning popular taste for intellectual and artistic symmetry. Part of the accommodation was an explicit emphasis on the Uniformity of Nature's Laws.

Germane, too, is Descartes' rejection of "atomism", this as a rider to his axiom that reality lends itself endlessly to enumeration. For you cannot then talk about the smallest possible number. From which it follows, so he was persuaded, that there is no limit beyond which you cannot further divide any of the myriad of minuscule particles of which the cosmos was composed.[34] Another perceived corollary was that one could never have a pure vacuum. That deduction struck chords the more sharply after, in 1643, Evangelista Torricelli had invented the mercury barometer.

In the French context, Blaise Pascal (1623–62) comes across as countervalent to Descartes. No less brilliant a mathematician, he early came to feel that rationalism – Cartesian or otherwise – was not enough. As much was made clear in his *Pensées* on religion and related subjects posthumously published in 1670. The proverbial "The heart has its reasons quite unknown to the mind" is one instance. Another can be "Atheism is the sign of a powerful mind but up to a certain point only." A salient trait is apprehension of the new cosmic dimensions, "The eternal silence of those infinite spaces frightens me".

Pascal and his dedicated sister sought to resolve their mutual doubts by enlisting in a big movement the Dutch Catholic, Cornelius Jansen (1585–1638) had launched. Its stress on Augustinian Grace, Pre-destination, personal holiness and hostility to Jesuit "casuistry" meant Jansenism could draw on the anti-papal tradition within French Catholicism. Through the turn of the century, this Gallicism would go out of popular and royal favour. Until then, however, it checked Cartesian influence considerably. In his lifetime, Pascal was its leading proselytiser. He thereby gained eminence in religious and literary terms. But to many, Pascal as a philosopher "was not one man but several . . . He never achieved mental unity".[35]

One initiative that confirms to my mind Pascal's ultimate commitment to truth and reason is his successfully organizing, in 1648, a demonstration that atmospheric pressure decreased with height. Needless to say, he used the new mercury barometer with its "Torricellian" vacuum or near vacuum in the enclosed space at the top of the mercury column. He thereby drew upon himself atavistic hostility, not from some *lumpenproletariat*, but from fellow intellectuals, Aristotelian and Cartesian. The Shakesperian revaluation of "nothingness" had not yet sunk very deep.

Two other leaders of this philosophic renaissance cannot be ignored in any such discussion. The Dutch Jewish apostate and skilled lens grinder, Benedict Spinoza (1632–1677), visualized immortality in terms not a few moderns would recognize. The German logician, Gottfried Leibniz (1646–1716), asked why anything existed. As a rule, his question got frozen out as unmanageable.

Rational or Romantic?

René Descartes certainly did all concerned a service by breaking the tradition whereby a European thinker was generally expected to declare himself either for Plato or for Aristotle, regardless of how whimsically or partially they might interpret their allegiance. Where he did disservice was through being committed so feverishly to mathematical reductionism even as applied to what could never be complete or completely quantifiable data bases. Acute polarization developed between those willing to proceed thus and those inclined towards a subjective and holistic approach, having due regard for aspects which little lend themselves to numeration. Undoubtedly, this polarity exacerbated the more profound divide we must now consider.

Our present mind sets are considerably legacies of the deistic/rationalistic "Enlightenment" which spread across the West (but especially Sweden, Scotland, France, England and America) in the century or so after 1675 It was epitomized by Voltaire's dismissal of Pascal as a melancholy misanthrope. Obversely, we are much influenced by the Romanticist counter-current which was surging through Europe by 1800. Technology was disdained for largeness and sameness as well as for grinding Nature down. Science stood condemned for its presumption that every problem could usefully be reduced to a batch of separate questions. It was an attitude epitomized by the laudanum-wracked poet Samuel Coleridge in 1801: "I believe the souls of 500 Sir Isaac Newtons would go to the making of a Shakespeare or a Milton".

Unfortunately, this unfruitful divergence coincided with or slightly preceded the onset of the Industrial Revolution. An ecological manifestation thereof by 1750, though unremarked awhile, was a secular rise in the atmospheric concentration of carbon dioxide, this due to rising coal consumption. The greenhouse effect therefrom contributed to a recovery in mean global temperatures.

A key to the big paradigm change the Industrial Revolution represented lies in the second of the two Laws of Thermodynamics enunciated by the German scientist, Rudolph Clausius in 1850. In effect, the First expresses the pre-Einstein understanding that energy can be neither created nor destroyed, merely redistributed. The Second tells how, in any self-sustaining process, redistribution will always be towards an even spread (locally and, in the ultimate, cosmically) in the randomised form we recognize as "heat". This makes it peculiarly hard to achieve and sustain energy concentration to the extent needed to apply it to mechanical work. There was no need to match the grandeur of, say, pyramid construction. But one had to achieve still higher, indeed much higher, mechanical precision than what the pharaohs remarkably attained.

Progress was slow for a long time, beautifully crafted though structures like the watermill were. But from the sixteenth century through the eighteenth, big gains took place in energy application just as they did in scientific instrumentation – Galileo, Torricelli, Spinoza, Wren, Newton, Harrison . . .[36] By 1960, say, the energy being generated across Europe by inanimate power will have been well over a thousand times what water power had yielded in 1800.[37]

Polarization

The philosophic divergence just identified did not connote a straight fight between progressives and conservatives – Left and Right – as they would, in due course, be styled. Let us compare two Leftists. The Anglo-American radical, Thomas Paine, wrote with abnormal urgency about everything from American Independence to the French Revolution of 1789 and to God. In many respects, however, his *The Age of Reason* (published in 1791–2) encapsulates the outlook of the liberal mainstream of his time. Take his averration that "our ideas, not only of the almightiness of the Creator but of his wisdom and beneficence become enlarged in proportion as we contemplate the extent and structure of the universe". It is a deistic pitch which contrasts sharply with that adopted by Jean Marat, the talented French medic who turned ultra-revolutionary only to be gruesomely murdered in the savage "Reign of Terror" days of 1793. His "anti-Newtonianism" was calculated to undermine the *Académie Royal des Sciences*, it having refused him membership. Against the background of wider pressures towards making Science democratic, if only in the sense of not being old elitist, the Academy did close in 1803.[38]

Marat evinced an amazing propensity for explosive resentment. Always, however, it was dangerous thus to scorn reason and deductive enquiry, claiming the while to set store by wholeness, organic unity, sensibility and naturalism. The dangers are all too evident in the impact of the amoral egotist whom Bertrand Russell dubbed "the father of the Romantic movement", Jean-Jacques Rousseau (1712–78). Writing from outside the confines of salon intellectualism, Rousseau envisaged a "general will" to which every member of a given polity should be required to surrender himself. However, he gave few clues as to its quality save that it is "always constant, unalterable and pure". Advice so nebulous left a wide flank open to totalitarian ruthlessness of the kind typified by Robespierre, Hitler, Stalin and Mao Tse-tung. Any emergent dictator can claim that he it is who represents the "general will".

Influential though Rousseau was, however, pre-revolutionary France was where Enlightenment rationalism burgeoned most distinctively. One measure of its progress will have been the acceptance of Newtonianism as the most up-to-date and sound account of how the deistic universe functions. For some time a big obstacle had been that Descartes had proposed an alternative theory of gravity which was based on the presumed effects of supposed cosmic vortices. Then come the 1740s, the Newtonians received decisive backing from the anglophile Voltaire.[39]

Herein can one see, too, the foundations for the remarkable ascension of French mathematics which occurred once the Robespierian "Reign of Terror" had given way to Bonapartism. In 1799, France pioneered the introduction of metrication, this to both civil and military ends. Among the distinguished applied mathematicians who came to prominence were Jean Baptiste Fourier, Joseph Lagrange, Siméon Poisson and – most famously, at the time – Marquis Pierre Simon Laplace (1749–1827). Together with Lagrange, Laplace underscored Newton's theory of gravitation. He also did seminal work on *a priori* aberrations in lunar, planetary and cometary motions.

Laplace contended that given complete information about the initial conditions (i.e. knowledge of the location and motion of effectively every particle), the future of the world should be predictable. Alas, such sanguineness (if thus it can be dubbed) was to be mocked by his experience advising Napoleon on climate, this ahead of his onslaught on Russia. His judgement was that an invasion begun in June would not be denied victory by the onset of winter. What he never realized was that, within the secular context of global warming, a transient reversion to earlier and colder Russian winters was under way. Come 1941, a German meteorological service headed by the similarly brilliant Fritz Baur, made just the same blunder for exactly the same reasons.

Middle Ways

Crucial to any attempt to build a new philosophic synthesis would be a broad endorsement, if not yet of Newton, at least of Copernicus. By the early eighteenth century, the latter had sufficiently been achieved. Not that you would think so from some of the evidence. Take Rome's declaratory position. Only in 1744 was the printing of Galileo's *Dialogue* allowed; and was, even then, subject to prescribed amendments. Not until 1835 did Copernicus, Kepler and Galileo come off the Index of proscribed works.

However, this was more a question of institutional inertia than dynamic resistance. Thus by 1700 the Jesuits in China were spreading the Copernican word much as the Dutch in Japan had.[40] In 1720, Shogun Tokugawa Yoshimune lifted the import ban on books about astronomy and medicine; and in 1744 founded Japan's first astronomical observatory.[41] So in the ever-eclectic Far East, the new cosmology was having tangible influence.

Awareness of the desirability of keeping reason and romanticism tolerably close together can be seen in the signal contribution made to sky science by Luke Howard of London (1772–1864), a practising chemist and amateur meteorologist. In 1802, he devised a Latinized cloud classification compounded from four basic forms: *cirrus, stratus, cumulus, nimbus* – i.e. curl, sheet, heap and storm cloud. It is very much the nomenclature still extant internationally. Among the near-contemporaries of Howard known to have used it were Constable, Goethe, Ruskin and Shelley.[42] That such interest did not lead to much improved artistic representation may show how deep this "two culture" divide readily became. One might add that the French evolutionary biologist, Jean-Baptiste Lamarck (1744–1829), came close to upstaging Luke Howard. But in 1809 Napoleon imperially advised him to forget about clouds, no doubt seeing them as a Romanticist fetish.[43]

A quest for synthesis on the grander scale (political as well as philosophic or aesthetic) inspired the idealism of the great German Idealist thinker, Immanuel Kant (1724–1804). Idealism is the tradition within Western philosophy of working in whatever manner from the premise that the realm of ideas is more fundamental than that of matter. Though often traced back to Plato, it enjoyed an early modern revival triggered off, obliquely one must say, by Descartes.

Neither then nor ever since has Idealism been the same as Romanticism. The former is more focused, esoteric and analytic; and the word its sole means of expression. Correspondingly there have been different pacemakers, Coleridge being one of the few with links to both these camps. Nevertheless, each community has been concerned to curb undue reliance on reason informed by a materialist understanding of reality.

Kant was another denizen of the Baltic shores. Reportedly, he never roamed more than sixty miles from his native Königsberg. He came to philosophy in part via a knowledge of physical science that was rather non-empirical and non-mathematical yet quite solid. In a 1755 treatise he endorsed a proposal Thomas Wright of Durham had made in 1750 which was that what we know as the Milky Way was a vast stellar population incorporating our solar system. He himself formulated a nebular theory of its origin. He also proposed that lunar and solar tides acting on the oceans would have a drag effect on the Earth's spin.[44] He went on to do as much as anybody to popularize the notion that all stars are massively grouped into such galaxies – "island universes" he called them. He had a holistic perception of the interdependence of everything within God's cosmos: "The starry heavens above me . . . and the moral law within me." He would have been refreshed by the astronauts' perception of the oneness of Spaceship Earth.

He might have seen it as an appropriate prelude to the somewhat utopian tract on *Perpetual Peace* he published in 1795.

Astronomy Consolidates

In astrophysics as in many other spheres, the nineteenth century begins in 1815 and ends in 1914. It was not a time of great leaps forward subject-wise, rather one of incremental gains. In 1817 Joseph von Fraunhofer (1787–1826), a brilliant Bavarian optical engineer, concluded that the numerous black lines he discerned on the solar spectrum were inherent in the light omitted. In 1863 it was finally confirmed that, not just in the Sun but in other observable stars, these "Fraunhofer lines" were absorption patterns characteristic of various elements in gaseous form – many of the molecules thereof constantly being created and disintegrated. It was a powerful indication, not gainsayed since, that the chemistry of the cosmos is homogeneous.

Meantime astronomy lost ground in public esteem before an Earth-bound positivism which professed boundless confidence in material and moral progress and tended, indeed, to conflate the two. The "Conquest of Nature" was pursued the Earth over with a relish akin to how some still enthuse about the "Conquest of Space". The Wallace–Darwin exposition of biological evolution through natural selection became, in due course, the vogue science. That the universe may have aroused less wonderment than previously will also have owed something to obscuration of the night skies by buildings, lights, haze and Dickensian fogs. Meanwhile, human progress afforded novel themes for celebration. Around 1850, the term "star" came regularly to be applied to somebody who "shines" with high visibility in the arts or, soon thereafter, sport.[45]

One somewhat negative respect in which astronomy can be said to have impacted, through the Victorian age and rather beyond, was via a parallel drawn between the eventual heat death of the universe and a similarly inescapable waning of the Pax Britannica. Among the active or aspirant statesmen who perceived this acutely were Arthur Balfour and the young Winston Churchill. It probably informed Gladstone's pessimism as well.[46]

From 1840 or thereabouts, the United States figured prominently in the construction and use of astronomical telescopes, the pristine star-field being part and parcel of the wilderness vision. In his first annual address to Congress (1825), President John Quincy Adams pointed out that the USA had no federal observatory whereas Europe had 130 such "lighthouses of the skies". His crusade after leaving office in 1829 for a national facility helped ensure the opening (in 1844) of the US Naval Observatory and the founding (in 1846) of the Smithsonian.[47]

Martians?

This discussion of the nineteenth century has been conducted as if Astronomy and Astrophysics were synonyms. In regard to really hard science, they effectively were. Astrobiology had no look in except that, in 1834, it was established that the Moon had no or extremely little atmosphere; and was therefore most unlikely to bear life.[48]

In the realm of unrestrained imagination, however, contemplation of extra-terrestrial beings intensified after 1800, perhaps in the hope that encounters with them could prove more agreeable than all the strife engulfing Europe. Among those who entertained the idea with some conviction were Lord George Gordon Byron, Samuel Coleridge, Erasmus Darwin, Immanuel Kant, Jacques Necker, Wilhelm Olbers, Thomas Paine, Percy Shelley and Madame de Stael. Of these just two, Kant and Olbers, studied astronomy closely. For the former, a belief in "pluralism" (i.e. life elsewhere) followed on from his conviction that the universe was galactic. For the latter, a belief in "pluralism" stemmed from, and outlived, his initial assumption that quite advanced creatures lived on the Moon. However, his 1825 agreement with Kant that it is "mostly highly probable" that "all of infinite space is filled with Suns" was flatly countered by the more original and celebrated axiom he soon propounded.[49] This was Olbers' Paradox, the contention that the Universe cannot be infinitely extensive. If it were, the night sky would be full of light streaming in from literally all directions. One might think that this would have stood as a compelling *a priori* argument against the later Steady State hypothesis. But that much was never conceded.

Through the turn of the century, the focus *vis-à-vis* extra-terrestrial life turned towards Mars. This was primarily due to advocacy by Percival Lowell (1855–1916). A very competent professional, he was one of the two American astronomers who predicted the existence of a planet beyond Neptune within our solar system. Furthermore, the observatory he had founded in Flagstaff, Arizona is where the said body – Pluto – was actually observed in 1930. Unfortunately, however, Lowell had allowed Mars to make him in later life the most extreme exemplar yet of astral hyperbole via press and radio.

In 1877, Giovanni Schiaparelli (1835–1910), an Italian field astronomer of very high repute throughout his life, claimed with high confidence to have identified a criss-cross spread of *canali* (i.e. water-courses) on Mars. In 1906,

Lowell brought to a head some years of fanciful speculation on this score with his book, *Mars and its Canals.* The nub of his interpretation was that Martian polar snows melted alternately causing water from them to be artificially impelled down constructed canals to the equator and beyond. These *canali* were supposedly evident from the vegetation on their banks.[50] Their global spread he believed bespoke high intelligence and peaceability on the part of their managers.[51]

A fierce disputation ensued. According to Carl Sagan, "It became so bitter and seemed to many scientists so profitless that it led to a general exodus from planetary to stellar astronomy, abetted in large part by the great scientific opportunities then developing in the application of physics to stellar problems."[52] By 1914, the sceptics had effectively won the professional debate. But in 1939, modish folklore around the world still allowed of some possibility that quasi-hominids were on the Red Planet, operating quite an advanced civilization.

Once the sceptics were ascendant professionally, conjecture began as to how far the *canali* may have had, like many more ancient myths, some factual foundations – meaning topographical fault lines or albedo boundaries. In 1975, Carl Sagan and Paul Fox concluded that such correspondences could be traced only for a low minority of the *canali.*[52] When in 2003 Mars came even closer than in 1877, webcam image acquisition congruent with eye–brain integration yielded enhanced albedo boundaries which may redeem to an extent Lowell's and, of course, Schiaparelli's observations.[53] Even so, the inferences Lowell drew were extravagantly dogmatic.

Fin de Siècle

The climacteric of 1914 was a terrible blow to the celebratory humanism that had suffused the West's literati the previous 150 years. Among the many causes adduced has been that "the little Eurasian peninsula that was Europe which had conquered the world and was its power-house, contained too much energy and power for the narrowness of its confines . . . There was in 1914, no more room in the world for fresh conquests".[54]

In some countries the excess energy had been expended, notably since 1880, in a more expansive nationalism. Yet soon a tightening of the global confines would be starkly evident. In 1909, Robert Peary became the first explorer to reach the North Pole[55] as, in 1911, Roald Amundsen did the South. In 1890, the "moving frontier" of the Old West ceased to figure in the US Federal Census. The formation of the Commonwealth of Australia in 1901 set the seal on that continent's being well and truly spanned. The

European encompassment of Africa moved from being peripheral to comprehensive between 1881 and 1911. As regards inner Asia, a British military expedition reached Lhasa in 1904.

Along with this political consolidation went a multiple revolution in communications that, by any previous standards, seemed epic. A world network of cable links stood complete in outline with the laying, from 1902, of a bifurcating line from British Columbia to Queensland and to New Zealand. Moreover, a dramatic pointer to the future was Marconi's wireless transmission from Cornwall to Newfoundland in 1901. Another was the beginning of powered flight in heavier-than-air aircraft, a departure most people allow came with the Wright brothers' transitting 260 metres in 1903.

At which point it is appropriate to pray in aid H. G. Wells (1866–1946) who has, after all, been described as "the last great literary figure" to exemplify "the profound effect that scientific developments had on the literature of the nineteenth century" in contrast with "the lamentable lack of cross-fertilization in the 20th".[56] Not that he was truly the prophet of the Space age *per se* he is often cracked up to be. In fact, he came to his scientific awareness via higher education in biology not mathematics or engineering. So while he did acquire a strong awareness of the role of astronomy in History, his insights into future Space travel are either banal or downright silly. Similarly, successive prognoses of trends in military aviation were regularly at variance with what actually happened.[57] His novel *The First Men on the Moon* is a political satire not much freer from Earthly mental furniture than Hergé's *Adventures of Tin-Tin.*[58]

However, where he scores heavily over the likes of Jules Verne, let us say, is as a social prophet. Much of his best writing is suffused by a compelling awareness of how finely we are poised by modern Science between triumph and tragedy. Occasionally he has flashes of uncanny prescience on this score. Witness the nuclear scenario he postulated in *The World Set Free*, published in 1913. After the requisite technical breakthrough in 1953, "Power after power . . . went to war in a delirium of panic, in order to use their bombs first . . . By the Spring of 1959 from nearly 200 centres . . . roared the unquenchable crimson conflagration of the atomic bombs . . . ". This scenario graphically anticipated the downside of the Einsteinian discovery that $E = mc^2$.

The Relativity revolution this formula epitomized was the most far-reaching of a clutch of fundamental Physics discoveries between 1895 and 1905: X-rays; radioactivity; the electron; the elements polonium and radium; and the granularity of Nature. This forward surge mocked the opinion, much subscribed to a quarter of a century before, that Science was within "a very few years" of the absolute limits to knowledge.[59]

5

THE COSMOLOGICAL REVOLUTION

THE SETTING FOR THE MODERN COSMOLOGICAL REVOLUTION was shaped by two genres of theoretical physics evolved during the first several decades of the last century. The one was Quantum Theory, the other Relativity. Though never yet properly integrated, they continue to form a reactive backdrop to on-going developments. At various points, metaphysics impinges on the said revolution. At many others, influence flows the opposite way. The interplay between particle physics and astrophysics has become basic to some crucial philosophic debates, not least that of Idealism versus Materialism.

A key tenet of Quantum Theory is that, at atomic level, particles can be seen as waves or waves as particles, depending on observer perspective. A prime mover in this formulation was Niels Bohr (1885–1962) of Copenhagen University. But central to quantum thinking, too, has been the Uncertainty Principle enunciated by Werner Heisenberg at Leipzig in 1927. The classic example has been that the more accurately we determine the momentum of an electron, the less precisely we determine its position and *vice versa.* The measurement of other pairs of conjugate variables must similarly be constrained. Contingently, this could be down to experimental or conceptual shortcomings or else to indeterminacy inherent in Nature. From Heisenberg onwards, most quantum theorists have stressed the latter, making it crucial to the whole Copenhagen interpretation. However, quantum effects in general are significant only at atomic level. Above it, countless interactions cancel out, leading to "common sense" results overall. This Bohr entitled the Correspondence Principle.

What quantum theory does not do is allow of human free will where Newtonian causation cannot. Ian Barbour has told of certain scientists imagining the mind acting on a responsive brain within a dualistic structure.[1] Apart from the question begging inherent in so Cartesian an approach, however, one must recognize that whatever pitch the human brain adopts will always have been mediated through a good proportion of its 100 billion neurons, the Correspondence Principle applied with a vengeance.

Einstein and Relativity

Conducting experiments or resolving mathematics, Albert Einstein (1879–1955) was talented by any mortal standards. But his most exceptional gift was being able, through what he termed his "thought experiments", boldly to draw radical conclusions from fresh data. His Nobel Prize was awarded in 1921 for one of several seminal papers published in 1905. Addressing seeming anomalies in the photoelectric effect, it showed how electromagnetic wave energy would always be organized into waves consisting of assemblages of energy quanta, later to be called "photons". Their individual strengths were an inverse function of wavelength. This breakthrough was evidently important for Quantum Theory.

In 1887, two Americans, Albert Michelson and Edward Morley, had found the speed of light was constant in whatever direction they measured it. This finding conflicted with the then received wisdom which was that light was a wave motion in a non-material medium called the ether. Accordingly, light's measured speed was expected to vary by virtue of the Earth's velocity relative to that of the ambient ether. But now the ether was a superfluous precondition. As for the speed of light, it was elevated to a ubiquitous principle by Einstein in his 1905 paper on Special Relativity. It became the keystone "invariant" in his universe of relativities.

He also followed up questions raised by H. A. Lorentz and others about whether Space and Time were respectively the discrete absolutes classically supposed. He proposed instead a four-dimensional Space–Time continuum in which any of the three dimensions of Space or that of Time might be foreshortened or extended by observer motion. But his General Theory of Relativity (formulated in 1916) particularly focused on gravitation, allowing as it did that the observer might be in irregular motion. It postulated that the presence in Space of matter caused it so to curve that a gravitational field was set up. Gravity waves were predicted; and now preparations to trace these are well-advanced. This is despite gravity actually being a strikingly non-intensive force. At the sub-atomic level, for instance, the reciprocal gravitational pull between an electron and a proton is estimated to be one part in 10^{40} of their electrical attraction.

Other Relativity results Einstein identified included the following. Mass increases with velocity although this effect counts for little below "relativistic" speeds – i.e. ones nearing the speed of light. At 90 per cent of that ultimate motion, an electron's mass is multiplied two and a quarter times.[2] Then again, mass and energy are, in principle, fully interchangeable attributes in accordance with the formula, $E = mc^2$ or energy equals mass times the square of the speed of light. That gravitational attraction can bend electromagnetic waves was shown adequately enough by Eddington and others during the total solar eclipse of May 1919. Two more effects, significant at relativistic speeds, are that (a) objects foreshorten axially to the direction in which they are moving, and (b) time as recorded passes more slowly with a platform in motion. In both respects, all measurements are with reference to specified rest frames. Relativity allows of no absolute frameworks of reference.

A concept destined to present big difficulties was the "cosmological constant": fields of energy within cosmic vacuo that Einstein initially felt obliged to propose in order to keep in balance his model of a homogeneous and isotopic cosmos which, he then wrongly assumed, was not expanding overall. He soon came to regard this "constant" as his "biggest blunder".[3] Yet the concept marched on partly because it was thought it might modulate the cosmic expansion which Edwin Hubble showed in 1927 was evidenced by radiation "redshift", a doppler effect.[4] Now some extra factor is being looked for in connection with an accelerated cosmic expansion the last few billion years, as confirmed in 1998. A form of "dark energy" may be the key needed.[5] But imponderables remain.

Einstein never got close to his cherished goal of formulating a Theory of Everything. On the other hand, Relativity *per se* has stood up well to theoretical and practical tests and as a source of concepts. It will continue to figure as we consider other relevant themes.

Entropy and Symmetry

Entropy is regularly cited as expressing how disorderly in heat energy terms a milieu or system is. The entropy change a particular system undergoes via a thermodynamic process is the quantity of heat taken up or sacrificed, gauged in relation to the prevalent temperature on the Kelvin or absolute scale. In other words, a given amount of heat absorbed at lower temperatures will represent a bigger entropy gain than if it were absorbed at higher.

The total entropy of a perfectly enclosed system can only rise or, perhaps in principle, stay static in the course of a thermodynamic process. The most basic implication thereof is that the universe is moving slowly but inexorably towards "heat death": a state of complete inactivity effected through the maximization of entropy all round. This must be the eventual outcome of the Second Law of Thermodynamics (see Chapter 4).

Important in modern physics (as was implied in Chapter 2) is the integration of symmetries effected, somewhat imperfectly, to help sustain the Standard Model for atomic structure and behaviour. But all that need actively concern us here is the customary view of simple symmetry as being the mirror-image correspondence of respective parts of a design seen in relation to a plane, line or point. Within European culture alone, one can see there has long been (Neolithic Man, Aristotle, the High Middle Ages, the Renaissance Continuum . . .) a disposition to take symmetry simply conceived as expressive of the soundness or sufficiency of a concept. Dante Alighieri came to see the cosmos as a huge sphere, with God at its centre, set in the very middle of Space and Time.[6] Admittedly, Copernicus and others were in error in assuming symmetry meant planets must revolve around the Sun on strictly circular paths. Even so, some of us remain prepared to be guided provisionally by naïve symmetry tests, as and when firmer guidance is still awaited. Some would say the whole universe seems symmetrical in the very general sense that the same Laws of Nature obtain throughout.

Much attention is these days paid to the often exquisite mathematics of "symmetry breaking". Sand dunes or convective shower patterns imposed on a flat open plain are examples. Symmetry breaking can also be seen in biological diversification and, indeed, the creation of atomic particles out of primordial nothingness. It has been well pointed out by a distinguished British physicist/astrophysicist[7] that many cultures from traditional Japan,[8] Pre-Columban America and elsewhere have well appreciated in the fine arts and music the creative beauty of what can be seen as symmetry breaking.

Why Anything?

The question of why there is anything at all has been addressed by philosophers episodically. Those who have raised it have included Aristotle, Leibniz and Heidegger. The point at issue is how any discrete forms could arise, with or without divine intervention. Genesis visualized initial moves being made on the face of deep waters. Buddhist texts talk of there having been a void. Given the possibilities opened up by modern physics, the distinction could be more apparent than substantial. Semantically, indeed, this is demonstrably so. Take the word "chaos", deployed by Hesiod. Its root means "to gape", maybe at boundless vistas. Then later, thanks to a false derivation from a word meaning "to pour", chaos came to mean the unorganized mass of basic material putatively scattered through Space to begin with.[9]

Still, the more searching challenge is why is there nothing whatsoever all throughout. One may turn for philosophic guidance to two twentieth-century French Nobel Laureates. The existentialist radical Jean Paul Sartre (1905–80) seems currently to be enjoying a centennial revival.[10] The other, Henri Bergson (1859–1941), made a strong appeal in the interwar period to agnostic opinion, religious and political, throughout the West. He did so by campaigning rumbustiously against constrictive utilitarian rationalism.

In his essay, *L'Être et le Néant* Sartre discusses Nothingness essentially in relation to Being, Consciousness, Ego and other parameters of the Self. However, he diverts to "naïve cosmogonies" for one involuted but arresting paragraph. He does so to make the point that to look back to the "nothing" obtaining before Creation is to accord it "a borrowed existence" within a line of causation. In other words, "non-being exists only on the surface of being".[11] Less cogently, Bergson argued in his most popularized work that "whether it be a void of matter or a void of consciousness, the representation of the void is always a representation which is full and which resolves itself on analysis into two positive elements: the idea, distinct or confused, of a substitution and the feelings, experienced or imagined or a desire or a regret".[12]

In essence, "nothing" has to be seen for what it is. But the seeing has to be by "something". Presumably, too, "something" has to be seen for what it is.

Cosmic Origins

By observing objects some 50 times fainter than ground-based telescopes have been recording, the Hubble Space Telescope has brought us to well within a billion years of an initial Big Bang. But closing to much within 100,000 years will be extremely difficult, not least on account of a profusion then of energetic and unattached electrons. Nevertheless, cosmological modelling now addresses what happened the first ten millionth or even first trillion trillionth of a second after Creation.

Certain aspects have been debated long and hard, especially since, in 1965, Arno Penzias and Robert Wilson discovered the microwaves residual basically from the Big Bang event, a legacy George Gamow had predicted – in outline at least – from 1948. Gamow was himself inclined towards a Big Bang/Big Crunch eternal alternation.

If the dimensions of Space and Time are all to be seen as granulated, it makes sense to think of everything cosmic originating as a quantum fluctuation over roughly the defined Planck Length, 10^{-35} metres – a fluctuation in accordance with the Uncertainty Principle. To adhere to mass-energy conservation laws, the resultant pristine universe should be homogeneous, isotopic and closed; and putatively consist equally of matter and anti-matter.[13] The present indications are that certain germane proposals may lend themselves to indirect testing at not too much above Planck-length dimensions.[14]

In 1981, Alan Guth of MIT proposed a phase of extremely rapid inflation incredibly early on in the life of the universe, this to account for various characteristics including the pronounced flatness of Creation as we know it. There seems now to be widespread agreement that this will have taken place between 10^{-32} and 10^{-36} seconds after commencement; and that it involved a concurrent expansion of Space, Time and Energy. The extent of the expansion then of the spatial region which now encompasses our universe was 10^{50} times – from one billionth the size of a proton to several centimetres. The expansion was taking place not within a true emptiness but into an excited state known as a "false vacuum". In the face of expansion, this released huge volumes of heat which raised the temperature of the embryonic universe to about 10^{27} Kelvin.

"Long after the era of inflation but still within the first millionth of the first second", [15] fundamental particles called quarks were combining to form protons and neutrons. Well within that first second, the Grand Unified Force existent at Creation separated into the fundamental forces we know today: gravity; the strong force of nuclear binding; the weak force of nuclear decay; and electromagnetism. After three minutes, the temperature drop allowed protons and neutrons to form atoms – 90 per cent of them of hydrogen and the rest of helium. Some 100,000 years later, it was cool enough (4000 K) for nuclei to capture electrons to form normal atoms. After another 280,000 years, the universe had thinned enough to allow electromagnetic energy wide access.

One's first impression of the Big Bang is of a tremendous but therefore randomized force. Instead it seems to have driven a built-in programme of amazing elegance, intricacy, precision and consistency. The number of protons had to match exactly or extremely closely what were pre-existing electrons. All protons and electrons carry an electric charge of exactly the same strength, "positive" in sign in the former and "negative" in the latter. The binding together of protons and neutrons as the basic constituents of atomic nuclei is facilitated by the neutron's being barely one part in a thousand heavier. Were this differential much greater, neutrons would be more liable to decay into protons than combine with them. And so on and so forth.

Matter and Anti-Matter

Moreover, it was indicated above that conservation laws would lead one to expect that, at the very beginning, the universe would have contained equivalent quantities of matter and anti-matter; that is to say, there would be the same mix of particles in similar quantities. On each side, the same type of particle would exercise the same gravitational force. However, the nuclear force charge would be opposite as would the electrical charge. If a matter particle and a corresponding anti-matter one collide, there is mutual annihilation with a proportional release of energy. So the advent of equal assemblages of matter and anti-matter as cited above would soon lead to the complete elimination of all material, the universe reverting to but an energy-rich false vacuum.

What was ascertained in 1964 in the Brookhaven National Laboratory, however, is that very early on there was an excess of matter of less than one part in a billion. An attempt to comprehend this was led from 1967 by Andrei Sakharov, backed by the Lebedev Institute in Moscow and using a very esoteric notion of Supersymmetry. Politically, this development was interesting in that it contrasted with how, in Stalin's time, Einstein had been condemned outright by the Kremlin for "bourgeois idealism" because his determining of an equivalence between matter and energy was held to compromise straight-down-the-line Marxian materialism. Not that this ever dissuaded Soviet science from availing itself of the practical fruits of Relativity.

Be that as it may, the matter/anti-matter imbalance still awaits elucidation, probably by going multidimensional.[16] Meanwhile, several other examples of structural fine tuning (mentioned in the following chapter) have become directly entrained in the debate about how life-friendly or otherwise our cosmos may be. These and other aspects of cosmic development suggest an organizing principle of singular complexity. A truly virtuoso unmoved mover?

Inconstant Constants?

What could soon be underlining this impression is confirmation that certain constant values actually change subtly across aeons of time. Take Alpha, the Fine Structure Constant which determines the strength and frequency of electromagnetic waves as well as the character of atoms and molecules. Using 147 quasars as markers over a two-year span, a team from the University of New South Wales has provisionally concluded that 11 billion years ago Alpha was weaker by seven parts in a million than it is today. Evidently a truly comprehensive Theory of Everything ought to explain *inter alia* how Nature determines the values of the Constants, be they changing or unchanging. However, one is advised this could require "many more dimensions of Space than the three we see".[17] But what, in such discussion, is meant by extra dimensions? Shrivelled vestiges introduced to accommodate mathematical loose ends? Or authentic geometrical constructs?

Black Holes

These phenomena, anticipated by Einstein, are today at the very heart of the perennial debate about the relationship between our universe and any others. The dialectic idiom is cast in terms of transfers of information as opposed to matter or energy, a convention reportedly initiated by John Wheeler to facilitate integrative analysis of big bangs, black holes, entropy and so on. From 1958, Wheeler was a pioneer of modern "black hole" studies and, indeed, endorsed this specific term a decade later.[18] He and others of like mind may now be leading us towards a situation in which all entities in natural science are seen as information assemblages.

In 1798, Pierre Simon Laplace had proposed that massive bodies might exist which would be quite invisible since their gravitational force would preclude light escaping. Current modelling of black hole compression usually indicates nothing escaping from what thus becomes the critical Schwarzschild radius. Were our Sun to become a black hole, however improbably, the radius thereof would be three kilometres.

Black holes have been understood mostly to form when large stars (not less than three solar masses?) collapse at the end of their life cycles. Much gravitational energy is released by material *en route* to the Schwarzschild horizon though not afterwards. However, other mechanisms may lead to black hole formation. Perhaps tiny black holes momentarily result from cosmic radiation in the Earth's upper atmosphere; and might be produced, too, by CERN's Large Hadron Collider. Most important, however, is the collapse into the centres of quasars (quasi-stellar radio sources) or of what may sometimes be several billion solar masses.

Some argue, too, that differential tidal forces would be less liable to tear apart things (humans perhaps included) within so big a singularity than might be the case within a smaller one. Reportedly, Igor Novikov, head of the Theoretical Astrophysics Center at Copenhagen, challenges the view that entering a black hole spells certain death. You might emerge in a different universe instead. But whereabouts, must one first ask? Only in the tiniest fraction of our present universe could a human survive unsupported even five minutes.

Lee Smolin of the Perimeter Institute in Ontario is among those tempted to speculate that black holes can give birth to baby universes, perhaps with certain constants of physics being just slightly altered. Michio Kaku, Professor of Theoretical Physics at City University, New York, believes sending a surveillance probe into a black hole could be a first step *en route* to a parallel universe. Akin for the purposes of this "multiverse" argument is an experiment by Martin Bojowald of the Max Planck Institute for Gravitational Physics in Golm, Germany. Applying loop quantum gravity, he has looked into the instant of Creation. He believes he has found therein a window to what had gone before.

If a universe can spawn others, there could be a commonality of information sufficient to open up the possibility of structural similarities within the family. But much would depend on how constant the Constants of Nature are within universes and between them. So might it on how much information was sent and how coherent it was. Scepticism seems to me to be in order so far as similarity is concerned. And how would we ever know?

String Theory

A way towards an understanding of such questions may unfold, these next two or three decades, through the fuller development of "string theory", a school of applied mathematics emergent from 1969. Its basic tenet is that the most fundamental constituents of the universe are not the elementary particles we familiarly depict but tiny vibrant strings or, more likely, loops. In this sector, too, the field work has to be considerably concept-driven because one is ultimately concerned with dimensions well inside the nanometric limits thus far attained. At the same time, one is reaching out to relativistic perspectives. One hypothesis still to be verified is that all fundamental particles will have a superpartner with a different though related spin.

Overall, pronounced symmetry is claimed for this string concept. Moreover, it holds out a promise of integrating quantum mechanics and relativity by operating in ten to twenty-six dimensions. What remains unclear, however, is what results can be obtained and corroborated; and whether they will simplify perceived reality rather than complicate it. Could one illuminate thus the inner workings of black holes? Or could one eventually find that the cosmological constant varied across Space? One of the progenitors of this field of enquiry has suggested that the "whole point . . . may be the huge diversity of environments that it may lead to. It, among all theories, is uniquely well suited to an anthropic theory of the constants of Nature".[19] But could it involve a pronounced departure from the cosmic homogeneity principle?

Many Universes?

Such discussion always takes more or less for granted that there will be many universes apart from ours. In the absence of natural breaks, indeed, one might ultimately be thinking of an infinitude of them. In practically all the recognized world religions, God is regularly equated with Infinity.[20]

Considerations of classic symmetry make our cosmos look too arbitrary to stand alone. Nor could a second one afford a "mirror image" complementality, thereby correcting imbalances. What, for example, can be the inverse of the speed of light as we know it? One must seek instead a spread of values. But that is one of the very many aspects that seem to proffer no natural breaks between two and an infinitude.

One ensemble of universes could be ancestors and/or descendants of our own. However, an alternative route to multiplicity has been proposed, notably by Hugh Everett and by Andrei Linde. It basically involves continual branchings of the on-going status quo. A fellow physicist, Bryce DeWitt, has told of the dramatically positive and lasting effect Everett's enunciation had on him: "Every quantum transition taking place on every star . . . is splitting our local world into myriads of copies of itself".[21] Linde has argued very similarly that the cosmos in its entirety is self-reproducing through "an extended branching of inflationary bubbles", the interior of each bubble being described by the Big Bang hypothesis.[22]

Addressing the Everett thesis specifically, Michio Kaku questions how meaningful would be thus linking worlds which, more or less by definition, could not communicate one with another. He believes the notion violates Occam's Razor[23] (see Chapter 3). William of Occam has been mobilized, too, by other critics either of Everett[24] or else of the whole "phantasmagorical" idea.[25] Paul Davies pitches thus against an infinitude.

However, a proliferation of universes in more prosaic terms is an idea which, although it does not figure in popular mythology the way Creation does, has a strong intellectual pedigree stretching back within Europe to the pre-Socratic Greeks. In particular, it emerges as a strong theme in the advance in philosophic awareness achieved in Latin Christendom in the thirteenth and fourteenth centuries. Ironically, both Plato and Aristotle had averred that just the one world was a physical necessity. The former had done so because only the one could derive from the mathematical perfection which logically preceded it. The latter did because the Earth was evidently *the* centre which natural notions proceeded either directly to or indirectly from.

Nevertheless, in 1277 Bishop Etienne Tempier of Paris challenged the received view that God could not create a plurality. He did so after consulting the Papacy and the Sorbonne. Paris schoolmen duly attacked Aristotle on this point. Among them were Nicholas of Oresme and Albert of Saxony. Oxford Franciscans who did likewise included Duns Scotus and William of Occam.[26]

A Plenitude of Stars

There is much on the more speculative side of modern astrophysical theory that this brief reconnoitre perforce ignores: negative energy, worm holes, loop quantum gravity . . . Even so, one is left with the sense of an entity evolved by a programme so elaborate and exquisitely honed that its ultimate explanation must be considerably metaphysical. In other words, one is looking for an Aristotelian "Unmoved Mover" or, if you like (see Chapter 17), a "Creator Now Dead". What we have to be very careful about, surely to goodness, is advancing *a priori* any claims that focal attention is paid to Humanity by any such entity.

The sheer scale of everything around us demands caution on that score. After all, the universe is basically organized into the agglomerations of stars we call galaxies. A galaxy may typically be several hundred thousand light years across and several million light years away from the neighbouring galaxy within a cluster. Our own Milky Way galaxy comprises 100 billion stars, our Sun being one of them. As regards the number of galaxies currently extant, one has to speculate about how many have died out since they were radiating light in our direction and how many are transmitting light which has not reached us yet. A range between ten and hundred billion has been mentioned. Even allowing that many may be less populated than is the Milky Way, this could imply a total of the order of a hundred billion billion stars.

Outside all of which is likely to be an infinitude we shall never know or, in mundane terms, need to know. One thing for sure, the cosmos we are a desperately tiny part of is possessed of exquisite majesty. This surely behoves us to curb the uglification of the sweetly pale blue planet we occupy within it.

PART TWO

The Life Dimension

OVERVIEW

The Life Dimension

The moral implicit in this section is that those who would write about the cosmos in the round should never ignore the presence within it of Life. We remain very uncertain how prevalent Life is throughout the cosmos as a whole. However, neither our terrestrial experience nor our astral observations suggest that Creation is any too Life friendly.

Yet there is in the relevant literature a disposition to suppose that the universe exists to sustain Life. We might do better to work from the opposite premise, namely that Life exists to serve the universe. It does so by virtue of its possession of Consciousness. For this allows Life a salient role in registering the universe, thereby securing the latter's existence. However, it could never observe anything like the whole of it. Therefore one is driven to the panpsychic perspective that all matter has some consciousness. It is a perspective with a rich historical lineage; and is amusingly illustrated by the parable of "Schrödinger's Cat" – Erwin Schrödinger (1887–1961), an Austrian Nobel Laureate in quantum mechanics.

Yet discussion is hamstrung by our continued inability to define consciousness satisfactorily, let alone gauge it. As and when this impediment is overcome, it will be through the blending of biological interpretations with metaphysical ones. To that extent, one will close the divide opened up in the seventeenth and eighteenth centuries (i.e. during the Renaissance Continuum) between Natural Science and Philosophy. This divide was the downside of the knowledge explosion.

6

ASTROBIOLOGY

NEXT ONE SHOULD ASK JUST HOW LIFE-FRIENDLY our universe may be overall. In 1978, Freeman Dyson, an elder statesman of Princeton astronomy, reflected that "The more I examine . . . the details of its architecture, the more evidence I find that the universe must in some sense have known we were coming".[1] As Dyson himself would readily acknowledge, further consideration must involve interweaving an overview of evolution with an evaluation of Consciousness, a twinning which reaches into the shadowlands between science and metaphysics (see Chapter 7). For the purposes of this chapter, it is best to work within a more conventional understanding of biological science.

THE ANTHROPIC PRINCIPLE

The microbiological revolution of the 1940s set the scene for the emergence, over the next decade, of modern astrobiology. Sir Fred Hoyle was well to the fore, not least in promoting the Anthropic Principle. Reportedly, the word "anthropic" was first used in the astral context in 1930 by the British theologian, F. R. Tennant. He averred that the intelligibility of the universe had, thus far at least, culminated anthropically – i.e. in the primacy of humankind.[2] As the terms Anthropic Principle or Anthropic Cosmological Principle gained currency, however, this semantically-correct usage was relaxed. A connotation more often favoured was that the cosmos was basically friendly towards life as a whole. Everyday words incorporated into professional vocabularies often evolve their meaning *en route*, sometimes quite radically. My own erstwhile specialism, strategic studies, is replete with instances.[3]

Much remarked of late has been how finely poised quite a number of basic values within astrophysics appear to be between starkly contrasting outcomes, not least as regards the prospects for Life. Had the initial fluctuations within the big bang fireball been a little smaller, the universe would have stayed unaggregated and dark; a bit larger, and it would have been dominated by black holes. Then again, although a proton and an electron bear equal electric charges (respectively "positive" and "negative", according to convention), the former has some 2000 times the mass of the latter; and so great an imbalance is what allows molecules with defined structures to form up. Other aspects of the proton, electron and neutron relationship have been addressed in Chapter 5.

Most remarked, however, is the rate of expansion of the universe. In 1986, John Barlow and Frank Tipler found this to be "irresolvably close" to the value at which expansion would not be sustained indefinitely.[4] In 1984, John Barrow and Joseph Silk had been still more categoric: "if the initial speed of the big bang had only been tuned to one part in 10^{29} rather than one part in 10^{30}, the expansion would either have reversed and recollapsed before any stars and galaxies and life evolved or would be proceeding so rapidly that stars and galaxies would be unable to form".[5] Now this reckoning looks too finessed, seeing how it was made from a data base so incomplete as to be misleading. But probabilities a few orders of magnitude less precise would still be remarkable.

Recent developments in field research since then have afforded important guidance about the outlook for the cosmos as a whole and therefore for life within it. Thus evidence adduced in 1998 confirms that expansion has accelerated these last few billion years. Meanwhile, data from the Kamioka research facility in Japan shows neutrinos to have rest masses. Neutrinos are minuscule elementary particles, stable and chargeless, which exist in their countless zillions. They were postulated from 1930 to conform with the laws of conservation in physics. Their cumulative rest mass is put at between 0.1 and 5.0 per cent of the cosmic critical mass – i.e. the value poised between allowing of indefinite expansion and ensuring eventual gravitational retraction.[6]

The latter outcome is defined as sustaining a "closed" universe or, if you will, an oscillatory one in which an ultimate Big Crunch becomes the next Big Bang. The question of what contraction implies for the laws of science and, above all, for the role of life has been beset with difficulty. In 1986, Stephen Hawking freely admitted to abandoning his belief that contraction connoted decreasing disorder.[7] Meanwhile, what Barrow and Tipler dubbed their "weak prediction" was that "on anthropic grounds . . . the universe is more likely to be closed than open".[8]

Even if so, vexing problems are still raised by the concept of Big Crunch/Big Bang cyclicity. One enunciated some 70 years ago is that, by the time the expansion of a cycle had given way to what would inevitably be an irregular contraction, considerably more disordered radiation will be present than earlier, this on account of transformation from ordered matter through nuclear fusion in stars. In other words, entropy (here gauged by the proportion of photons to protons) will increase cycle to cycle. Therefore each successive cycle will become spatially larger, a prospect which would contradict sustained cyclicity.

However, it is now proposed that such ratios will be determined *a priori* (as one might say) within each successive Big Bang. What is liable, in any case, to curtail things is lambda: a refinement of the cosmological constant proposed by Einstein. As the universe expands, this "vacuum energy" should become progressively stronger relative to gravity, the implication being that at some point there will be an escape to limitless expansion. But if, after all, there is a reversion to Big Crunch in our present cosmic cycle, it is unlikely to be completed for another 50 billion years.[9]

Then again, the standard Big Bang model was formulated without provision for Black Holes. Yet these would proliferate (and oftentimes amalgamate) during expansion and, the more so, contraction. Therefore any sequence involving the transfer of information through a Big Bang/Big Crunch event would become more "holey" with each successive cycle.[10] Ultimately, everything could be thus engulfed. Also mooted occasionally is the possibility that a contraction phase would involve reversal of the "arrow of Time". It is a prospect which would hardly make sense in terms of effects following causes or, indeed, of memory. Closely related is the argument that if the universe is to effect gravitational collapse, it must by definition have a centre of gravity. Yet the theories of relativity militate against such Newtonianism. Any and all of these propositions can have implications for the place of Life in the scheme of things.

Even so, it is too tempting to single out Life in this regard. One student of such questions, Vahe Gurzadayan of the Yerevan Physics Institute in Armenia, speaks of the relationship between Space curvature and the unidirectional "arrow of time" as the "curvature anthropic principle."[11] But all else is affected too.

Carbon and Silicon

Within the mystery of why a cosmic setting emerged which allows Life to develop is the enigma of how this actually happened. It has to be addressed via chemistry more than physics. For at the nanometric dimensions of individual molecules, the complexity of the inanimate or "inorganic" part of Creation palls into insignificance in comparison with what we know as the "organic" or life-giving fraction. As of 2002, the most intricate molecule definitively identified anywhere in circumstellar or interstellar Space was one of 13 atoms – namely, $HC_{11}N$. Had it been inorganic, it could have been described as large and complex. Being organic, it would have to stand comparison with some comprised of hundreds of atoms.

Central to the tree of life of ancient legend is carbon, an element with a marked capacity for molecular bonding – not least as between rings or chains of carbon atoms. It is the core feature of very long molecular chains or polymers, the component molecules of which tend to be stable yet available for further reactions of specific kinds. These polymers may store energy, impose structure on protoplasm, catalyze complex reactions, and hold genetic data. Moreover, carbon dioxide is unique among the common gases in being able, at representative terrestrial temperatures, to diffuse stably in water to about the same concentration as in air. Therefore it readily makes transitions between these two fluids; and hence between living forms and their surrounds. Also its forming with water a weak and rather unstable acid gives it a role in modulating acidity within life forms.

All the elements are grouped, according to their Atomic Numbers, in the Periodic Table: an incomplete grid, the original version of which was formulated by the Russian chemist Dmitri Mendeleev in 1869. Next to carbon in the vertical alignment is silicon, reflecting the fact that the two can have very similar properties. However, silicon cannot match fully the bonding capabilities of carbon. Besides, at anything like Earth surface temperatures, silicon dioxide (SiO_2) is no gas. It is quartz, one of the hardest of rocks.

By early 2002, some 120 types of molecules had been identified as "free floating" in circumstellar or interstellar Space. A high proportion are ones we describe as "organic". Thus 85 contain carbon and nine silicon.[12] Meanwhile, silicon is – after oxygen – quite the most abundant element on the Earth's exterior, nearly 150 times more so than carbon within the outer 24 miles of the lithosphere. Granted, the idea of silicon-based life gets short shrift these days from the astronomers. But can its occasional occurrence in certain cosmic locales be entirely excluded? Might the information-rich crystalline silicates in clayey mixtures sometimes at least facilitate the transition to what we know as organic life?

Among the further possibilities to conjure with are life forms using fluids other than water to bear substances in solution.[13] Mooted, too, have been life forms on or in the Sun or neutron stars. In the former, the interaction between the magnetic field and moving electrical charges might generate information patterns. In the latter, gravitation might fuse electrons and protons, creating a neutron sea within which bonding patterns might emerge.[14]

The Primordial Situation

At all events, the prevalence of certain germane materials only shows that life's emergence cannot be ruled out *a priori.* It does not *ipso facto* rule it in. Other ambient preconditions have to be met. After all of which, the chances of life actually appearing at a given place and time can be but one in many orders of magnitude.

Working within the ambit of familiar biochemistry, a crucial prerequisite might be a water-based "primordial soup" with biotic ingredients, a situation affording a measure of protection but, above all, fluid transport. In 1924, a rising star of the pristine Soviet academe, A. I. Oparin proposed origination within a swampy regime. In 1929, the British pioneer geneticist, J. B. S. Haldane, envisaged some relevant ingredients accumulating in the atmosphere (as vapours or droplets) while the remainder dissolved in the oceans.

Impetus was given to this realm of enquiry when, in 1953, Stanley Miller showed how an energy source (such as a spark discharge or ultra-violet radiation) applied to gaseous mixes of methane (CH_4), ammonia (NH_3), water vapour, etc. could yield a range of organic compounds. His findings still bear on the important Murchison analysis (see below).

At the time, however, this work came under stringent scrutiny. L. G. Sillen believed thermodynamic constraints meant that organic accumulations in the ocean would have remained so dilute as to make talk of "prebiotic" soup unreal. Meanwhile, P. H. Adelson objected that, had methane been present appreciably in the primordial atmosphere, as much would have been evident in sediments.[15] Yet even today confirmation is lacking of the major involvement, early in this Archaean era, of methanotrophic microbes.[16] In the fossil record, sparse and flimsy as it is, cyanogenic materials rich in nitrogen as well as carbon are dominant.[17] Thus cellular fossils of cyanogenic character have been dated at *c.* 3400 million years BP in Western Australia.[18] Meanwhile, isotopically-light (i.e. organic) carbon from whatever source has been found from 3850 million years ago.

All the same, the warm global climate of the later Archaean (under a Sun much weaker than today) might best be explained in terms of abundant methane acting, as it does, as a greenhouse gas. Moreover, it could have been a vital link in the process whereby much more oxygen eventually entered the atmosphere. Intense heat (solar or volcanic or hydrothermal venting) can separate the hydrogen from the oxygen in water or water vapour. The free hydrogen then joins with carbon to form more methane. But its molecules are disaggregated in the upper atmosphere by Ultra-Violet C solar radiation. Hydrogen atoms, being exceptionally light, may then readily escape into Space, leaving oxygen ones still unattached.[19] However, the free oxygen in the atmosphere did not rise all that markedly until two billion years ago.

Cells, RNA then DNA

Not that the argument can rest there. In the world we know, all living organisms are comprised of cells: discrete, membrane-bounded units which are usually filled with protoplasm in two distinct forms – the nucleus and the surrounding cytoplasm. Within the former reside the chromosomes while the latter constitutes an ambience within which scores of distinct chemical reactions may be taking place. Many micro-organisms (e.g. bacteria and protozoa) consist of just the one cell. But a human being is basically made up of a million million. A cellular structure facilitates reproduction and growth. It also helps preserve genetic stability.

Today methanogenic archaea (i.e. micro-organisms) adjust variously to local situations which (being airless and chemically impoverished) closely replicate the general conditions on the early Earth – no free oxygen and few, if any, pre-existing organic compounds. Among the features which may distinguish them from bacteria are aspects of cell wall composition.[20] Cellular development goes back to or nearly to the beginnings of life on Earth (see below).

One must also allow for the advent early in geologic time of RNA then DNA, each essentially composed of "nucleotides" – compounds with a nitrogen, sugar and phosphate base. Both are long thread-like molecules, RNA being a single chain while DNA is characterized by its now-famous "double helix" configuration. Each form bears genetic information of remarkable complexity. In the human genome, there are three billion DNA building blocks.

DNA is a constituent of some viruses as well as of the chromosomes found in the cell nuclei of all plants and animals. RNA is the corresponding constituent of the other viruses. Also it acts as a transmitter of DNA instructions in many other species. Neither these modes nor the principle of genetic regulation could have evolved easily. Should we not look to an impersonal Unmoved Mover? An Immanent Intelligent Design?

The Emergence of Terrestrial Life

The geophysicists now calculate with confidence that the Earth coalesced from disaggregated matter 4500 to 4600 million years ago, round about the same time as the other solar planets. But how long did it then remain a liquid magma? Early in 2001, an international team reported that a zircon crystal from Western Australia had been shown by isotope analysis to be 4400 million years old, 600 million more than the previously confirmed oldest specimen. In other words, prospective life forms were proffered a solid platform much sooner than previously thought.

However, conditions thereon were anything but salubrious. There was much vulcanism. Worse, heavy bombardment by extra-terrestrial objects would not be starting to die down for another 700 million years. Studies of the uneroded lunar surface provide an instructive parallel. Their gist is as follows. For 3200 million years, the cratering rate has been effectively constant. But between 3200 and 3850 million years ago it averaged a thousand times more than now. Before that, it averaged 4000.[21] For the Earth, the time profile will have been similar, save that the Earth's much stronger gravitational pull might have led to the variations over time being greater than on the Moon.

There could also have been exposure to severe Ultra-Violet C radiation with damaging consequences for life. This will have been in the absence of the organization at high altitude of a protective ozone layer. Since ozone is oxygen in molecules which are triatomic as opposed to the usual biatomic, it will have been little present in the early atmosphere because oxygen in general was still not very prevalent. But an important caveat to enter is that thinnish dust veils from the impacts may have afforded cover instead. Very occasionally, too, the UV-C may have effected a beneficial genetic mutation. The "panspermia" factor could also have been operative (see below).

But conditions for any living creatures already *in situ* will have been extremely tough. In fact, there does seem to be a close concurrence between the scaling down of impact incidence 3850 million years ago and the first signs of life we can confirm. Whether by "close" one should understand a few thousand years or a few million is not resoluble, not yet at any rate.

The Resilience of Life

So was the emergence of life on Earth triggered by inputs from elsewhere? A reality which tantalizingly strengthens both sides of this argument is the astonishing tenacity and adaptiveness *in extremis* of life on Earth as we have known it. This could mean that organic compounds or even organisms might survive transits through Space. Yet it could otherwise mean, additionally or alternatively, that life might have developed autonomously even in the highly hostile environments of early Earth.

Life, some of which is decidedly advanced by standard definitions, thrives today immediately above and around the vents of volcanic mid-ocean ridges, vents which discharge sulphidic water superheated to perhaps 400°C. Most of the animal species are blind though a few have either embryonic or vestigial eyes – e.g. eyespots with no lenses. These creatures seem to have an evolutionary connection with species inhabiting the tidepools and reefs bordering *terra firma*. Then again, the freshwater Lake Vostok lies four kilometres below the surface of the ice sheet of East Antarctica. It is 14,000 square kilometres in area and up to 500 metres deep. It hosts a variety of microbes and microfungi.[22]

Likewise cyanobacteria have been identified which can live within Antarctic rocks or else on Jerusalem roof-tops at 80°C. Meanwhile, cyanobacteria ensconced within Antarctic desert sandstones receive so little moisture that their carbon turnover time (from fixation to release as carbon dioxide) is 10,000 years.[23]

Slow metabolism due to ambient cold can be a factor in the achievement of longevity within many microbial species. Work at the Russian Academy of Sciences on bacteria within Siberian permafrost indicate that some can remain alive, if quiescent, for three million years. Other microbes may live in ice for 160,000 years and more. A more general manifestation of this tenacious resilience is the propensity to proliferate new species.

Panspermia

Ideas about life being seeded around the cosmos can be traced back to classical times. But interest in this possibility received a strong fillip when Louis Pasteur's microscope-led studies finally demolished what had originally been Aristotle's counterproposal – namely, that moderately advanced life could generate "spontaneously" from inanimate matter. Thus in his 1871 Presidential Address to the British Association for the Advancement of Science, Lord William Kelvin aired the possibility that meteorites might carry seeding organisms from planet to planet.

More influential in this regard was Svante Arrhenius, the versatile Chemistry Nobel Laureate from Sweden who *inter alia* was reinvigorating the debate about "greenhouse gases". A core consideration behind his belief in global seeding was his expectation that free microbes with diameters not above one micron (a millionth of a metre) would resonate with Infra-Red radiation pouring forth from stars. The radiation pressure the microbes would thus be subject to might impel them outwards.

One objection was that many microbes are susceptible to damage by shorter wave radiation. Another could have been that any radiation pressure would not be exercised consistently. Once a microbe was outside its parent stellar system, it might be impelled all over. Gravitational influences could be irregular, too. Nor was this problem just one of planet to planet or star to star. It surely applied orders of magnitude more to intergalactic travel. In the febrile post-Sputnik years, there may have been a disinclination to explore these issues lest a perceived biological threat from Space complicate our approach to the one latent in the Cold War.[24]

The lead expositors of panspermia in the more modern context, often publishing together, have been Chandra Wickramasinghe and the late Fred Hoyle. In one of the last of their joint statements, they summed up the current understanding as they divined it.[25] The gene set coding for 256 different proteins can be taken *pro tem* as a minimal information content for cellular life. The chances of this being achieved randomly on a given occasion may be as low as one in 10^{5120}. But there could be, in our galaxy alone, 10^{33} tons of interstellar dust, quite a bit of which seems, on preliminary spectroscopic sampling, to be closely akin to freeze-dried bacteria in that it is hollow and organic.

The source of the organic particles could well be cometary. After all, some analysts believe there could be over a 100 billion comets in the outer solar system, their spread extending up to a light year away. All or nearly all of the comets we terrestrials encounter are ones from this assemblage, having been nudged or tugged at some point into highly eccentric orbits. Yet there may over time be a few derivative from other stellar systems. In which connection, it has been claimed that something approaching one per cent of the dust grains that enter our atmosphere from without do so at over 100 kilometres a second; and this is seen as indicating they have come in from beyond the solar system.[26] Moreover, comets may expel organic particles. Halley's did so in great bursts during much of its close passage in 1986; and was adjudged to have a highly carbonaceous nucleus.

An alternative rendering of the Panspermia thesis, discussed by Hoyle and Wickramasinghe, is that bio-material is exchanged between the inner planets of our solar system. Meteorites have been recovered on Earth which have originated on the Moon or Mars. Early but still contested evidence to date of life extraterrestrially is a meteorite of Martian origin found in 1984 in the Alan Hills region of Antarctica. The question outstanding is whether the "very ordered structures" visible on cut sections are micro-organisms.[27]

Studies have suggested that organic materials contained within bodies 200 metres or more across (e.g. asteroids or large meteorites) would be unlikely to withstand the heat and shock associated with transit to impact through an atmosphere of the same sea-level density as at present (defined as "one bar"). For smaller bodies, the impact velocities would be lower though the thermal stresses sometimes more acute.

However, models have suggested that the atmosphere 3800 million years BP, say, could have been of 10 bars surface pressure; and this could have allowed comets 200 metres or more across to deliver biogenic substances tolerably intact. Additionally or alternatively, disintegration to fragments at high altitude may have been the pattern. Another possibility mooted is that high-velocity impact could sometimes have triggered organic synthesis in the terrestrial materials effected.[28]

Otherwise biogenic interstellar clouds deriving from comets might ease the transport problem somewhat. After all, stellar systems revolving within galaxies do pass through interstellar clouds occasionally. Questions remain, however, about how microbes survive a lack of air, water, warmth, radiation immunity and so on. Talking about species definition when reproduction is asexual is not easy because sexual pairing is itself defining. But it may not be unreasonable to talk of there being millions of microbial species on this Earth alone. Many of them are specifically adapted to cope with particular extreme conditions but maybe not some vast assortment thereof. On the other hand, going into deactivation or suspended animation is a generalized microbial response which can be extremely effective. Besides, Hoyle and Wickramasinghe had a point when they suggested that, in dense clouds, panspermia might still be viable if only one bacterium in a trillion (10^{-12}) remained sufficiently intact.

But a further issue is how far organic molecules observable in deep Space are produced by life processes as well as potentially being contributive to them. Asymmetric molecules such as amino acids assume two modes, a right-handed R and a left-handed L. Biological syntheses on Earth yield only L molecules whereas non-biological lead to a mix in equal proportions. The latter circumstance has regularly obtained to date with meteoritic organic matter. Nevertheless, in the debris of the Murchison impact north of Melbourne in Australia in 1969, a small excess of L was recorded (after contamination tests) in several amino acids.[29] This finding seems now to be quite widely accepted, notwithstanding a previous history in this field of results being compromised by contaminants.[30] Additionally, Chandra Wickramasinghe is persuaded that many of the scores of morphologies identifiable within the Murchison residues resemble those of known microbial species.[31]

Relevant, too, is the pioneering work by Stanley Miller, then a doctoral student at the University of Chicago working under Harold C. Urey. One outcome of his subjection of gaseous mixes to electrical discharges was the formation of some 18 amino acids which we find are also present in the Murchison remains. Moreover, the relative abundances show surprisingly close concurrence.[32]

Now interest is burgeoning in comets transporting organic materials, these eventually to be distributed either by simple impact or else (as per Halley in 1986 and Hale-Bopp in 1995–7) via explosive outbursts. In 2006, we should receive on Earth material extracted from Comet Wild 2 in January 2004. Meanwhile, imagery from that extraordinary foray "has provided startling evidence of a seething hot cauldron of organic material bubbling beneath a frozen crust".[33]

Prospectively, one of the most promising lines of panspermia enquiry is the use of sterile collection systems carried by balloons into the Earth's high stratosphere. In 2001, a collaborative venture took place involving the Tata Institute at Hyderabad and Cardiff University with follow-on support from the University of Sheffield. This found hard evidence of bacterial clumps at 41 kilometres, well above the local (i.e. Hyderabad) tropopause at 16 km – the very highest any bacteria ascending there from the Earth's surface would normally reach.[34]

What then if panspermia can or once did work? What if one stellar system may seed a thousand others? How radically could this qualify the one chance in 10^{5120} Hoyle and Wickramasinghe spoke of? Must this whole debate hinge on statistical procedures more finely honed than any yet to hand?

Advanced Evolution Deferred?

May this layman air a further thought. It is striking how soon, geologically speaking, microbial life emerged once the heavy bombardment had peaked out. Yet even more striking is its taking another three billion years for hard skeletal remains (very largely invertebrate) to feature firmly in the fossil record, this being the development which defines the Pre-Cambrian/Cambrian boundary.[35] Is this because more softly structured life forms (bacteria, algae, plankton, worms . . .) were so adaptive to a not too unsteady environment that natural selection afforded little scope for progress to what we see as higher life? Did the RNA–DNA transitions prove difficult to make early on? Had the advent, rather over two billion years BP, of eukaryotic cells (i.e. ones with a nucleus) likewise been hard won? Had extra-terrestrial imports been imperative at any stage? How cosmically representative may this Earth-bound experience be?

The Cosmic Density of Life

Disk-like configurations aligned with the equatorial planes of spherical bodies are a feature of our universe. The giant planets in our solar system have rings. Already disks are observable around many young stars; and are often called "protoplanetary" because of their broad similarity to the one out of which our Sun's planets formed. In some binary star situations, gas escaping from the one star is being captured by the gravity of the other. Accretion structures can likewise exist around the huge black holes which seem often to be at the centres of galaxies.[36]

Within a consolidating rotating star, loose material in the equatorial plane will tend to move inward ever more slowly while other such matter around the rotation axis will fall more and more quickly towards the equatorial plane. Accordingly, a high proportion of young stars exhibit disks. More needs to be ascertained about such disk characteristics as turbulence patterns and magnetic fields. Elongations of the latter may help explain the long narrow jets of gas sometimes emanating from youthful stars. Yet despite such persisting uncertainties, one can feel confident that a sizeable fraction of stellar disks will consolidate into the planets and moons on which life might most readily thrive.

A planetary blob orbiting a star other than our Sun was first detected in 1995. Since when, over 100 such sightings have been made. Thirty years hence, say, there should be thousands as efforts intensify and technology advances. So far detections have almost all been via a largish planet or protoplanet's gravitational influence on the star it orbits. But Space telescopes operating in the infra-red (where a planet's light is less swamped by that of its star) promise to broaden optical discrimination considerably.

The fact that many planetary assemblages will not as yet have consolidated need not totally preclude microbial life. A useful analogy could be that microbes may flourish profusely in water clouds in our terrestrial atmosphere. But the basic assumption still is that the emergence of ecosystems more advanced in terms of the standard criteria would require firmer platforms.

A Mediterranean Syndrome

To bring the story briefly much closer to the present, there is a geophysical aspect which must have influenced the emergence of the hominid family to which our *homo sapiens* species belongs. This was a drying up of the Mediterranean on several occasions, some 5 million years ago. This is liable to have had major implications ecologically for the hominids' locale in East Central Africa.[37] [38] Natural selection will duly have called forth qualities of adaptiveness able to compensate for shortcomings thus highlighted. Certain of these positives could have brought those concerned closer to godliness. No matter that we must here be talking about random concurrences, not the deliberate intervention of an Unmoved Mover. How much closer such outreach could bring us is something to ponder.

The Tsiolkovsky / Fermi Question

Since 1959 astronomers in the USA and several other countries have been quite actively engaged in the Search for Extra-Terrestrial Intelligence (SETI). This has involved scanning the heavens for electromagnetic emissions signifying intelligent origination. So far, nothing has been confirmed. However, scanning capability is now advancing radically. So if SETI continues several more decades without results, many people will conclude there may be no advanced life elsewhere after all.

Popular interest in the West in this matter is sometimes traced back to a lunchtime chat at Los Alamos in 1950 involving Enrico Fermi (1901–54), the Italian–American refugee from fascism who gained a physics Nobel Prize in 1938 and then in 1942 generated the first self-sustaining fission reaction in uranium. Flying saucers glided into the conversation and extra-terrestrials too. At which point, Fermi asked, "Where are they?" Why have we had no authenticated visitations? He went on to argue that these might have been expected from life forms which had had time to evolve to a higher level than ourselves.

Konstantin Tsiolkovsky had posed the same enigma perhaps a decade before. In an undated essay in Russian (retained in Kaluga in the Tsiolkovsky State Museum of the History of Cosmonautics), he had written "Millions of milliards of planets have existed for a long time; and therefore their animals have reached a maturity which we will reach in millions of years of our future life on Earth. This maturity is manifest by perfect intelligence, by a deep understanding of Nature and by technical power which makes other heavenly bodies accessible . . .".[39] What is nowadays more apparent, alas, is that, whereas such reasoning presumes that a superlatively intelligent consciousness must be the apotheosis of evolutionary progress, it could as well be the route to precipitate oblivion through ecological destabilization and intra-specific conflict. Frank Drake, the SETI pioneer, has taken this consideration on board. He has inferred that throughout our galaxy there could be 10,000 civilizations able to communicate across Space, assuming *inter alia* that such advanced social orders averagely last 10,000 years.

Also to reckon with is the possibility that, however tough life on Earth can often seem to its sentient inhabitants, our "pale blue" planet is actually even more of a life-friendly oasis than usually envisaged. Does the proximity of a relatively large moon – the Moon – keep the Earth's spin axis sufficiently stable to avoid unbearable temperature swings? Have the orbital locale and gravitational strength of Jupiter moderated, the extra-terrestrial bombardment of Earth? If so, we might play down further the prevalence of advanced life elsewhere.[40]

A further factor to feed in is that broad swathes of any galaxy are outside its Galactic Habitable Zone (GH) for any of a number of reasons. Heavy metals may be in too short supply to allow of planetary formation or else so abundant that unsatisfactory giant planets appear. Supernovae may have been too near. Not enough time has passed for sentient life to evolve. And so on. Research in this field has a long way to go. But the tentative indications are that not more than a tenth of the stars in our Milky Way are within its GHZ.[41]

Not that one is entitled to expect life elsewhere always to follow an evolutionary progression closely akin to our own. As and when it does have thought processes, these may not be directed the way ours are. Nor may life elsewhere be individuated to the extent it is on Earth. Nor may it invariably differentiate between plants and animals as sharply as does ours. How cosmically widespread, for example, are the chlorophyll twin molecules: chlorophyll-a ($C_{55}H_{72}O_5N_4Mg$) and chlorophyll-b ($C_{55}H_{70}O_6N_4 Mg$)?

My own hunch is that (with ancillary support from panspermia) life has emerged, albeit rather thinly, throughout the cosmos. However, much of it may yet be in quite lowly forms. Let us recall that the late Stephen Jay Gould insisted that even on Earth we still live in an Age of Bacteria. This is as judged by biochemical diversity, range of habitats, and resistance to extinction; and maybe, "if the deep hot biosphere of bacteria within subsurface rocks matches the upper estimates for spread and abundance, even in biomass".[42] It is a perspective not incompatible with his notion of the continual "punctuation" of evolution. Cosmic bombardment may be a big factor here.

Though usually spreading mayhem, it may also facilitate fresh starts, sometimes assisted by panspermia. Still, specialists have debated the utility of the "punctuation" concept ever since its enunciation in 1972.[43] Most distractive are the endeavours of certain Creationists to appropriate punctuation.[44]

A Weak Anthropic Principle?

All of which could be consonant with the Weak Anthropic Principle currently modish among astronomers. It may have been best defined thus. The observed values of "all physical and cosmological quantities are not equally probable but they take on values restricted by the requirement that there exist sites where carbon-based life can evolve and by the requirement that the universe be old enough for it to have already done so".[45] The Strong Anthropic Principle, as defined by the same authors, says that the "Universe must have those properties which allow life to develop within it at some stage in its history".[46] This could seem like a distinction without too much difference. In any case, should we not be talking more subjectively about a spectrum of relatively weak or relatively strong Anthropic Principles?

Stephen Hawking has instead turned the spotlight back to "intelligent life".[47] It is a focus evidently open to the same objection as the Fermi/ Tsiolkovsky contradiction. Besides, some might see this as a first step towards re-establishing the pre-Copernican notion of human paramountcy: as the Psalmist rendered it, "lower than the angels" yet still crowned with "glory and honour". Surely the vastness and complexity of Creation is enough to put the kibosh on that. And is there not a deeper issue? We may have spent too long across the generations asking what God and Fate do through the universe for us. Perhaps the time is ripe to ask what we (as a distinctive life form, albeit among countless others) do for the universe. True religion is about Grace received through giving.

7

CONSCIOUSNESS

TWO PROPOSITIONS MUST HERE BE VENTILATED. The former says that the twinned concepts, Life and Consciousness, should regularly be integrated into accounts of the cosmos and its ways. All too often even now, overviews of mainstream cosmology either ignore Life or treat it as incidental. How can this be? Is not Science suffused with the Aristotelian precept that Nature does nothing in vain? And must not Life development be no exception?

In the Britain of the interwar years, the two great exponents of cosmology and pioneers of its development were Sir James Jeans and Sir Arthur Eddington. The Augustan elegance of Jeans' *The Mysterious Universe* inspired me in adolescence and does so still today. The fact remains, however, that it considers extra-terrestrial life but briefly and dismissively, the disposition being to underline human uniqueness. Only one star in every 100,000, Jeans opined, would be likely to have even one planet revolving within a life-friendly zone.[1] Similarly, in 1924 Eddington felt "inclined to claim that at the present time our race is supreme; and not one of the profusions of stars . . . looks down on scenes comparable to those which are passing through the rays of the Sun".[2] Each was subscribing to an "exceptionalist" perspective too emphatic to be warranted scientifically, then or ever since.

Which brings one to the second of the propositions alluded to above. This is that Life is spread across the cosmos sufficiently extensively to play a special part in its registration. Since "nothingness" has to be recognized for what it is, so too does "somethingness" (see Chapter 5). This precepts resonates closely with the Idealist tradition in Western philosophy considered in Chapter 4 in the context of its early modern revival. It is a metaphysics-driven top-down approach. How far is it corroborated by bottom-up empiricism?

The Kantian Perspective

Of no little import for today is how notions currently emergent about a living cosmos were anticipated by Immanuel Kant (1724–1804), currently the best regarded of the Idealists. He saw his "island universes" (i.e. galaxies) as all containing many stars, as per our Milky Way. A high proportion of stars would have planets, a very high proportion of which would bear life. How "things of different natures . . . seemingly worked for such excellent co-ordination" bespoke to him "an infinite intellect in which all things were designed". That he was envisioning, too, the prevalence of Life with advanced Consciousness comes out in his expecting Mercurians to be dullards compared with ourselves, whereas any Saturnians would be much keener intelligences. The reason given, in this 1755 text, was that coarser materials would tend not to be thrown so far from the Sun centrifugally.[3]

A Lag in Comprehension

Time was when the sardonic might say of cosmology as a whole that, since its data was so scanty, its theories could evolve unconstrained. Needless to say, this has never really been so, not since Classical Greece at least. Today it would be invalid except (and it is a big exception) on various conceptual frontiers. Among these may be the nature and role of whatever smacks of cosmic consciousness.

The study of the human mind is germane. By 1914 there was dour altercation between the "Behaviourists" and the "Stream" literati, to deploy a term the social philosopher William James (1842–1910) coined *c.* 1890. Behaviourism was long dominated by a triad: Ivan Pavlov (1849–1936) of Leningrad; J. B. Watson (1878–1958) of Johns Hopkins; and B. F. Skinner (1904–90) of Harvard. It has always played down the cardinal tenet subscribed to, in one form or another, by the Austro-Swiss community of analysts (Adler, Freud, Jung . . .). This was that each individual's outlook largely derives from underlying thought patterns and emotions which may lend themselves to redirection but which can never be discounted. Instead, a high premium is placed by Behaviourism on the objective (preferably numerical) recording of responses to external stimuli.

Conversely, the Stream of Consciousness takes full account of the *melée* of words and images endlessly passing through anybody's mind. William James effectively launched it as a *modus operandi* for professional psychologists. His brother Henry is seen as its trail-blazer within the literary fraternity. Others usually deemed to have been on board include William Faulkner, Gertrude Stein, Virginia Woolf and William Yeats.

Against this talent galaxy were pitched some of the most eminent philosophers: Bertrand Russell, Gilbert Ryle, Ludwig Wittgenstein . . . Also through the 1930s, Behaviourism drew much support from the broad Left, an ambience in which Pavlov enjoyed great repute. Where it became ascendant in academe, Consciousness along with Freewill virtually dropped out of psychology and philosophy syllabi. Little or no quarter was given, perhaps because Skinner himself had long struggled to deny or forget how tempted he had felt, during a troubled early manhood, to try his hand at Stream of Consciousness literature.[4]

Consciousness Defined?

A computer plays chess with a theology student. The latter is widely understood to possess Consciousness, the former usually thought not to. Does this difference bear on the prospects for this contest? Is the metaphysics relevant more broadly?

In 1995, David Chalmers, a prominent figure in the Consciousness debate, identified as the "hard problem" the subjective reaction left in the brain's response after every behavioural and cognitive aspect has been counted in.[5] But in 1998 and 1993 respectively, Roger Penrose at Oxford and the late Francis Crick agreed it was premature to try and define Consciousness at all precisely.

Nevertheless, Crick advanced what he dubbed the "astonishing hypothesis" that one's joys, sorrows, memories, ambitions, sense of identity and free will are "no more than the behaviour of a vast assembly of nerve cells and their associated molecules".[6] Together with Christof Koch of Caltec, he was seeking to understand visual awareness via the modelling of neuron links in our brain. His averrations that this could be the entrée to a general comprehension of brain function were matched by his scorn for any suggestion of bringing in philosophers and theologians, given their general lack of scientific expertise.

Similarly esoteric was an approach adopted by Penrose around this time. Interacting with Stuart Hameroff of the University of Arizona, he focused on the role within neurons of microtubules: micro-porous protein structures, some form of which is widespread in living cells. Penrose felt these might lend themselves, in the neural context, to quantum gravity fluctuations large enough for Consciousness discernibly to emerge.[7] Such an outcome would bear on the relationships "between the physical world, our mental world and the Platonic mathematical world". In which context, he cautioned against seeing the mental world as so rooted in the physical that "you cannot have souls or so on floating around without any physical basis" or that "the whole of the Platonic world is accessible to our intellect".[8]

However, his microtubule hypothesis was soon subject to astringent criticism, notably from near colleagues of Crick's at San Diego. Is human thought not more algorithmic than Penrose suggests? Are microtubules really adapted to the role envisaged? Where is the evidence of their being thus engaged? Might Platonism be a myth?[9] Roger Penrose's latest major study, though billed as "a complete guide to the laws of the universe", has no indexed reference to Consciousness *per se.*[10]

More generally, something of an impasse prevails. Consciousness studies have barely taken off. Neuroscience has progressed a long way yet still has far to go. Though interaction between it and the metaphysical approach has scarcely begun, many of those active in the quest for explanation do profess confidence that within decades one will be forthcoming. However, some of us are too awed by the qualitative uniqueness of Consciousness to believe in a resolution so near-term if, indeed, ever. But all concerned can at least agree with Susan Greenfield that progress on Consciousness has been retarded by an obsessive desire to avoid subjectivity on the part of scientists plus a chronic failure to achieve multidisciplinary collaboration to the extent required.[11]

Engaging Metaphysics

Lately a bold response to current quandaries has come from the Centre for Mind and Cognition at San Diego. It depicts a cosmic milieu characterized by three ontologically equal domains. There is Space–Time which contains all physical matter. There is Phenomenal Space, the mind's understanding of reality. There is the "real time" an observer records.[12] The panache notwithstanding, one is bound to wonder whether such metaphysical physics is yet ready for the Consciousness debate; and whether, in any case, that debate is ready to receive it.

Whether such elaboration is, in any case, compatible with the Aristotelian–Occamite precept of keeping explanations as simple as may be viable could depend on what one believes the role of Consciousness may be. Is it just to confer on any living thing thus endowed a tactical advantage in the survival battle? Or does it have a wider purpose within a process of registration which, in the final analysis, is what makes Creation for real?

An exploration of these issues needs to involve both the recognized branches of the Philosophy of Science, the epistemological and the metaphysical. The former asks how valid knowledge acquisition may contingently be. Can given methodologies unearth permanent truth? How far can competing theories be objectively resolved? Are experimental results being stultified by theoretical assumptions? Obversely, the metaphysics considers those attributes of the natural world as described by science, which could be philosophically challenging. Do all events have identifiable natural causes? Is everything reducible to physics? Does Nature have wider purposes?

A Great Divide

One is here looking towards the progressive easing of the schism between Natural Philosophy (alias Science) and General Philosophy that became so pronounced during the Renaissance Continuum of the seventeenth and eighteenth centuries. Natural Philosophy waxed strong early on in physics and astrophysics while General Philosophy was characterized by the development of political thought but also a revival of philosophic Idealism.

Benjamin Franklin (1706–90) and, of course, Immanuel Kant were signal exceptions to the divergence of interests. So by dint of their mathematical zeal were René Descartes, Pierre Laplace and Gottfried Leibniz. Otherwise those who shone in what then was called "natural philosophy" (e.g. Boyle, Galileo, Harvey, Hooke, Laplace, Linnæus, Newton . . .) were not much disposed to venture beyond it. Obversely, most who have stood the test of time as philosophers in the strict sense (Berkeley, Hobbes, Locke, Spinoza . . .) were amateurs at best regarding what still we refer to as natural or pure science.

Having well identified this dichotomy in respect of those two centuries, Steven Fuller relates it less successfully to the nineteenth. Two of the aspirant bridge-builders he identifies as scientific "losers" are Jules Henri Poincaré (1854–1912) and Ernst Mach (1834–1916).[13] But Poincaré would widely be seen as the leading mathematician of his generation, not least by virtue of his work in such applied areas as mathematical physics and celestial mechanics. Not a few scholars have been, or are, persuaded that he might well have enunciated Relativity theory had not Einstein done so first.[14]

As for the polymathic Ernst Mach, to this day aerodynamicists endlessly use "Mach numbers", measures of speed whereby the velocity of sound in air at the given altitude is taken as one. At a more profound theoretical level, there was a taut and prolonged dialectic between Mach and Einstein. The latter in due course reflected that "even those who think of themselves as Mach's opponents hardly know" how much of his views they have imbibed.[15] The fact that Mach so obtusely denied, on philosophic grounds, the reality of atoms and molecules no more reveals his whole self than did a taste for alchemy in the case of Isaac Newton. The chief reason why the gap between Philosophy and Science remained wide was that very many scientists were persuaded that Philosophy, as institutionalized in established religion, was obtusely antipathetic to their enquiries, not least in regard to sentient life. Bishop Wilberforce's verbal extravagance did much to engender this view.

From early in the twentieth century, the Logical Positivists sought to reforge links between Science and a Philosophy which they strove to make pertinent and, hopefully, godless. However, the prophets of Logical Positivism (Carnap, Russell, Wittgenstein . . .) were to be dismayed by the facility with which Science was harnessed to the juggernauts of 1914–18. Then two decades later, Logical Positivism was rocked by the collapse of the nodal "Vienna circle" (in which Mach and later Carnap had been prominent) hard upon the 1938 Anschluss occupation of Austria by Nazi Germany. After 1945, its influence diffused anew but much more in the humanities than the sciences. Meanwhile, not a few scientists (cosmologists, nuclear physicists, biologists . . .) keenly communicated to the public at large what they saw as the philosophic implications of their respective subjects.

Not that the links between the scientific and the metaphysical ever sundered completely. One perspective which militated against this was philosophic Idealism: the disposition to treat, in one way or another, concepts or ideas as more basic than substance. Not infrequently, Sir Arthur Eddington has been dubbed an Idealist. He urged that, while recognizing "that the physical world is entirely abstract and without *actuality* apart from the linkage to Consciousness, we restore Consciousness to the fundamental position instead of representing it as an inessential complication occasionally found in the midst of inorganic nature at a late stage of evolutionary history".[16] According to this approach, obtaining anything like a synoptic view of the Cosmos must depend on treating Idealism and Consciousness as twinned.

Likewise Sir James Jeans advanced the proposition that our cosmos is fundamentally a "universe of thought", the creation of which was an "act of thought". He quoted this passage from George Berkeley (see below): "All the choir of Heaven and furniture of Earth . . . have not any substance without the mind" of some eternal or created spirit. Also cited is this opinion of Plato: "Time and the heavens came into being at the same instant in order that, if they were ever to dissolve, they might be dissolved together. Such was the mind and thought of God in the creation of time."[17]

Philosophic Idealism Today

A signal example of contemporary bridge building in this domain is afforded by Roger Penrose, someone who can fairly be seen as *primus inter pares* among cosmic theorists and who also adopts a neo-Idealist perspective. He has often extolled Plato for his belief in concepts existing in "a timeless ethereal sense", a belief Penrose equates with the "absoluteness of mathematical truth" and its primacy over the material world.[18] Tentatively he discerns something of a prevision of "Big Bang" theory in Plato's notion of a point "before the beginning of years" with an Earth "without form or void", a point at which the *demiurge* [19] resolves to generate the Cosmos out of Chaos[20] – meaning, in Plato's day, a boundless emptiness. Certainly this great thought experiment was quintessential Idealism as well.

Otherwise Penrose has cited, as exemplifying mathematical "absoluteness", the "Last Theorem" enunciated (in 1637) by Pierre de Fermat. Until 1995, this intricate proposition lacked a published proof. Yet ever since Creation, it must have been a reality.

No less interesting is John Wheeler of Princeton, a Project Manhattan veteran and quantum cosmologist *par excellence.* Shortly before reaching his ninetieth birthday in 2002, he resolved to revert for the rest of his career to figuring out how the universe came to exist and what this connotes. He is reportedly intrigued by a 1984 experiment at the University of Maryland which, with an architecture of mirrors and slits, indicated that whether the constituent photons of a shaft of light appeared as wave forms or particles depended on whether or how the shaft was being observed.

Wheeler has lately been conjecturing that an ability thus to modulate the present connotes a considerable ability to shape the past too. He is said to believe that seeing "the universe as a vast arena containing realms where the past is not yet fixed" could hold the key to explaining how anything has come to exist. Meanwhile, his close colleague, Andrei Linde of Stanford, seems even more strongly persuaded that conscious observation is essential to the universe being: "The moment you say the universe exists without any observers, I cannot make any sense of that. I cannot imagine a consistent theory of everything that ignores consciousness." He reportedly regards as a virtual corollary the proposition that the most we can say about the universe is that "it looks as if it were there ten billion years ago".[21]

Meanwhile a disposition is abroad to view the whole cosmos as a quantum computer. Every photon and every sub-nuclear particle stores some data. Entropy relates to the situations and velocities of the molecules in a given substance.[22] It all seems closer in spirit to the philosophic Idealism of three or four centuries ago than to the prosaic scientific realism/materialism modish in the pre-Einstein world.

George Berkeley

Evidently René Descartes, by drawing a radical distinction between Mind and Body, was magisterially influential in energizing the debate out of which philosophical Idealism emerged. Nevertheless, it was the Anglo-Irish George Berkeley (1685–1753), Bishop of Cloyne from 1734, who best spelt out an Idealist philosophy free from mathematical or scientific distractions.

His seminal contribution, virtually complete by the time he was 28, is well summated in a three-part dialogue he crafted in 1713 between Hylas (a scientifically-trained sceptic) and Philonous (Berkeley himself).[23] The primary issue he tackled at two levels. The one was recourse to common or garden experience which he felt vindicated "mentalism" as opposed to realism. We humans do not all appreciate equally the same tastes (p. 271), heat sensations and so on. We subjectively measure Time (so he thought) by the rate at which ideas flash through the mind. The quicker they do, the longer any external event appears to take (p. 281). With "regard to the Copernican system" (p. 384), we do not perceive the Earth to be in motion. But we would, were we detached from it.

On the higher plane, he mused on the cosmos and "its boundless extent with all its glittering furniture" (p. 303), a scene neither human "sense nor imagination are big enough to comprehend . . . " (p. 303). Berkeley clearly belonged to the large genre of thinkers who have seen proof of an overriding "Mind, Spirit or Soul" not in arbitrary or miraculous events but rather in holistic comprehension and in orderliness, in how "the vast bodies that compose this mighty frame . . . are . . . linked in a mutual dependence and intercourse with each other, even with this Earth . . . " (p. 303). Noting, too, how steadily a fountain of water rises, breaks and falls due to the "principle of Gravitation" (p. 360), he averred that similarly "the same principles which, at first view, lead to Scepticism, pursued to a certain point, lead men back to Common Sense" (p. 360). He, as Philonous, taunted Hylas with not knowing what might be meant by matter existent (p. 317) nor what form the overriding cosmic force took (p. 335). Berkeley saw himself as the purveyor extraordinary of plain common sense.

Most modern philosophers would probably receive Berkeley well. This they would do in spite or maybe because of the engaging frailty of much of his reasoning. Our focus of concern for now must be how he renders the role (within the Cosmos) of Life and the Consciousness associated with it.

If one admits of overall control by an extra-cosmic power which is unalterable and "indivisible" (pp. 306 and 326), what special contribution can Life make to the registration of existence?

Must one not postulate that Life functions within a registration process which is entirely within our Cosmos, not at all as an extrinsic god-like influence? But what then happens to the registration of substance and process when these are out of contact with Life, the "Schrödinger's cat" dilemma? It is hard to imagine how, even on our little life-sustaining Earth, even a thousandth part of the total mass is being sensed by Life at a given time. So what if one endorses the notion of existence itself slipping into abeyance when it is not being consciously observed? How can such observation be everywhere sustained? Should we say, after all, that all matter has some degree of Consciousness? Can we otherwise surmise that organizational complexity will achieve more for a given mass? Might momentum or kinetic energy do so? Are there threshold values?

Then do not certain higher life forms, ourselves included, achieve a unitary and in-depth Self-Consciousness which represents a decided qualitative advance? In what was the former's last academic paper, Francis Crick and Christof Koch proposed that, in human beings, the requisite integration might take place in the claustrum, a thin sheet of grey matter at the base of the cerebral cortex.

Bounded Consciousness

After Matter versus Energy, the sharpest qualitative divide throughout Creation appears to be that between the living and the inanimate. From the molecular level upwards, many of the materials directly involved in the life process are strikingly more complex. Theoretically, the contrast can be delineated in terms of entropy, a truism which was well defined by Erwin Schrödinger when *c.* 1950 he digressed awhile to reflect on the life phenomenon. Any reaction a living organism is involved in cannot but result in entropy increases overall, as per the Second Law of Thermodynamics. Nevertheless, the organism can thereby reduce its own entropy, thus underpinning its chemical complexity and its accentuation of Consciousness as well as prolonging its life.[24]

Manifestly we humans are much inclined to see our own Consciousness as being, for intensity and refinement, in a class by itself. An inclination to self-celebration may well be something we share with chimpanzees, elephants, porpoises or whoever as well as, retrospectively, with Neanderthals, mammoths, dinosaurs, trilobites . . . Maybe an elephant is mighty proud of being so elephantine. In our case, however, a justification on the Consciousness front is afforded *a priori* by the big boost our very exceptional linguistic ability gives to our comprehension. One has only to observe an infant talking to itself to appreciate the contribution language is incipiently making to the ontogeny of that individual and so in the longer haul, to the phylogeny of the species into which she or he has been born.

However, it is one thing to acknowledge exceptionalism. It is something else to endorse immoderate claims for it. Sometimes one is told we are the only animals on this planet who know we have to die.[25] But are we really alone in this? The animals in one's garden continually devote far more time and energy than do we to deferring the inevitable. Besides which, we as a species find death hard to face squarely. The source just cited remarks how some psychologists believe that the mechanisms our minds evolve to repress a dread of dying "are at the root of how we construct our societies . . . treat others . . . and see ourselves". It is a two-edged interpretation of what heightened Consciousness is about. It surely suggests the absence of any very basic divide between ourselves and the other mammals.

Susan Blackmore's reading of the specialist literature on snakes has given her to understand that one will track, kill and ingest a mouse through stages marked by sharply contrasting sensing modes: light or heat; then smell; then touch.[26] These may constitute a sound operational sequence. One does wonder, however, whether these sensing disjunctions rather cut across the standard notion of advanced animals attaining a peculiarly unitary Consciousness. Maybe not.

Moving on outwards, as one might say, the first Consciousness divide one encounters which looks at all radical may be as between the animal and plant kingdoms. Granted, a big problem at this juncture is the sparse discussion in the Consciousness literature of plant life. Nevertheless, the reactive sensitivity whereby plants generate volatiles in response to a whole variety of ambient stimuli is well known. One instance is orchids which will attract pollinators by simulating the female sex scent of certain bees.[27] Tobacco plants within a colony may give off methyl salicylate to warn their neighbours or unaffected parts of themselves of disease incidence.[28] Dare one talk of Consciousness here? How consciously does the sunflower, in the words of the old refrain, "turn to her god when he sets, the same face as she turned when he rose"? Does tightly-specific genetic programming make talk of Consciousness superfluous?

More critically to be evaluated on this score, however, are the viruses. For they cannot be said to have the measure of autonomy one otherwise associates with Life. After all, they depend on host cells they have requisitioned in other living beings "for the raw materials and energy necessary for nucleic acid synthesis, protein synthesis, processing and transport, and all other biochemical activities that allow the virus to multiply and spread".[29]

Still, they can do much over and beyond multiplying themselves by binary fission. Suppose a host cell has died due to the destruction of its DNA. A virus may induce the cellular mechanism within the remaining cytoplasm to use its (the virus's) genes as a guide to assembling viral proteins and replicating the viral genome. Then again, some viruses incorporate or encode enzymes which repair host molecules. And so on. Viruses cannot be dismissed as passively inert though neither can they be unreservedly regarded as autonomously alive. Some consciousness may attach to them.

Artificial Intelligence

Outside the biological realm, on any current definition, is the dilemma presented by Artificial Intelligence. If that computer has played chess with the said theology student and trounced him, which party is the more conscious? And does not a microchip implanted in a damaged brain make it more conscious?

Debate has raged extensively but thus far inconclusively. Take the proposition (inferable from the interpretation here enunciated) that a tailor-made "thinking machine" may have heightened Consciousness albeit a sight less than its specific mathematical capacity could have led one to expect. Nothing said above rules this out. But does anything rule it in? Writing in 1950, Alan Turing cited acquiring a sense of humour as a good test of whether a machine was progressing towards true Consciousness. In humanoid terms, it could be the best test of all. But we have to face the fact that almost all the advanced life forms known to us show little sign of this faculty.

Benedict Spinoza

The emergence of Life within the Cosmos seems to me extremely hard to account for unless one does assume the latter requires a certain immanent self-awareness in order to gain registration. Yet it also seems to me that this requirement is hard to meet in any case unless one predicates that all matter down to a low, probably molecular, level of organization can be defined as Conscious to at least some small extent. In other words, one has drawn close to a "panpsychic" perspective – to the belief that all matter has, or maybe all phenomena have, this elusive mind-like quality. It is to be distinguished from "animism", the more tribal mind-set which has natural features possess defined local spirits. But panpsychism readily conflates with "pantheism", the notion of God as a ubiquitous unifier. The latter concept also has old roots though the said term entered general currency only from 1720.

The perspective here enunciated is close to that adopted by Benedict Spinoza (1632–1677), the scion of an Amsterdam Jewish family with an Iberian ancestral background. His novel recasting of Cartesian modes of thought was appreciated by Einstein and characterized by Bertrand Russell as "brief, dense and deeply obscure" yet of "unparalleled breadth, consistency and beauty".[30] This thoroughly unassuming grinder of lenses tried, through what he wrote in the 1660s, to help *Homo Europa* emerge from a gloomy one-and-a-half centuries of violent religious strife, recent events in England being much in his mind. His writings cover an assortment of themes. But his understanding of God is what attracts most attention today and is of salient interest in this context.

The essence resides in his celebrated expression *Deus sive Natura* – "God or Nature". This is to say, God is a conscious entity who or which is synonymous with the whole of Nature. Spinoza's actual definition was that God is a "substance constituted by an infinity of attributes, each of which expresses eternal and infinite essence".[31] Here "infinite" should probably be read as "infinitely many".[32] However, he would have eschewed any attempt to particularize as being an affront to an holistic Divine.

None the less, there are problems with Spinoza's metaphysics as presented in his *Ethics* – mainly in the cumbersome form of numbered propositions with demonstrations, corollaries and commentaries attached. Thus Proposition XVII of Part I states that "God acts by the sole laws of the Divine nature and is constrained by nothing". But He is evidently constrained by those laws. Thus He cannot prevent the internal angles of a triangle summating to two right angles. He remains the overseer who stands above everything although His divine immanence is ubiquitous.

Then again, Spinoza dismissed personal salvation after bodily death as illusory (Part III, Proposition III). Yet he did suggest something in the mind/soul was eternal. Witness its "adequate knowledge or cognition of the eternal and infinite essence of God" (Proposition XLVII). Moreover, his concern to have people free themselves from destructive passions by becoming more aware of them may not leave due scope for righteous anger as a worthy and contributive motivation. Nor can one be entirely happy with how his linear Cartesian logic leads him to treat individual advantage as always our overriding motivation. Also, like others who treat the future as predetermined, he was curiously keen to help shape it, to curb the Evil he otherwise saw as a God-given mystery.

Still, his confusions must have owed something to the difficulty of writing in a Netherlands in which formal republican toleration could be belied by popular religious bigotry – Christian and, the worse in his personal experience, fellow Jewish. At all events, his equating of God with Nature can fairly be seen as tantamount to seeing Consciousness in everything, albeit at varying levels of intensity (Part II, Proposition XIV).

Panpsychism to 1900

Among the pre-Socratic Greek intelligentsia, the *psyche* was seen as the attribute which engendered motion. Thales credited a loadstone with a soul since it could move iron. Heraclitus averred that all things were fired with divine spirits. Nor did the philosophic revolution launched by Socrates break with the past decisively in this regard. Plato's rendering, though piecemeal, was emphatic. Defining "soul" (in his *Timæus*) as "the source of motion", he deemed not just the cosmos *per se* to be ensouled but its component parts to be so individually as well.

Aristotle was less categoric. True, he famously adjudged plants to have "nutritive souls". More generally, he associated souls with, though only with, life forms. Since, in accordance with the said Greek tradition, he related life and souls to motion, he got into difficulty over the revolving heavenly bodies. He could hardly depict them as inert. Yet considerations of symmetry inhibited him from viewing them as potential venues of life within his geocentric cosmos. David Skrbina concludes there could perhaps be "something of a subconscious panpsychism in Aristotle".[33]

The chief development during the Hellenistic era or Late Antiquity was in regard to the atomic theory of Democritus then in vogue among Epicureans and Stoics. The concept was that atoms falling through a void might swerve "at uncertain times and at uncertain places", thus effecting atomic collisions. Such events were considered basic to human free will and also to assemblages of life forms. But any consequent tendency towards panpsychism was soon to be curtailed by the ascendancy of monotheistic Christianity. Nevertheless, a revival was to come during the Italian Renaissance, with several of its leading philosophers adhering to this perspective. One of them, Francesco Patrizi, is understood to have coined the term "panpsychism".

With this genre, Giordano Bruno stands out, his combative personality proffering strong affirmation. The Earth occupied no privileged position in the universe; and humankind enjoyed no prerogative *vis-à-vis* possession of a soul. "For in all things there is spirit, and there is not the least corpuscle that does not contain within itself some portion that may animate it." To Skrbina, the importance of this lies in its anticipating the thinking of Leibniz as well as of Spinoza.[34]

Come the French Enlightenment, religious interpretations of natural phenomena are pretty much discarded. Yet such writers as Denis Diderot and Julien La Mettrie also found uninviting any notion of mind as emerging from nowhere. They therefore proposed that a mind-like quality was inherent in all matter, a perspective which came to be known as "vitalist materialism". Between 1751 and 1780, Diderot's writings reflect the dawning of a conviction that sensitivity is what one should be talking about: "from the elephant to the flea, from the flea to the sensitive living atom, the origin of all, there is no point in Nature but suffers and enjoys". In the century following the French Revolution, thinking within or close to the bounds of panpsychism developed most strongly among German philosophers. The names of Johann Herder, Ernst Mach, Arthur Schopenhauer and Wolfgang von Goethe come readily to mind. To Schopenhauer, the unifying "mental" theme is not "consciousness" which he sees as confined to ourselves and the other animals. It is "will". This is "manifest in every force of nature that operates blindly, and it is manifest, too, in the deliberate action of man; and the difference between these two is only a matter of degree . . . ".

Ernst Haeckel (1834–1919) can be seen as the last representative of this Germanic school. By 1892, he was commenting that "One highly important principle of my monism seems to me to be, that I regard all matter as ensouled, that is to say as endowed with feeling (pleasure and pain) and motion." Nearly twenty years earlier, however, William Kingdom Clifford had written that "along with every motion of matter, whether organic or inorganic, there is some fact which corresponds to the mental fact in ourselves". Thus the panpsychic baton was passing to the Anglo-Americans. Over many years, William James, the eminent psychologist and philosopher, was repeatedly drawn to panpsychism but never took the plunge because of its discordance with various of his preconceptions.[35] Still, in due course the whole theme would become subsidiary to the great debate about human consciousness, a debate which left Behaviourists ascendant.[36]

A Dichotomy Persists

Taking the discussion of Consciousness within Britain alone, one finds it still split between holistic notions about Cosmic Consciousness and the

Earth-centred neurological approach. The former seeks scientific provenance no less than the latter. Speaking now as the Gresham Professor of Divinity at Gresham College, Keith Ward feels that the scientific quest for a "theory of everything" resonates "with the religious idea of God as a cosmic mind". He further avers that if "mathematical realities exist only when conceived by some consciousness, we can frame the idea of a consciousness in which all mathematical structures, all possible states, and all moral and aesthetic values exist". This "ultimate consciousness" can be spoken as an omniscient, omnipotent Supreme Good. We thus bring "ourselves very close to the classical Christian, Jewish or Muslim idea of God".[37] Perhaps one can do so by passing through the last of successive thresholds on an ascending scale of Consciousness. But an "idea of God" may or may not correspond at all closely to any divine reality.

Depending on precise interpretation, there may be an important rider to the notion of Cosmic Consciousness. It does look as though the peculiarly complex phenomenon we know as Life emerges within the universe in order to bolster its self-awareness as it expands and cools. What this would mean is that, as and when Life dies out because the universe has become too cold and disaggregated to support it, Cosmic Consciousness will fall below a critical threshold and the whole of Creation will therefore disappear. In other words, it cannot expand forever. Nor will it ever contract unless, rather improbably, this process is then already under way. Thus will our universe undergo the "heat death" so often talked of.

However, Susan Greenfield has defined Consciousness strictly neurologically as "an emergent property of nonspecialized and divergent groups of neurons (gestalts) which is continuously variable with respect to, and always entailing, a stimulus epicentre. The size of the gestalt, and hence the depth of prevailing consciousness, is a product of the interaction between the recruiting strength of the epicentre and the degree of arousal".[38] The Oxford Centre for the Science of the Mind (of which Baroness Greenfield is the first Director) will endeavour to blend these and other relevant perspectives. How far it succeeds will be quite indicative of, and may be quite critical to, the general prospects in this field.

PART THREE

Utopia Lost

OVERVIEW

Utopia Lost?

In many ways, being a human being is veritably a cosmic privilege. We occupy what feels like a paramount position on a planet that must be exceptionally equable and salubrious. For it is partially shielded from cosmic bombardment by the mighty Jupiter while its own gyrations are held in check by our relatively large moon. Between three and one million years ago, the challenging circumstances of our prehuman existence were such that average brain size increased by 150 per cent, an exceptionally fast rate of evolution which ultimately would endow us, their descendants, with marked virtuosity.

Even so, the current reality is that we are dangerously close to making a hellish mess of everything. Our incessant erosion of Nature is depriving us of an invaluable context for meaningful existence. The urbanized mammon so many pursue so avidly does have manifold attractions short-term but is increasingly associated with a general flight from reason. Biological warfare looms as the biggest military threat humankind has ever faced.

All these challenges have what we may term their heavenly dimension. Light pollution of the night skies is denying us starfields our forebears found to be richly inspirational. Irrationality can find religious expression at least as readily as reason or spirituality can. At every level from operational efficacy to philosophic import the weaponization of near Space would be a terribly inept response to terrestrial insecurity.

8

AN END TO NATURE?

In the twentieth century, the political process was besprinkled as never before by single-word war cries, iconic or demonic. "Freedom, Democracy, Justice, Tyranny, Racism . . . " come readily to mind. As the twenty-first unfolds, "Natural" may eclipse them all. Should material advance be allowed to distance us ever further, by default or design, from what our forebears saw as "natural"? Should we avail ourselves willy-nilly of whatever new options science and technology present in respect of security, comfort, leisure, health, longevity, reproduction, information exchange, landscape modification or settlement? Can we help ourselves doing so? But can pollution be conquered without economic growth being limited? Might failure in this domain undermine our liberty, welfare, peacefulness, sanity . . . ?

What is Natural?

For centuries, the basic concept informing Western political thought was "natural law". It was understood to be a higher form of law derivative if not from divine disposition then from how things truly were. Therefore its ethical authority was ubiquitous and everlasting.

The origins of Natural Law are generally seen as Classical Greek. Still the lineage is not easy to trace. Plato's episodic contributions on the subject point to the tautological conclusion that what is natural is what is best and *vice versa.* With Aristotle, natural law is close to customary justice and, as such, a *sine qua non* for a stable political process.

The themes become more organized with the Stoics. To them Natural Law proper was, in effect, a deterministic account of how the cosmos works.

Within that context is *ius gentium*, the human regime where individual freewill obtains. A wise and virtuous person will seek fulfillment by living in accordance with Natural Law, *ius naturale*. Somewhat similarly, in the High Middle Ages, St Thomas Aquinas deemed it feasible, nay imperative, to harmonize Natural Law with (a) Divine Law as laid down in the Bible (not least the last six Mosaic commandments), and (b) Eternal Law, the principles whereby God rules all Creation.

Come the seventeenth and eighteenth centuries, Natural Law was the formalized expression of the State of Nature, the primordial setting (historical or mythic or just inferential) which might proffer a framework of reference for current political goals. In their diverse ways, Hobbes, Locke and, in due course, Rousseau claimed State of Nature provenance. Its influence can further be traced in the disquisitions of the classical Marxists (and especially Engels) about "primitive Communism". All such visions drew succour from the "noble savage" interpretation put on reports about such forays as the serendipitous visit to Tahiti in 1768 of two French naval vessels led by Louis de Bougainville, a classicist of high repute as a mathematician and military veteran.[1]

The trouble was that those who wished to extol noble savagery could never attain the best Newtonian standards of information and deduction. So, being decently anxious to dissuade the forces of imperialism from expelling, enslaving or annihilating the native peoples they came across, they romanticized the latter. The whole mind set had its roots in the Classical Graecian legend of a departed "Golden Age".

Noble or Otherwise?

What needs be guarded against is a revival of this tradition via averrations that a simpler, plainer life "closer to Nature" could almost spontaneously prove more gratifying for "ordinary people" than what they know today. These expectations may considerably derive from the Behaviourist cultural anthropology so much in vogue last century.

Take the study of Samoa (as observed in 1925) by the very popular American anthropologist, Margaret Mead (1901–78). To her, Samoan society appeared far less constrained and more spontaneous than it actually was or, might one say, could have been. Take by way of example her understanding of punishment in infancy. She concluded that, within the family, infants received nothing more than tellings-off and "occasional cuffings".[2] But in his comprehensive 1983 critique of her study, Derek Freeman adduced solid evidence of infants not infrequently being savagely beaten as well as forced to assume submissive postures over extended intervals. He saw Mead's many misconceptions as sustained by her minimal knowledge of the local language and inadequate familiarity with previous studies as well as by her decision to base herself in an expatriate compound.[3] Today his critique would be quite widely endorsed.[4]

The primary moral of the Mead affair is that one ought not to try and reconnoitre another culture without requisite skills and with a headful of presuppositions. Meanwhile Freeman's materials also reminded us once again how much humankind remains the creature of its evolutionary past. Since the severe shrinkage ten million years ago of Africa's Miocene rain-forest, this past has repeatedly turned extremely tough. Our human and pre-human forebears survived through belligerent loyalty to pyramidal tribal groupings basically several hundred strong. We their progeny are ineluctably legatees of this *longue durée*, inimical though it must be to genial co-existence within the global mass society now taking its rather formless shape.

Nor dare we assume our adaptiveness could be decisively improved by systematic conditioning in school and elsewhere. Remoulding individuals is unlikely to erase the cumulative experience of 100,000 life spans. Besides, however good their intentions, the apostles of Behaviourism will be uncertain what to aim for. Nor will intentions always be good. Sometimes, too, bad intent may achieve results more readily.

Innate Aggression

Since 1860, social philosophy has been much affected by a quasi-Darwinian concern about human aggressiveness. Through the late 1960s some were especially exercised by the roots of violent strife, surprising levels of which were lately to be observed. Vietnam was an overriding influence. But opinion was swayed, too, by the Cultural Revolution in China and the Warsaw Pact invasion of Czechoslovakia. So was it by open warfare in the Middle East and Nigeria; American urban unrest; and a dreadful pogrom against the Chinese minority in Indonesia. All was in sharp contrast with a decade or so earlier when it had been so modish to look towards steady convergence between the West and the Soviet bloc. So was it, indeed, with the euphoria prevalent for a year or so after the resolution of the Cuba crisis in 1962.

The new mood was well articulated by Anthony Storr, a British consultant psychiatrist. He noted how an inherited sense of unremitting struggle found expression in our ready recourse to pugnacious language: "We *attack* problems or *get our teeth into* them. We *master* a subject when we have *struggled with* and *overcome* its difficulties. We *sharpen* our wits, hoping our mind will develop *a keen edge* in order that we may better *dissect* a problem." He further warned that, certain rodents apart, "no other vertebrate habitually destroys members of his own species. No other animal takes positive pleasure in the exercise of cruelty on another of his own kind . . . The somber fact is that we are the cruelest and most ruthless species that has ever walked the Earth."[5] These judgements may be overly categoric but one can recognize large grains of truth within them.

As to where lethal aggressiveness stems from and what may aggravate it, some emphasis was put through the late sixties on thresholds of overcrowding, comparisons with other members of the animal kingdom being freely drawn.[6] A modulation of this theme was afforded by Robert Ardrey. Noting the appeal that war and territorial acquisition have for *Homo sapiens*, he related it to our special needs for identity as opposed to anonymity, stimulus rather than boredom, and security instead of anxiety.[7] Through atavism to fulfilment?

Lorenz and Koestler

A more singular perspective had been afforded by Konrad Lorenz (b. 1903), the Austrian zoologist and Nobel Laureate dubbed by Sir Julian Huxley "the father of modern ethology". He sought analogies between human behaviour and that of other animals. He proposed that, if the physical attributes of the members of a given animal species are such that they could kill one another at a stroke, then their psyche will very generally have incorporated a mental block against their so doing: "A raven can peck out the eye of another with one thrust of its beak, a wolf can rip the jugular vein of another with a single bite. There would be no more ravens and no more wolves if reliable inhibitions did not prevent such actions." Conversely, doves, hares or chimpanzees feel no such inhibitions about belligerency towards individuals of their own kind because a single strike will not be fatal.[8] A modicum of intra-specific conflict helps to determine leadership and to distribute the population evenly.

The trouble with us humans is that we, too, lack any very formidable natural armament. Therefore we have no in-built mechanisms to limit our violence against human aliens. Yet we have acquired artificial aids to combat far beyond the broken branch a chimpanzee might grab. This could presage big trouble for ourselves and the rest of terrestrial Creation.

This theme was taken up by Arthur Koestler in *The Ghost in the Machine* published in 1967, an admonitory anti-Behaviourist tract that draws eclectically on Lorenz. Koestler boils his own argument down to there being a major disjuncture between the older part of our brain which provides our emotional drive to service and self-sacrifice and the "all-too-rapidly grown" modern part which generates intellect, military inventiveness included.[9]

Koestler's empathetic but piercingly stringent biographer sees good reasons why this text has not "worn well" after a reception initially favourable, at least among the young. To David Cesarini, it "represented the bankruptcy of Koestler's political and social thought; and the beginning of a move away from empirical realities into mysticism and the occult in the quest for a solution to the predicament of mankind".[10]

Specialists may long debate how far one should look to other animals for confirmation or otherwise of the naturalness of our hierarchical group bonding, the structural basis of organized aggressiveness. However, Koestler went well beyond that. He also prayed in aid the structure of books, symphonies and the cosmos itself.[11] Nor was it sensible to declaim that "The only periods in the whole of Western society in which there was a truly cumulative growth of knowledge are the three great centuries of Greece; and the last three centuries before the present."[12] None the less, Arthur Koestler showed courage in taking on board the scholarly pessimism of the likes of Lorenz at a time of intellectual turbulence in which not a few continued to extol innate human virtue with a veritably millenarian zeal. Konrad Lorenz himself was subject to verbal abuse because, under Nazi duress, he had briefly allowed that his ethology gave some validity to racial supremacism. His contemners were unwilling to acknowledge the agony of choice presented in such a situation. Make a Galileo Galilean concession or have the Gestapo trash your facilities and maybe deport self and family.

Huxley and Krishnamurti

Regarding the celebration of humankind, Julian Huxley (who, 1946–8, had been the first Director-General of UNESCO) extolled in 1988 the "uniqueness" of achieving Earthly dominance as a single species not a genus.[13] He mainly attributed this to how speech had facilitated the pooling, integration and retention of knowledge along with conceptualizing, concretely or abstractly. Complementary to it was the development of the forepaws. This helped in the definition of nearby objects. More importantly, sophisticated tools could be made and utilized.

Also contributory has been a propensity to rear small families and to do so actively across some 30 per cent of a progeny's expected life span. Laughter, too, is seen as another valuable trait, not quite uniquely but well to the fore. A further attribute Huxley believed at least some humans possessed in some measure was Extra-Sensory Perception (ESP), probably not peculiarly human but maybe another asset in our panoply.

Only a thirtieth of this long essay was given over to human frailties as Huxley acknowledged them. A rather singular propensity to be subject to several kinds of motivation proffers nuanced responses but may too often engender confusion instead (p. 409). Likewise, mental plasticity can give scope for waxing "nonsensical and perverse" instead of astute (p. 408). Then again, the range of variation in individual minds can lead to mutual incomprehension: "The difference between a somewhat subnormal member of a savage tribe and a Beethoven or a Newton is assuredly comparable in extent with that between a sponge and a higher mammal" (p. 400), a rendering which surely rates as Whiggish condescension at its most extravagant. Unfortunately, Huxley's failure to develop any of his reservations lent credence to Koestler's contention that he was altogether too dismissive of the dangerous proclivities residing within the human psyche.[14] He could have added within those forepaws, too.

Impatience with doubts about human perfectibility was all too typical of Huxley's generation of Western progressives. The essay here cited first appeared in 1937 in the *Yale Review*. In November 1938, the Nazis in Germany launched their *Kristallnacht* nationwide pogrom against their Jewish fellow citizens. The same month, Jiddu Krishnamurti (an Indian social philosopher by then assuming some prominence on the world's religious Left) responded to questions as to how he – a professed pacifist – proposed to respond to this persecution of the Jews. In reply, he agreed that "the poor Jews are having a degrading time. It's so utterly mad the whole thing". Still, "one dominant race exploits another . . . all over the world". Witness the Brahmin loss of "all sense of humanity with regard to the untouchables" in parts of South India. The answer, he felt, was for "one to be an individual, sane and balanced, not belonging to any race, country or particular ideology. Then perhaps peace and sanity will come back to the world".

This widening of the argument reflected a concern on Krishnamurti's part to justify his pacifist stance in terms of a kind of original sin. He said it was "so easy to curse Hitler, Mussolini and company but this attitude of domination is in the heart of almost everyone". He further opined (as Mussolini was wont to) how "having grabbed half the Earth, the British can afford to be less aggressive".[15]

All the same, the crux here is not pacifism. It is individualism *à l'outrance.*

Admittedly, Krishnamurti's desire to interpose nothing between the upright individual and the eternal verities might elicit some support in the Elysian fields of modern Western culture – from the likes of Bruno, Descartes,Voltaire, the "free trade" political economists or, indeed, the Behaviourists. Yet this inclination might soon lapse into a very anarchic State of Nature in which life was "nasty, brutish and short", to use the imagery famously invoked by Thomas Hobbes in the melancholy aftermath of the English Civil War. Such an ambience of insecurity could only be precursory to uncompromising authoritarianism.

In short, supposed States of Nature never took proper account of Human Nature – in particular, of the desire to bond ourselves within smallish close-knit communities, the trait identified by Lorenz then accented by Koestler as a dangerous manifestation of altruism. Recognition of how basic this need is can be a prerequisite for suitably restraining it.

Climatic Bouleversement

Still, the great counterbalance to Human Nature has to be Nature in the round. Since the mythic Dawn of History, the latter has been cherished as the setting for the former. Some 15 years ago, Bill McKibben, a youthful Anglo-American columnist, interwove stern indictment with solemn prophecy on this score. The gist was that by depriving "wild nature" of any separate existence, we have unhinged ourselves. Above all, by "changing the weather we make every spot on Earth man-made and artificial. We have deprived Nature of its independence and that is fatal to its meaning". A likely consequence will be the divorce of religion from naturalism, this exposing us to a "siege of apocalyptic and fanatic creeds".[16]

More prosaic warnings about "changing the weather" through the "greenhouse effect" (the way mounting fuel demands were altering world-wide the gaseous composition of the atmosphere) can be traced back to the early nineteenth century, almost to the delineation of the spectral absorption patterns (see Chapter 4) on certain of which greenhouse analysis rests. Yet through 1975, received wisdom still strongly was that secular cycles in the Earth's rotation and revolution were sustaining a natural cooling trend. Soon, however, informed opinion would be shifting. Monitoring the increase in the Earth's atmosphere of carbon dioxide was one influence. Another may have been comparison with Venus. Probes to that planet (and especially the Venera soft landings by the Soviets from 1967) had indicated that carbon dioxide (known from 1932 to be abundant in Venusian skies) comprised no less than 94 per cent of a dense planetary atmosphere. Surface temperatures therefore reached 750 Kelvin or nearly 500°C. Pre-1932, Venus had been much favoured for manned landings.

A turning point in public attitudes to global greenhouse warming (in the USA but more widely too) came with the great American drought of 1988. Over much of the country from the Great Plains to the Atlantic seaboard, the heat and aridity were prolonged, all this hard upon what had widely been the driest Spring for 50 years. Professional circles were understandably concerned about unduly trite media correlations between this aberrant season and global warming.[17] All the same, the UN's Intergovernmental Panel on Climatic Change (IPCC) was established that year.

At the turn of the century, the IPCC agreed its Third Assessment. In the absence of strong mitigating measures, mean global air temperature was predicted to rise somewhere between 1.4 and 5.8°C by 2100. This prognostic spread bespoke persisting geophysical uncertainties, the biggest perhaps being the extent and character of cloud cover under given circumstances.[18] However, it also reflected varying assumptions about the pace and character of economic globalization. Cited estimates for annual economic output worldwide at the next turn of century ranged between seven and twenty-two times the aggregation of Gross Domestic Products in the year 2000.[19]

Nor was total production by any means the only indeterminable aspect of future energy consumption. What gains can still be made in energy efficiency? How long will (a) oil and (b) natural gas reserves hold out? What balances can best be struck between these fluid hydrocarbons and coal, bearing in mind that coal generates more carbon dioxide per quantum of combustion than oil while natural gas generates less? What balance may be struck between the hydrocarbons as a whole and (a) nuclear power or (b) "renewable" forms of energy – e.g. wind power? Will not the Kyoto Climate Change Protocol of 1997 still be a valuable departure despite the refusal of Bush to come on board? But is Washington still too disposed to play down the urgency of the climate problem?

A range of values for air temperature rises this century apprehended by the IPCC was given above. Their median value is 3.6°C. This is more than enough to induce acute alterations in rainfall patterns and other climatic parameters, regional and global. It is important to recognize how rapid, pronounced and therefore disruptive a change this would be by comparison with virtually any other alternation in climate, global or continental, the past two thousand years.

Take the "little climatic optimum" which climaxed in and around Europe though also various other parts of the world in what we know as the High Middle Ages. In West and Central Europe, there was a gain of 1°C between 700 and 1275.[20] Closely concurrent with this quite insistent but agreeably gradual progression came many signs of enhanced well-being. Across much of Northern Europe the treeline ascended a typical 100 metres during those several centuries while arable margins widely did likewise. Vineyards were created at 54°N in Yorkshire. Granted, technological progress and capital accumulation also govern such adjustments as do cultural evolution and demography. But just how critically certain limits depended on the climate being favourable was shown when it turned inclement from the late thirteenth century.[21]

A truism that emerges from studying the historical impact of climate change is this. When societies or regimes are well founded in all other respects, they can cope with a considerable measure of climate variation. But when either are fragile for other reasons, climate alteration may pitch things critically, especially if it is sudden or rapid.

Take Europe in the aftermath of the Second World War. In 1946–7, much of the continent experienced its coldest winter since the 1840s. This event influenced the ramification of the Cold War, not least in divided Germany. Witness the launch that summer of the Marshall Plan for European economic recovery. Obversely, the winter of 1962–3 was widely the most severe since 1740. Yet on the world political stage it was absolutely a non-event. The reason was that both sides of Europe were at a peak of post-war stability thanks, first and foremost, to sustained economic growth. What some of us apprehend is that, as this century progresses, the world will become very prone to regional crises brought on by a conjunction of causes. A particular influence in the climate domain will be shifting rainfall patterns on desertic margins.

The North Atlantic Gyre

The circulation of the oceans is so massive a process that any modulation thereof could have adverse consequences regardless, this continentally or even globally. A case in point could eventually be the North Atlantic gyration, especially its Gulf Stream/Gulf Stream Drift sector. For this oceanic circulation, like others, is considerably driven by latitudinal temperature gradients in the sea though also in the lower atmosphere, notably as gauged between sub-tropical and sub-polar latitudes.

Under conditions of global warming, these gradients are liable to diminish. This is partly because ice reflects sunlight strongly, especially when the Sun is low in the sky. It acts as its own coolant, therefore responding the more readily to temperature trends in either direction. But it is also because polar marine ice (being but a metre or so thick) will extend or contract in area the more readily on that account too. A lessening of South-North temperature gradients in the Northern Hemisphere when greenhouse warming is under way will mean less propulsive energy in the oceans themselves and in the surface winds above.

Moreover, the fact that sea water is non-compressible and, in addition, much more viscous than air, means that its eddies and currents can be orders of magnitude larger and more enduring. Correspondingly, the oceanic gyres are a lot larger and far more persistent than atmospheric depressions or even most anticyclones. Clearly the inertia that keeps existing features going will also inhibit the emergence of new ones. However, a corollary is that when the pattern does basically alter, this adjustment can be rather acute – vertically as well as horizontally. Along with colleagues, Christian Pfister – an eminent Swiss climatologist – has proposed that upwellings of North Atlantic deep water may have brought on the extended spells of savage bad weather associated with secular cooling in fourteenth-century Europe.[22]

According to a 2002 report by the Aberdeen Marine Research Laboratory in Scotland, the Gulf Stream seems to have slowed 20 per cent since 1950. But another secular effect now being observed in the USA (by the National Oceanic and Atmospheric Administration and the Woods Hole Oceanographic Institution) is a big redistribution of water from the tropics towards the respective poles via the atmospheric segment of the water cycle. Warming at low latitudes considerably increases evaporation then condensation; and a good deal of the resultant precipitation occurs at higher latitudes, boosting there the supply of fresh water.[23] Not that climate prediction is ever straightforward. More salinity at low latitudes and less at higher would result in a density gradient conducive to surface flows polewards. Also less salty sea water may lose ice cover more readily. On the other hand, it has been surmised that extra water from the northerly skies could constrain a little the pre-existing circulation of the North Atlantic.

An acute curtailment of the North Atlantic gyre seems unlikely too much before the twenty-second century. But if and when it does occur, it could lead to quite sudden mean temperature drops of as much as eight degrees in North-West Europe – a paradox which well highlights the complexity of the global warming prospect. This is one of many matters about which terrestrial reconnaissance from Space has already been seminally instructive. An influential 1943 rendition of weather forecasting cited a comment from an ex-president of the Royal Meteorological Society: “If it were possible to divide the Atlantic into sections . . . thus preventing any flow of water, it would cause very little change in the climatic conditions of North-West Europe.”[24] Since when, energy emissions measured by satellite show sea currents to effect no less than 40 per cent of all heat transport between the equator and 70° N.[25]

To which one should add that the need to look well ahead in this whole field is the greater on account of the inertia inherent in the whole ocean/atmosphere system. Were a comprehensive counter-greenhouse regime to come into force tomorrow, it could take a good half century to be fully effective.

The Arctic Basin

Apprehension that catastrophic adjustments could already be in train is heightened by the current signs of warming around the Arctic. A report led by Matthew Sturm, a veteran of the US Army Cold Regions Research and Engineering Laboratory in Alaska, maintains a Bush-style agnosticism about causes. But the evidence he and his colleagues intensively deploy leaves little room for doubt about effect. Arctic air temperatures are at their highest for 400 years; and their rise is accelerating. The rate of shrinkage of Alaskan glaciers has increased threefold the last ten years alone. Some models suggest that by 2080 the Arctic Ocean will be ice free in summer.[26]

This syndrome is pre-eminently one which needs be conjured with from the McKibben perspective that, by denying Nature autonomy, we unhinge ourselves. Through the nineteenth century, those the Amerindians scorned as "eskimos" (eaters of raw flesh) assumed a "privileged place in the imagination of a West fascinated by their position amidst the ice".[27] Still one can feel in the far North vibes from an ancestral response to glacial vicissitudes. Hence the dour celebrations thereof in modern prose and poetry: Vilhjalmur Stefansson, Robert W. Service, Fridtjof Nansen, William Morris, Jack London, W. H. Auden . . . Admittedly, evocations of a mythic Nordic past can be silly or even sinister. Witness the Thule Society of Munich, a group linked in the 1920s with the infant Nazi party. However, deviant attitudes towards the natural and primeval may be more easily curbed if pristine landscapes remain authentically in place. Around the Arctic today, the prospects of their so doing is further diminished by the way the on-going accumulation of pollutants (radioactive kinds included[28]) is being malignly favoured by its geography: the encompassing continents, the ice, the low stratosphere, susceptibility to greenhouse warming . . .

One would like to hope this situation will bring together to good effect the riparian nations themselves, plus other parties actively interested. However, two factors militate against this. The global nature of the underlying menace is one. The other is a dawning realization that, by 2040 or thereabouts, the historic North-West Passage could be ice-free much of the year, thereby proffering a sea-route from, say, Europe to Asia 7000 kilometres shorter than the one through Panama. Various issues arising seem bound to exacerbate long-standing differences between a Canada anxious to maintain managerial authority over those waters and other countries (above all, the USA) concerned to preclude or circumscribe this. Meanwhile Moscow has been evincing interest in a revitalized North-East Passage.

Yet here one has further to insist that the question of comprehensive Arctic protection cannot be reduced to econometric managerialism, however reckoned. We are talking of the preservation of a landscape (sea and sky emphatically included) which is an invaluable part of our mythic and spiritual heritage.

Regardless of Greenhouse

What should be recognized, too, is how serious the Man-induced ecological crisis would be worldwide even if climate change had no part to play. As much is apparent from even a cursory review of the 2003 evidence concerning wildlife in the oceans. A feature in *Nature* on the catching of big fish reported that "Since 1950 with the onset of industrialized fisheries we have rapidly reduced the resource base to less than 10 per cent . . . for entire communities of these large fish species from the tropics to the poles."[29] Moreover, a study by the Netherlands Institute for Fisheries Research had found that, for a mix of reasons, the population of European eels was but one per cent of what it had been in 1980; and that their American cousins were doing little better.[30] Furthermore, the ambient threat posed to marine well-being by rising levels of man-made noise was highlighted by persuasive evidence of whales (and maybe porpoises and dolphins) dying as a result of dysfunctional behaviour due to close exposure to naval sonars.[31] Meanwhile, the evolution of a comprehensive strategy for whale survival was being hagridden by an impasse between the "anti-whaling" majority and the "pro-whalers" represented chiefly by Iceland, Norway and Japan. A celebrated bird species considered in danger of extinction (especially through injury from fish-hooks) was the Wandering Albatross.

Celestial Erosion

Johan Huizinga, prominent within a veritable galaxy this last century of Dutch historians, observed in 1924 how in the fourteenth century, say, a small town could experience a nocturnal peace more profound than was ever the case currently: "The modern town hardly knows silence or darkness in their purity nor the effect of a solitary light or a distant cry."[32] Nowadays the contrast would be much starker almost anywhere. In such respects, indeed, the rural Western Europe of 1924 will have been closer to 1324 than it would be to 2004. Closer also to our inner nature? Closer to tapping our spiritual reserves?

There has been a collateral decline in what one may term primordial knowledge of the night. Researchers at Japan's National Astronomical Observatory recently interviewed 348 schoolchildren in the 11 to 14 year old bracket. Forty-two per cent were persuaded that the Sun revolved around the Earth. As to where it sets, 73 per cent said the West but 15 per cent said the East and 2 per cent the South.[33]

One factor in this casting adrift is constricted horizons. Another, much discussed of late, is "light pollution". The year 2001 saw the publication of the Italian–American *First World Atlas* on this subject, the registered luminosity under cloudless skies being corrected to sea level, A most striking feature was the very stark but even division of the USA effectively down the 98°W line of longitude, the more lit area to eastward contrasting sharply with that mainly less lit to westward.

The broader conclusions were as follows. Two-thirds of the world's people live in places where night sky luminosity exceeds the defined pollution level. For the European Union and the continental USA (i.e. excluding Alaska and Hawaii), the reckoning was 99 per cent. Working from average optical acuity, for a fifth of the world's people the Milky Way is no longer visible to the naked eye. In the continental USA and the EU, the proportion exceeds two-thirds and a half, respectively. Moreover, a tenth of the population no longer views the heavens through eyes adapting to the night.[34]

An added complication are condensation trails or "contrails", lanes of ice crystals pumped out in the engine exhaust of jet liners. These fairly soon assume forms akin to cirrus cloud. They then impair considerably starfield visibility. But they also diminish the day–night temperature range. When all commercial aircraft were grounded in the USA for three days from "9/11", the average temperature range diurnally proved to be 1.3°C more than immediately before or after. In the mobile weather situation then obtaining, this was a good indication that contrails were otherwise contributing considerably to cirrus cover.[35]

With world air traffic predicted to increase between 2 and 5 per cent per annum over the next half century, that sounds ominous. However, the general outlook on this particular score may not be too bad. Condensation trails tend to form within a zone a couple of thousand feet or so deep at an altitude (over Britain at least) typically around 20,000 feet. Above that the air is too dry for them to form the way they do, while below it is not cool enough. The contrail altitude spread ought to be predictable on a given occasion and pilots enjoined to stay clear. Many pilots will have little interest in any case in spending time at the given height. However, a lot may depend in any given locality on how crowded the skies are overall, in relation to the control facilities available.

Limitation of light generation cannot be easy for communities. Even so, many steps are being taken. A sample from a recent global survey may be illustrative. Canada has a number of "dark sky preserves" where lighting ordinances are aggressively enforced. In the USA, the number of state ordinances on the subject is increasing. Italy has applied lighting legislation in nine of its twenty regions. Australia has national standards for the restraint of obtrusive lighting.[36] It is hard to believe such remedies are as yet reversing the global increase in light pollution. But we seem to be in with a chance. Success would bring much comfort and reassurance to many forms of life, not just ours. One's hopes for some such consummation are fortified by a burgeoning literary interest in the darker hours.[37]

9

RETREAT FROM REASON

A MODERN SYNDROME

IN HIS ACCEPTANCE SPEECH AT THE 2004 REPUBLICAN CONVENTION, George Bush alluded to inspiration from "beyond the stars". Yet astral perspectives, as nowadays understood, square ill with the religious fundamentalism the President looks to for core support and personal inspiration. Ronald Reagan's televised address after the 1986 *Challenger* Shuttle disaster told to effect of how its astronauts had "slipped the surly bonds of Earth" to "touch the face of God". But he had come out of a less closed conservative genre, theologically speaking.

Still, the Bush allusion did confusedly acknowledge that our appreciation of the majesty of Nature, of Creation, is enhanced by cosmic perspectives. So might humankind afford itself a metaphysics cast in such a setting? Could this become for many individuals within today's pell-mell society a prerequisite for staying sane? And might such underpinning also help societies and regimes to craft mental maps indicating where best to head longer term? At first sight, all three questions invite affirmative answers. In practice, it depends.

MATERIAL WELL-BEING

At first sight, too, the notion of modern society being that insecure seems out of kilter with advances, in the West especially, the past half-century:

income per head more than doubled; much better health care; nobler physique; much extended educational access; bigger housing stock; better working conditions; cleaner rivers; cleaner urban air; more leisure and leisure facilities; far more travel; vast consumer choice; marked social and occupational mobility; far less prejudice in various directions; and, informing everything, much stronger data flow, often in real time. We should not become too inclined to take these gains for granted.

A source of gratification, too, has been a levelling-up in the recorded distribution of income worldwide.[1] A caveat to enter throughout, however, is that, across a large part of the developing world, half the work force is primarily within an "informal economy", much of which may not be particularly illicit but all of which is beyond government ken.[2] This vitiates comparisons. A further complication is that, this last decade or so, inequality has worsened within various countries, richer as well as poorer. Nevertheless, one can still look with fair confidence to more global equality in real income over the next two or three decades. One can also look with high confidence to the lessening of absolute poverty as lately defined. Yet how well such progress can assuage rising expectations is another matter, not least because these never lend themselves well to calculation.

Happiness or Satisfaction?

Hard upon the youth revolt of the sixties, came an upsurge of concern about "fulfilment". With the onset of the dot.com revolution, this anxiety has returned, though with the accent on "happiness". It is a criterion which seems sometimes to lend itself to physiological explanations and tests. Yet difficulties will long persist, conceptually and operationally. There are transcultural differences about how "happiness" is perceived. Does one mean "contentment" or "pleasure" or "joy"? Endeavours to calculate "satisfaction" exacerbate the inherent difficulties.

The upshot (as adumbrated in the *New Scientist*) has been that, while all bar one of 32 countries selected from World Value Survey returns showed a higher rating on "satisfaction" than on "happiness", the differential varied widely. In Mexico 58 per cent of respondents declared themselves to be "very happy" and 81 per cent "satisfied". In more backward China, only 12 per cent were reportedly very happy yet 64 per cent well satisfied.

Beyond a certain point, little correlation was found between average income levels and general happiness. In the USA between 1957 and 2002, the proportion of declared "very happies" stuck around 32 per cent. Yet average annual income (2002 prices) rose from 8 to 22 thousand dollars.[3]

Swedish studies have similarly found that, above quite lowish thresholds, extra annual income per head does not generally enhance feelings of well-being. A 1997 study put the said ceiling at $13,000 p.a.[4] It is also to be remarked how literati interest in "happiness" burgeoned in the eighteenth century, just as technical progress was proffering quantum advances in material well-being.[5]

Democracy

Attention was also drawn in the *New Scientist* to a lack of correlation between recorded happiness and the vitality of democratic institutions. Considerable reservations though one must have about such enquiry, this outcome does call into question any presuming of "democracy", as one perceives it, to be a positive regardless. In the present context, the most compelling argument in favour of parliamentary democracy is actually a rather negative or, at any rate, precautionary one. It is that everywhere the information explosion is presenting all regimes with a starker choice than ever. Become either more open in whatever ways are appropriate or else a sight more closed.

One must also be wary of the precept that democracies are not disposed to make war on each other. Revived by President Clinton in his 1994 State of the Union address, it is today a mantra. But take the First World War. Among the big European powers, the two most cautious about embroilment were actually the two least democratic, Austria-Hungary and Russia. Likewise, other wars have involved, on opposite sides, polities with plausible credentials, democracy-wise.[6] In a shrinking world, a steady strengthening of democracy has to be a necessary condition for peaceable progress. Yet it is by no means a sufficient one.

Since the sixties of the last century, the notion has burgeoned that a people can be persuaded to trust government by being accorded more democratic access. However, this gain may be hard to make convincingly. When the histories of New Labour in power in Britain are written, the verdict may well be that its biggest failures concerned constitutional reforms essayed with the declared intent of making the governed feel at one with governance.

Thus early on, Tony Blair's administration devolved political power to Scottish and Welsh elective assemblies. It did so extensively in Scotland and considerably in Wales. It also created the elective post of Mayor of London. Then forthwith it did whatever it legally could to obtain preferred electoral outcomes in Wales and London. As for Scotland, it remains far from clear whether devolution will prove to be an alternative to outright independence or its precursor.

Likewise, New Labour's forays into Lords reform, Proportional Representation and English regional assemblies have thus far led only to undignified hiatuses. Routinized postal voting is the latest issue to engender confusion. Meanwhile a diametric policy switch has meant there could some day be, after all, a British referendum on a revised constitution for the expanded EU; and this might test Scottish–English solidarity to the limit. So eventually might one about replacing the pound with the Euro. The 2005 proposals for new legislation essentially to curb "Islamic terrorism" raise issues more fundamental than Whitehall cares to admit.

Nor have unedifying altercations about the reasons proffered for invading Iraq done public trust any good. Nor have policy *volte faces* on domestic issues. Nor has the failure of successive formal enquiries into official *modus operandi* to reach trenchant conclusions. In short, a government which came to power committed to reducing the "democratic deficit" has found that mission hard to progress with or even to gauge.

Ultimately, the supreme tests of good government have to be avoidance of corruption and of overly short-term thinking, these together with the encouragement of imaginative innovation. The twentieth-century signs have been that parliamentary democrats rise to such demands better than do the authoritarian alternatives. Whether the former can do so sufficiently remains to be seen. Much may depend on whether a new metaphysical consensus can proffer inspiration and guidance.

H. G. Wells Revisited

A good point of departure for consideration of the further outlook could be *Anticipations*, the 1901 essay in which a youngish H. G. Wells surmised how England and the world might look the next turn of century.[7] One parallel between then and now is quantum gains in information flow. Another is the transoceanic projection then of military force – Mafeking, Manila, Peking, Tsushima . . .

Wells foresaw how the diffusion of settlement would blur the distinction between countryside and town. Yet he was incongruently complacent about the collateral impoverishment of wildlife. He saw no reason why "the essential charm of the countryside should disappear . . . the lane and hedges, the field paths and wild flowers" (p. 63). Since when, the fabled English hedgerow, viewed as a wildlife habitat, has all but disappeared.

He better realized how greater mobility and data flow would lead to cultural uniformity, evidenced not least in the erosion of the vocabulary, idiom and inflection of the local dialect (pp. 223–5). Before 1914, dialectical singularities were often discernible across a mere several miles in city or country, even in the English south-east. Throughout England today, local and indeed regional linguistic differences are disappearing fast in the face of an "Estuary English" emanating from about the Lower Thames.

Wells expected a residue of "unemployable" paupers always to be around (p. 82) though he seemed unsure on what scale. Of more moment, in any case, would be "the reconstruction and the vast proliferation of what constituted the middle class of the old order" (p. 82), considerably due to the expansion of service industries. This prognosis would be hard to fault.

Above all, however, he apprehended democracy would prove to be "but the first impulse of forces which will finally sweep round into quite a different path" (p. 146). Political philosophies were deemed to have left the very word "Democracy" as but "a large empty object in thought" (p. 145); and never to have made a case for the "elective government of modern states that cannot be knocked to pieces in five minutes" (p. 146). No less pertinent was "the confusion of moral standards" (p. 132) to be expected from the intermingling of cultures.

Insecure governments might thereby be impelled towards demagogic belligerency, drawing the "voter to the polls by alarms seeking ever to taint the possible nucleus of a competing organization with the scandal of external influence" (p. 167). Periodically, such posturing could bring those concerned to the brink of major war.

Belligerent Resolution

All to soon, Britain herself would be teetering on that brink. Her measure of immediate responsibility for what happened in 1914 stemmed more from geopolitical equivocation than anything. But there was a background of internal tension which, for her as for other participants, dissolved – one could almost say miraculously – once the gauntlet was down. The social dynamics were and are a sobering commentary on the mechanics of mass society, democratic or otherwise.

In his *The Strange Death of Liberal England*, George Dangerfield analysed the United Kingdom pre-1914 at two levels. The one was everyday life in the Westminster village. The other encompassed, on the nationwide canvas, the successive surges of angry impatience with Respectability, "one of the chief articles in the Liberal creed . . . unwritten deep in the heart".

The anger had multiple causes. No doubt the organization in 1910 of the

Union of South Africa had rekindled the bitter cleavages over the Boer War. Worse, a "triple alliance" between miners, dockers and railwaymen had been spearheading industrial unrest. Meanwhile, women's suffrage was being militantly pursued with hunger strikes and self-immolation among the tactics employed. The Liberal government had forced through in 1911 a Parliament Act to curb the powers of the House of Lords, this hard upon the second chamber's blocking of the 1909 government budget. A mass resignation of officers at Curragh camp in Ireland in March 1914 signified how close general unrest had been brought by the Home Rule question.

Yet once war had been declared that August, "one nation" was reforged remarkably. Dangerfield, who was writing 20 years later, was more mystified than he need have been.[8] Basically, the energy of anger had been redirected outwards. Although the particulars varied country-to-country, the basic mechanics of mass psychology were very similar throughout. Nor were the doyens of dissent outside the redirection process. Witness, in Britain, H. G. Wells, David Lloyd-George and, indeed, much of the Suffragette leadership.

A danger now extant is that, in various parts of the world, there will be a sharpening of the contradictions between religious fundamentalism and the advance of Science, not least cosmology. Affected regimes may be tempted to control this spiritual gridlock by outward posturing. That social groups are perennially inclined to define themselves in terms of the enemy without, "the Other" is an axiom of modern psychology.[9]

The Urban Syndrome

At the heart of the discourse about the human condition today is urbanization. As recently as 1970, barely a third of the world's population were urban dwellers. Now, on any reckoning, over half are, even though world population overall has nearly doubled in the interim. Undeniably, the glittering image of the city long endemic in many cultures still allures, especially in late adolescence and early adulthood.

As evident, however, is the susceptibility of those either side of that narrowly resilient "age of man" to the congestion, eco-denudation and social rootlessness of many town and city environments. Early on in the current debate, a British student of drug abuse rendered thus one aspect of that problem: "Few modern teenagers have ever been physically extended. Walking under a hot sun till exhausted, swimming in a cold river instead of a heated pool, cycling home soaked to the skin, running across country on a cold winter's day."[10]

Not that the subject area is spared facile generalities. Even megacities are not entirely novel. At its Late Classical peak, Rome, the imperial metropolis, numbered a million free people plus 150,000 slaves.[11] But modern cities no longer threaten to mushroom population-wise as unmanageably as seemed in train some decades back. Nor was the cause of good analysis then furthered by intimating strong ineluctable correlations between urban growth and the cult of the gigantic in technology and bureaucracy.[12]

Nobody could say, in any case, that the tension associated with city life has been entirely uncreative. It has by no means always found expression in crime or insurgency. Take the Japanese scene. Present-day Tokyo–Yokohama is a concentrated conurbation some 25,000,000 strong. Yet it remains remarkably free of such individualized malpractice as mugging and vandalism. Furthermore, there and throughout the Japanese archipelago encounters between strangers are characterized by cheery politeness of a kind seen less and less frequently in, say, the South-East of England.

Still, there are downsides. A big one has to be the million or so Japanese youngsters below the age of 25 (the great majority male) who have been living as recluses, sometimes for years on end, usually in their parents' home. One can easily imagine this asocial *hikikomori* flipping into customary Japanese "groupism" (*shudan shugi*) managed by *yakusa* crime syndicates.

Another downside is that discourse throughout the corridors of national power and influence is too mannerly and restrained – too "respectable", as Dangerfield might have said. Perhaps this is why Japanese government has failed of late to mobilize properly the social energy latent in the people. Meanwhile, the extent to which China's urban implosion is accommodated in mega-blocks of flats may bode ill for the medium term.

A Social Paradigm Change?

What is going to test social stability severely, more or less everywhere, is the passing from the scene of the generation which came to maturity before 1950, before the ever freer flow of people and information afforded by motor cars, jet liners, transistor radios, television, fax machines and latterly the internet. For it is this generation who have done most to preserve a sense of extended families operating within local communities: a milieu which, for all its frictions and constraints, is much to be preferred to *anomie* within a lonely crowd.

A prevision of what this could mean in terms of spiritual starvation can be gleaned from a searing assessment of current British attitudes to newsworthy ill-fortune or demise. It tells of transient surges of mourning sickness acting as substitutes for old time religion. The said study lacks historical perspective; and was taken up too cavalierly within the political Right. As much can be seen in the healthiness of so much of the immediate British response to the London bombings of July 2005. Nevertheless, it does identify reactions which may too often be symptomatic of an *anomie* malaise,[13] in Britain as elsewhere.

In seemingly diametric contrast is how blatantly people fail to interact as they mingle in public places day by day. Already this syndrome affects various parts of Europe severely. "In Stockholm negation will be expressed in a formalism which is too studied. In London, it is the informality which will be too studied. Yet in neither direction will formal correctness be the real point. It will be instead a lack of natural graciousness in word and movement, avoidance of eye contact, generally cheerless demeanour . . . ".[14] There is little scope here for statistical collations but much for sombre reflection. Are we entering an Age of Isolation as per, let us say, the art of Edward Hopper?

The Work Station

An ancillary though quite important question concerns trends in job satisfaction. As of today, the received wisdom probably is that, for most people in the richer world, work has become more fulfilling as various of the more routine operations are either computerized or outsourced abroad.[15] Yet already this is debatable. Reduced scope for physical prowess is as evident in the workplace as anywhere. The same often goes for the customary practical skills. Nor can these tendencies be masked much longer by an expansion of the service sector. There, too, routinization is gathering pace.

Nor can the way the internet facilitates horizontal information exchange be expected to undermine at all decisively the new pyramids of decision taking. Granted, it does challenge certain of the prerogatives of the pyramid builders. All else apart, it makes youth more articulate across the board. But at the same time it furthers, in a myriad of respects, the cultural melding which so readily makes centralization more feasible. All in all, we are entered upon an era in which the gateways to truly challenging occupations will be much narrower than the upsurge of youngsters aspiring to practice them. To my mind, this challenge is rendered the more acute by its near total neglect by the commentators, in the media and academe.

Indeed, it obliges one to square up to contradictions at the very heart of the idea of progress. One recalls how George Orwell, in his *The Road to Wigan Pier*, mocked Left-wing visions of utopias peopled by god-like heroes. H. G. Wells, typecast as a kind of "socialist planner", is scorned for the evasiveness inherent in his opining that "once you have got this planet of ours perfectly in trim, you start out on the enormous task of reaching and colonizing another". The plain truth was that "in a world in which nothing went wrong, many of the qualities which Mr Wells regards as god-like would be no more valuable than the animal faculty of moving the ears".[16]

Therapeutic Violence?

The frustrations liable to build up as our urban civilization evolves do not lend themselves well to articulation within established linguistic and thought patterns. But they may wax all the more dangerous on this account, generating at various levels an intra-human belligerence that may be rationalized in morbidly fantastical ways.

In the final analysis, of course, one has to reckon again with armed violence between states as well as within them. In which connection, one continues to reiterate as if by rote the received wisdom of the Gladstonian–Wilsonian era (1868–1919). Its gist was that, while military force is bound to have a role to play deterring or resolving armed conflict, this must be underpinned by the political redress of grievance – particularly of any denial of basic liberties. Yet nowadays one has the uneasy feeling that not a few people the world over are ruminating to the effect that, for Washington and its allies, political remedies will soon be superfluous to requirement. After all, advanced weaponry can be so precise and lethal; and modern surveillance so multi-mode, voluminous, synoptic and integrated that a Pax Romana might always be imposed more or less regardless.

However, if through over-reliance on such analysis, the West contrives a raft of inequitable solutions, it could be sowing the "dragon's teeth" of a terrible legacy perhaps three decades hence. For such could be the time span over which genetically-engineered biowarfare could start to be a proliferating option. It could become the ultimate weapon of the alienated, their preferred way of exacting terrible vengeance.

My own evaluation of this specific menace has lately been spelt out elsewhere.[17] Suffice for now to adumbrate the main tenets. The military applications of genetic engineering could likely be contrived in laboratories as small and outwardly nondescript as household garages. Breakthroughs would largely take the form of the Genetic Modification of existing microbial diseases in order to create mutants which no immune system (animal or plant) would previously have had a chance to adapt to. This surprise factor is acutely accented by how microbial forms typically go through a quarter of a million generations while human beings are progressing through one.

The assumption once was that germ warfare was an improbable departure, given the risk of accidents during preparation as well as of any epidemic operationally contrived then going out of control. However, the recent upsurge in several parts of the world of suicide strikes reminds us how readily human beings become so resolved to harm their antagonists as to be careless of their own welfare as well as of proportional response. Suicidal genocide could be the end game.

Conflict Avoidance

Naturally, the avoidance of violent solutions will widely be read as upholding a modicum of "rationality". But what does this mean? Though "rationality" may come across as a precept which is axiomatic to the point of platitude, it has long been open to varying interpretations, not all of them positivist. May it be but the trite playground logic of military deterrence theory? Does it represent a narrowly reductionist thrust which looks for simple linear causation regardless of wider influences? Or is full, factual truth what one must be concerned with? Then again, should one be seeking objective pragmatism, the "smile of reason" extolled by the eighteenth-century *philosophe*? Can "rationality" be taken to subsume, in their due proportions, truth, reason and reasonableness?

While "rationality" is hard to delineate, its *alter ego* "irrationality" is easier at least to identify. Yet here, too, loose usage can divert one. Through the early eighties as the Cold War approached its *dénouement*, the militant Left and the hard Right within the West reciprocally accused one another of blanket irrationality. The former adjudged the latter to be hooked on a "bomb culture". The latter saw the former as gripped by utopian fantasies. Today this typecasting continues across very similar battle lines with civil nuclear energy and global warming now salient bones of contention.

Yet it ought to be obvious to all concerned that genuine irrationality cuts deeper than transient political opinions one has personally felt unable to endorse. In April 2002, the US National Science Foundation found that 30 per cent of American adults thought UFOs are, indeed, vehicles from Outer Space; 40 per cent saw astrology as scientific; 32 per cent believed in lucky numbers; and 70 per cent accepted magnetic therapy as good science. Alongside all of which should be placed declared beliefs about two concepts which either are not or may not be as palpably absurd as those just cited but which ought to be subject to astringent scepticism at this stage in our understanding, not least since each covers a broad field. A reported 60 per cent believed in Extra-Sensory Perception as did 88 per cent in Alternative Medicine.[18]

The Sirens of Fascism

Faced with multiple patterns of tension and threat, many could eventually turn yet again to the contrivances of aggressive authoritarianism: the sadistic attractions of exercising total power over others coupled with the masochistic one of dissolving oneself in an overwhelmingly strong collective identity. Feed in a few external guidelines and you have a political religion. A narcissist image of one's own group is raised to a high while the demeaning of those without reaches a new low.[19]

In the twentieth century, fascism and communism were the chief political religions, Hitler's Germany and Stalin's USSR being taken as archetypal.[20] Each had arisen from a mass minority movement hard to delineate in terms of customary ideas about social groups and their mores.[21] Each could achieve considerable feats of social mobilization, especially in wartime.

Also, both credos attracted awhile considerable support from outside their core domains. As Hitler came to power in Germany, not a few countries in Europe and beyond were swinging more towards the authoritarian Right in the wake of the Wall Street crash of 1929. As for the USSR, there was the disposition long evident within the Left intelligentsia to view it oneirically, much as medieval Europe had the Indian Ocean - an "exotic fantasy" where "dreams freed themselves from repression".[22]

To an extent, this aberration was expressive of desperate wishful thinking about Soviet liberality. Yet in some measure, too, it bespoke an inner predilection for strong-arm rule.[23] Pertinent is a comment by G. K. Chesterton (1874–1936) – English literary virtuoso and Roman Catholic convert. He cautioned that "when men cease to believe in God, they do not believe in nothing, they believe in anything." But a complication in this instance is that many fascists, at least, did continue to profess belief in God or – in Japan – the Gods.

ASTRAL FANTASIES

The notion that a resurgent awareness of the cosmos may influence considerably whether civilization can survive draws succour from the extent to which this awareness is suffusing society at large. After all, space research and exploration may soon be at least as much a part of popular awareness as aviation became after the First World War. Indeed, it has proved able to embrace a utopian aspect aerial flight rather missed out on. Televisual treks through the galaxy may be interwoven with previsions of extra-terrestrial urban milieux encompassed by airtight domes. These may be alluringly unpeopled and noiseless.

In the wake of the final Apollo flight in 1972, enthusiasm for space fantasies flagged awhile, even in the USA. But it was never dead. The *Star Treks* of the Spaceship Enterprise, first viewed in 1969, remained a television staple. Within six months of its American debut in May 1977, *Star Wars* was proving "the most popular film ever",[24] with success in the USA being repeated round the world. Soon, *Close Encounters of the Third Kind* was registering comparable triumphs.

In 1977, too, a Princeton physics professor, Gerard K. O'Neill, deployed arithmetic projections to argue that "perhaps by the middle years of the next century" a start could be made with reducing the Earth's population through emigration to space. Within the first decade, 290,000 might have been settled thus; and after the third, the total (boosted by natural increase) could be 631 million.[25] This prognosis left one uneasy. All else apart, these target levels were incongruously exact, bearing in mind the logistical imponderables.

More vexing still is the question of what social values would need to be cultivated for any such exodus and what other political adjustments made. In the Anglophone tradition, the "moving frontier" of colonization is archetypically envisaged as a matter of rugged homesteaders staking out their individual independence. Yet historically the pattern of colonization around the world has quite often been martial and authoritarian. My own suspicion is that the *modus operandi* of any extra-terrestrial settlement *en masse* would have to be more Spartan than Sparta. All in all, it is hard not to consign the O'Neill proposal to realms beyond the policy arena.

Other contributions have been more blatantly escapist. Buck Rogers made his debut in 1932 as the great slump was troughing out. *Star Trek* came along just as South Vietnam was being abandoned. The 1999 rendering of

Star Wars was released in the USA against a background of anxiety about the disorderliness of the "new world order" and one year ahead of a world economic downturn.

Not that space fiction as televised or broadcast has always imparted reassurance. A famous exception was on 30 October 1938 when CBS broadcast a hastily modernized version of H. G. Wells's *The War of the Worlds.* Orson Welles was director and also a lead actor. The radio audience was put at six million.

Unfortunately, the "breaking news" of a Martian invasion was too convincing. There were switchboard and highway jams plus a myriad of other overreactions. Unremarkably, a Princeton study found people of low educational attainment more prone to take it at face value. Two million listeners nationwide at first believed it was for real. Welles thus became a celebrity at the age of 23.[26]

False Prophets

A number of works of astral fiction have lately been presented as well-founded hypotheses. Among those thus attaining lucrative prominence have been Immanuel Velikovsky and Erich von Däniken. Among the acute professional critics has been E. C. Krupp from the Griffith Observatory in Los Angeles.

Velikovsky brought out his *Worlds in Collision* in a Cold War crunch year, 1950. Its modest spread of references are mainly from or about the Old Testament but do include other sources tantalizingly obscure to most of us. He asserted that, around 1500 BC, a comet issued from Jupiter. After several near passes by the Earth, it settled down as the planet Venus. The close encounters had affected our planet in divers ways, nearly all of them inimical to human well-being.

Krupp leveled a score of separate objections to this historical rewrite, virtually any one of them clinching. To my mind, the most elementary yet conclusive was his comparison of the mass of a cometary nucleus with that of a middle-range solar planet. The latter he put at anything between a million and a hundred billion times the former. As for a cometary tail, all its atoms would fit "comfortably into a suitcase".[27] Yet to such protestations, Velikovsky was all but impervious. He told how his hypothesis came to him in the summer of 1940. That at least was believable.[28]

Meanwhile, von Däniken's books were selling in their tens of millions. He told how, between 40,000 and 592 BC, the Earth was visited, recurrently and influentially, by a majestic breed of astronauts. Krupp demolished this notion even more thoroughly. Then he generously allowed that "if scientists can learn to communicate their discoveries with the same enthusiasm as von Däniken, and if we can learn to partake of the tree of knowledge without losing the innocence and curiosity that prompted our first questions, the human spirit will evolve".[29] With respect, however, astronomers have often displayed these very qualities, not least this past century. So, for that matter, have their peers in contiguous areas like particle physics and biological evolution.

Another aberrant contribution came in 1974 from John Gribbin and Stephen Plagemann. Their advice was that in 1982 "the Los Angeles fault will subjected to the most severe earthquake known in the populated regions of the Earth in this century".[30] The nub of their argument was that a close alignment of almost all other planets would trigger under the Earth's crust a surge of molten rock liable to be catastrophic.

Evidently nothing of the sort transpired. However, the more essential point is that in no way could it have. Like tides in water or air, those in molten rock do not depend on the overall gravitational pull exerted on the Earth by particular heavenly bodies. They derive instead from the contrasting strengths of the pull on points on and/or within the Earth short distances apart. This puts a premium on the operative heavenly body being close by. The reason why the sea tides induced by the Moon have several times the amplitude of those the Sun causes is that the latter is some 400 times further away. Never mind its being millions of times more massive.

Gribbin is a fine interpreter of modern science, not least through his comprehension of the philosophy which informs it. Neither he nor, indeed, Plagemann needed phantasmogoria to keep moving on. Moreover, astral phantasy does have a role to play in entertaining us but also in extending our conceptual horizons. Arthur C. Clarke, Fred Hoyle and, indeed, Douglas Adams are among the many who have demonstrated this. But a distinction must be preserved between science fiction and hard science. Sometimes this will hinge on getting orders of probability right. In 1983, Gribbin and Plagemann admitted they had "been seduced into putting all our eggs in the planetary alignment basket".[31]

In his first offerings (1978–9), Adams advised us that the Earth was actually a giant computer. Its original human dwellers had died out only to be replaced by a coterie of hairdressers and management consultants. His cosmic travelogues had a Swiftian flare for blending social satire with high comedy. Nobody could mistake them for pure science. Yet neither could they deny his understanding of science nor his ecstatic reverence for how a cosmos of "inordinate complexity and richness and strangeness" had arisen "not only out of such simplicity but probably absolutely out of nothing".[32]

Modern Astrology

In evolution studies, it can be misleading to identify a particular group as having been "dominant" at a given time; and to dwell overly upon its fortunes. In cultural history, the same applies. Astrology has often been drowned out by some ascendant belief system yet has never really drowned.

Take the Enlightenment "Age of Reason" as experienced in Britain. The first edition of the *Encyclopedia Britannica*, published in 1771, dismissed it as a science which "has long ago become a just subject of contempt and ridicule". Yet come the turn of that century, a number of astrological almanacs were still selling well. In the case of Moore's, this meant between a quarter and half a million copies a year.[33]

The British Library history of astrology covers incisively the sequence from the Enlightenment to the Second World War. Suffice now to focus on Germany's experience. Astrology did appear almost to die there before 1914; and, in every country where it still thrived, failed to predict the cataclysm.

Nevertheless, in the fraught atmosphere of the Weimar republic it was renascent. However, its mien was now more individualist, the accent being on how someone's personality was supposedly pre-ordained by the heavenly situation at their moment of birth. Cast thus, astrology attracted the attention of Carl Jung, himself a German Swiss. He related it to what he termed "synchronicity" – the concurrence of events which are understood not to have common causality but which, none the less, have like connotations.

Within a year of gaining power, the Nazis had suppressed astrology publishing. However, Whitfield sees a parallel "with the Roman emperors' decrees against astrology while still resorting to it themselves".[34] Granted, Hitler himself was not seriously engaged despite the "war by horoscope" legend surrounding him. But Hess, Himmler and – opportunistically – Goebbels were. Peter Whitfield asks what connection there might be between the imprisonment in 1941 of many astrologers and Rudolf Hess's bizarre flight to Scotland that May. The question is a very good one.

Today astrology survives more or less worldwide. Indeed, it is adjudged lately to have surged up anew. It had no entry in the first (1983) edition of Japan's *Kodansha Encyclopedia.* Nevertheless, it currently looks quite strong across East Asia, most notably perhaps in South Korea, Taiwan and among the Tibetans. In India, it figures considerably especially in relation to marriage plans. In Britain, the astrology business ticks over steadily. What never can be confidently gauged is how much solid conviction attaches. At all events, attitudes anywhere could readily gel in favour in circumstances of acute stress, this as part and parcel of a wider flight from reason.

Unidentified Flying Objects (UFOs)

UFOs – alias "flying saucers" – have shown a pronounced tendency to proliferate at times of climactic world tension. Indeed, their modern lineage can be traced directly to strange projectiles recorded over Sweden, 1943–5; and to the "foo fighters" (i.e. accompanying lights) Allied bomber crews reported seeing over Germany around that time.

Early on in the Cold War sightings resumed, notably in the USA though with considerable incidence in Europe and elsewhere. The frequency (as tabulated by the United States Air Force) underwent a ninefold increase in 1951–2 with hot wars raging in Korea, Indochina and – less intensively – Malaya. Then sightings rose by a half in 1957 as against 1956, no doubt with the advent of Sputnik I. As a year of international *détente*, hard upon the Cuba missile crisis, 1963 was to be "remarkably quiet" UFO-wise. The sightings in 1968, that year of protest and violence, were twice those for the more tranquil 1969.[35] By 1986/7, with the USSR sliding into terminal crisis, UFOs were a source of fascination in the Soviet media.[36]

Belief in "flying saucers" as phenomena sometimes more substantial than optical aberrations or mental flips was shown in 1978/9 American polls not to profile along the lines one might have expected. Now the elderly or ill-educated were not the more credulous. The 57 per cent or so of the public said to regard UFOs as not "just people's imagination" rose to 70 per cent in the under 30 bracket. Over 80 per cent of applied scientists and engineers under 26 thought UFOs probably or definitely existed.[37] These spreads of declared beliefs explain why scientists of eminence (e.g. Carl Jung and Carl Sagan) felt constrained to join the authenticity debate.

Photographic evidence of UFOs is strikingly sparse and ambiguous. Nevertheless, the attributes accorded them by the putative observers exhibit a marked, though not total, consistency. They are usually lens-shaped though sometimes oblong. They are silent; and have little or no mass, judging from their capacity to hover unsupported and also to accelerate abruptly and fast. UFOs can be manned or unmanned; and vary in diameter from one foot to 1500.

Give or take sundry allegations of abduction, their reported behaviour is not actively harmful. Nevertheless their perceived visitations could bespeak an intrusive curiosity especially towards the USA and, above all, airfield and nuclear installations. Otherwise there is no evidence of an alien master plan of the kind synoptic surveillance would require.

Something even UFO buffs generally concede is that well over three-quarters of all putative sightings lend themselves to mundane explanation – meteorological balloons, aerial night flights, castellated cloud . . . Never mind that the UFO stereotyping adumbrated above hardly relates well to any such objects. What then of the 5 or 10 per cent which cannot be so readily explained away? In his special UFO study, Carl Jung discerned the mandala symbolism which readily wells up within the human psyche in many cultures in times of general stress.

Unfortunately, the comparison was drawn too casually. Jung depicted the mandala as a protective circle in the same genre as the prehistoric "sun wheels" and magic circles or in much of the architecture of authority in more modern times.[38] However, what is generally recognized as a full mandala has a square which is set exactly within a circle. The square stands for the "four corners" of the world; and the circle for the enveloping cosmos. The whole may symbolize the inner peace an individual may gain through mystic unity with Creation. Arguably, though, the square matches well enough the flattish cupola that normally surmounts a perceived "saucer".

The circle within a square Heaven–Earth imagery appears in *Jirang* ji, a mandarin poetic anthology by Shao Yong (1011–77); and can be traced back to the Book of Changes,[39] a millennium and a half earlier. Likewise mandalas have customarily played a large part in Tibetan Buddhism, especially as elaborated to take some account of a large pantheon of quasi-deified spirits. In Japan, the mandala is seen as "the central symbol of Esoteric Buddhism".[40] In the period 975–1025, stars figured exceptionally (by Japanese standards then) in mandalas designed by Heian Esoteric Buddhist artists.[41] Hinduism also features this motif. Indeed, "mandala" is a word of Sanskrit derivation. The influence is also evident in the art of medieval Christendom.

That summation cannot be exhaustive. But perhaps it says enough to indicate how deeply embedded is the mandala image (perhaps, most basically, as a foetal symbol?) within the human psyche. It is not hard to foresee its finding UFO expression. If there is a problem, it may be that the world-wide reporting of UFOs seems not to coincide well with the historical geography of mandala portrayal. But mismatches are not easy to confirm and may not be material.

Thus far, this particular appreciation has proceeded as if UFOs belong to the recent past rather than the on-going present. Undeniably the theme does not currently command public attention as once it did. Yet like astrology, it does not die. In March 2004, a Mexican military aircraft flying over Chiapas and Campeche states photographed in the infra-red a mysterious set of lights. These were afterwards shown to emanate from offshore oil wells. Nevertheless, certain of the less scrupulous UFO promoters deployed *à l'outrance* verbal extravagance and fabrication, hoping to make a bigger story stick.

Nor should we forget how readily UFO legends phase into broader space-oriented fantasies. Alien abduction can currently lay claim to over 200,000 dedicated web sites. In 2003 some notoriety attached to a sect rooted mainly in Canada and calling itself the Raëlians. Raël is the name they accord their founder/messiah. He is Claude Vorilhon (b. 1946), a French motor-racing journalist and singer.

One reason why the movement attracted attention was its plans to host an Earth visit from the planet Elohim, a heavenly body which has thus far eluded all non-Raëlians. Things were put on hold until a landing site with consular facilities could be built in Ireland to accommodate the peaceable Elohims, a quest for a site in Jerusalem having proved fruitless. Raël claims to be of human-cum-Elohim descent. His movement has a vague interest in building utopia, H. G. Wells-style. However, their greatest object of enthusiasm is the scope for human cloning.

The Falun Gong

Now very comprehensively suppressed as subversive by the Beijing regime, this Chinese popular movement has bestowed on itself a bizarre cosmology. Yet when founded in 1992 by Li Hongzhi, born *c.* 1952 and previously a minor official, it was simply devoted to *qigong* ("energy cultivation"), an ancient system of breathing exercises believed to channel bodily energy so as to enhance health, prolong life and enrich spiritual experience.

Initially it enjoyed as much official toleration as a voluntary association could hope for in the wake of the suppression of the pro-democracy movement from 1989. But in 1996, with the authorities now exercised by the soaring popularity of the Falun Gong and other *qigong* groups, an official crackdown began. In April 1999, some 15,000 followers protested in central Beijing against the constraints. At the time, the government estimate was of 30 million adherents across the country while the Falun Gong abroad was claiming, far less plausibly, 100 million worldwide.

That July the sect was banned and total repression ensued. According to Falun Gong reports, some 100,000 people were arrested within three years. Twenty thousand had been sent without trial to labour camps; and a further thousand detained in mental hospitals.

Meanwhile, Li Hongzhi (who had gone into reclusive American exile in

1998) had formulated a metaphysical perspective. Suffice for now to adumbrate its cosmology. Our universe, like countless others, comprises billions of galaxies within an outer shell. Three thousand such cosmic entities form just one layer of a megauniverse. Beyond which, the narrative gets more confused. But outside the megauniverse there is a true vacuum; and then, eventually, a still larger heavenly agglomeration where matter, life, space and time are quite different from how we know them. It is populated by buddhas and gods. Moreover, the whole business is replicated endlessly in microcosms of worlds within worlds. As Li puts it, "one grain of sand contains 3,000 worlds".

Regarding life and consciousness, a variety of tenets are proffered. Everything in our universe is alive. Everyone was born simultaneously in many dimensions. The humans we know of were created by the gods at "high levels" but, thanks to their depravity, have slid to where they are today. Yet all that now exists on Earth somehow derives from them.

The Yale-based study on which this resumé is based does much fuller justice to this admixture of bowdlerized hypothesis, pure concoction, and maybe cosmic consciousness.[42] But enough will have been said to indicate that Li has hereby done no service to the many thousands of Falun Gong who have lately struggled in isolation for their human rights. Ask people well up in the Beijing hierarchy of power about the Falun Gong, and the stock answer you get is that "They are all mad". The disaggregated metaphysics of Li Hongzhi cannot but feed the illusion that this response was adequate, even when the movement was at its height. As of 2003, the worldwide membership of the Falun Gong was authoritatively put at well below a million.[43]

The harnessing of cosmology to political legitimation has a long history. Broadly speaking, the two themes evolved together in pre-modern society (see Chapters 1 and 2). Not infrequently, cosmology/metaphysics was mobilized in short order by a newly-found regime seeking the "mandate of heaven". About half a century into the Han dynasty in China, in fact, a twin track approach was adopted to the underpinning of legitimacy. A legal code was formulated concurrently with a cosmological overview.

Political Religion, Soviet Style

A much later example of such concurrence was how the infant Soviet Union aligned itself (albeit querulously) with astrophysics. The legitimacy quest behind this might be expressed thus. An uncompromising "dictatorship of the proletariat" is warranted *pro tem* because it will usher in the era of "scientific socialism": the word "scientific" here connoting both "rational" and "science-based". The ultimate demonstration that this can fully meet both the emotional and the material needs of the masses is afforded by its ability to reach out to the heavens.

The person long recognized as the "father of Soviet space science" is Konstantin Tsiolkovsky (1857–1935). Excluded by early deafness from laboratory or factory management, he developed a flair for conceptualizing aerospace. After being pretty much sidelined by the Czarist scientific establishment, he was taken up avidly by the Soviets. Through the 1920s he was able to give vent to previous seminal work (notably on interplanetary travel) and to follow it up with further studies – e.g. on multistage rockets; jet propulsion; and air cushion vehicles. He was steadily gaining international recognition as well.

Between 1917 and 1923, scores of thousands of Soviet literati were involved in a Protelkult movement. Influenced by Tsiolkovsky, many of them sought to hitch their futurism to the stars via Russian Cosmism. This cult, which had flourished through the recent turn of century, combined philosophic tendencies as disparate as theosophy, panslavism, Russian Orthodox religion and technocratic optimism. Its chief ideologue (and a big influence on the young Tsiolkovsky) was Nikolai Fedorovich Fedorov (1828–1903). The nub of his philosophy was that "everything in the universe from the tiniest grain of matter to the gigantic suns of distant galaxies *was alive and had some degree of consciousness* (my italics). As beings of the highest consciousness, humans had a special role in introducing design and purpose in the chaotic workings of Nature, here on Earth, in the solar system and throughout the universe".[44]

However, Stalin apparently saw Cosmism as essentially retrogressive. Leon Trotsky, the great prophet of utopian Leftism, certainly did. He expressed himself thus:

> we have conquered all Russia recently and now we are going on towards world revolution. But are we to stop at the boundaries of "planetism"! Let us put a proletarian hoop on the barrel of the universe . . . we'll cover it all with our hat.
>
> Cosmism seems, or may seem, extremely bold, vigorous and proletarian. But in reality, Cosmism contains the suggestion of very nearly deserting the complex and difficult problems of art on Earth so as to escape into the interstellar spheres.[45]

In 1928/9 Trotsky was forced into exile, leaving the field to Stalin. Since 1925, straight-down-the-line atheism had waxed very militant. In 1929, the constitution was changed to omit clauses permitting religious advocacy. Over the next year or so, many churches were closed and their books and icons burned. This pitch was relaxed in 1943 to enable the Orthodox Church to help sustain a year or two longer the commitment to the Great Patriotic War of a desperately traumatized and weary Soviet people. But this concordat, such as it was, did not provide for constitutional revisions.

Late Stalinist Science

By now, Soviet science and technology was receiving oneiric treatment in the West. Take a comprehensive review first published in 1936 by the one-time science correspondent of the *Manchester Guardian*. Alluding to dialectic materialism, he averred that the "activities to be described in this book could not have been conceived and brought to the present level of development in 18 years without the inspiration of some remarkable doctrine".[46] This makes all the more surprising there being (in the 1942 abridged edition) no reference to Space, save for the "outstanding" research being done on cosmic rays.[47] Instead, the top aerospace priority was aviation, civil and military. The civil side did figure prominently in the development of the Arctic coast and Siberia, areas then subject to climate amelioration and seen by Stalin as akin to "high frontiers". Soon industry and the military were to respond to the aviation demands of the Great Patriotic War in a manner which can be seen, through the frightfulness of it all, as impressive by any standards.

Then in the doleful aftermath, the true spirit of science wilted badly. In the broad drive towards even stricter conformity led by Andrei Zhdanov in 1947–8, Soviet physicists were enjoined to study Einstein's specific conclusions but discount the "bourgeois idealist" philosophy which supposedly permeated their presentation. The late Stalin era was also characterized by such practical absurdities as regional climate control by means of endless arboreal shelter belts; and Trofim Lysenko's short cuts to biological mutation. Neither Space nor long-range rocketry were on the overt agenda. Biological evolution, Lysenko-style, was even more anti-scientific than Creationism. In 1956, against the background of Khrushchev's anti-Stalin drive, Lysenko resigned as president of the All-Union Academy of Agricultural Sciences.

Sputnik Fantasy

A fair number of German weapons scientists were now in Soviet hands. This circumstance was to be instrumental in the USSR's launching, in the autumn of 1957, Sputniks 1 and 2 – the Earth's first artificial satellites. To many people the world over, it indicated that Soviet scientists, being "ten feet tall", had won a crucial measure of military superiority via a "missile gap". After all, a rocket sufficiently powerful and accurate to launch Sputniks would be well able to deliver a thermonuclear warhead on an American city. Yet the broader inference was unsound for several reasons (see Chapter 10) which in themselves reflected badly on Soviet Communism's claims that "scientific socialism" suffused its political religion. Even so, until perhaps the Cuba crisis of 1962, the "missile gap" notion was widely enough believed to acquire a certain validity. The Soviet Star was in ascension awhile.

10

CONFLICT IN HEAVEN?

MISSILE DEFENCE

THE PROPOSITION THAT SOCIAL TENSIONS MAY ENGENDER WAR-LIKE CRISES does not relate easily to emergent global perspectives on security. Yet the latter also contrast starkly with the mental maps which sustained military campaigns from 1914 and from 1939, not to mention countless other historical encounters. Thus one emergent reality is that the qualitative ascendancy now being achieved by the leading Western countries (and especially the United States) in the assize of modern arms makes it next to impossible for developing countries or insurgent movements within them to present the West with a serious challenge on the time-honoured field of battle. Therefore the angry underprivileged of this planet may soon feel obliged to look more to Weapons of Mass Destruction, nuclear or – failing them – biological.[1] Already this prospect has implications for the military uses of Near Space.

The debate about space militarization effectively took off with President Reagan's keynote speech of 23 March 1983, announcing what soon would be officially entitled the Strategic Defense Initiative (SDI) and popularly dubbed "Star Wars". The gist was that the United States would strive insis-

As noted in the Preface and the biography, space weaponization is a topic this author has been immersed in. See the *Fundamental Issues Study* (especially Chapters 3 and 4) and *Global Instability and Strategic Crisis* (especially Chapters 7 to 10). What follows is an updated overview.

tently to replace "the threat of instant US retaliation to deter a Soviet attack" by an ability to destroy ballistic missiles (i.e. rockets) "before they reached our own soil or that of our allies".

From the outset, the Strategic Defense Initiative Organization (SDIO), founded in April 1984, had space-based weapons centre stage within a broader remit. Indeed, James Abrahamson (the quietly charismatic USAF general who headed SDIO from its inception until 1990) would soon speak of SDI as the prospective

> "nucleus for a Space renaissance for the twenty-first century. It will spur the growth of much new technology. Our programme is forging an alliance with scientific researchers in industrial and academic circles; and interdisciplinary research will continue to be important.
> I believe that SDI will enhance and increase the civilian uses of Space, rather than inhibit them . . . like the shuttle programme, SDI will be viewed as a symbol of national pride and will pay for itself."[2]

The general's upbeat attitude will have owed something to a disposition long characteristic of the American space community. Promote a capability; and then look for useful applications. In the middle sixties, the Pentagon had been allowed to proceed with a proposal for a Manned Orbiting Laboratory (MOL) despite much scepticism within the Johnson administration as to what useful purpose it might fulfil.[3] Between 1958 and 1965, Project Orion had been extant. This was an officially-backed paper study for a fleet of nuclear-propelled spaceships weighing (when out of orbit!) several thousand tons apiece. Orion was successively accorded a variety of missions (ranging from deep space exploration to military bombardment[4]) which ought never to have been accommodated within the same specification, even on a modular basis.

The First Climax

For three tumultuous years from 1984, a raft of interceptor techniques were reviewed by SDI. Among those actively considered for orbital installation were mini-rockets, lasers, sub-atomic particle beams and electromagnetic launchers. Yet by 1987, operational viability seemed far from assured, certainly for the last two genre just mentioned. Moreover, the year in question saw the departure from the Reagan administration of the Pentagon's five pro-SDI stalwarts: Frank Gaffney, Fred Iklé, John Lehman, Richard Perle and Caspar Weinberger. Though each in turn cited personal grounds for exiting, there was visibly some backlash against the SDI school.

Technological unease will have been a prime reason, unease about interception feasibility but also such other aspects as early warning and command-and-control. There was, too, a dawning recognition that SDI had already made its distinctive contribution to the winding down of the Cold War. The managerial panache its *modus operandi* betokened had convinced powerful circles, military and party, in Moscow that they could no longer compete because they lacked not only resources but also the right entrepreneurial ethos. Never mind the critiques of SDI techniques emanating from within their own Academy of Sciences,[5] critiques markedly more factual and incisive than previous disquisitions on modern military science published by the Soviets.

Allowing that a definitive historical judgement is still awaited, evidence for an interpretation much along the following lines built up. By 1986, Raymond Garthoff had concluded, as the senior Brookings analyst of Soviet military affairs, that central to Moscow's concern over SDI was a perception of it as a fount of technological spin-offs throughout the panoply of theatre war.[6] Meanwhile, Westerners in contact with Gorbachev and his entourage at the Reykjavik summit and elsewhere were receiving intimations to the effect that SDI had convinced the Kremlin the Cold War was unwinnable. Lady Thatcher has since advised us that this was just what Ronald Reagan had predicted, two years before.[7] Moreover, this SDI effect was affirmed at Oxford in the Spring of 1992 by Roald Sagdeev who through the mid eighties had headed the Institute for Space Research at the Soviet Academy of Sciences.[8] He came over as eminently reasonable and trustworthy.

By 1987/8, Vice-President George Bush was among those persuaded of how unwise it was to push SDI too far, both for intrinsic reasons and on wider political grounds. Correspondingly, in his inaugural State of the Union address in 1991 he announced as President that he had "directed that the SDI programme be refocused on providing protection from limited ballistic missile strikes whatever their source". This revision led to the Missile Defense Act of 1991, the first Ballistic Missile Defence (BMD) legislation to mandate actual deployment, albeit for limited purposes. Its prime stipulation was the single-site installation by 1996 of 100 ground-based midcourse interceptor missiles, the aim being comprehensively to protect the continental United States against limited strikes by "rogue states" such as North Korea. Shortly, Grand Forks, North Dakota was nominated as the single site. With the passage of the 1991 Act, SDIO made way for a Ballistic Missile Defense Organization (BMDO) which has now given way to the Missile Defense Agency. But by 1990 the initially intense public debate about the principle of SDI was markedly dying down, in the USA and around the world. It has yet to revive.

The Clinton Era

With President Clinton assuming office in 1993, another strategic sea-change was effected, a change likewise reflected in successive Missile Defense Acts. The nub was a switch in emphasis away from National Missile Defense or, in the 1991 parlance, Global Protection Against Limited Strikes. Now the accent was heavily on Theatre Missile Defense in locales like the Persian Gulf or the Korean peninsula. This would likely be surface-based and/or airborne. In 1993, too, the Clinton administration enunciated a comprehensive Counter-Proliferation Doctrine. All the recognized stratagems were incorporated: diplomacy, export controls, bilateral or multilateral safeguards, arms control pacts, BMD, deterrence through retaliation, economic sanctions and security guarantees. Pre-emptive elimination was duly included as well. Indeed, in September 1993, Secretary of Defense Aspin confirmed that "an important category is attack on buried targets because in many of these countries, proliferators are using hardened underground structures to build or operate special weapons arsenals from". This confirmation contrasts ironically with how heavily pre-emption was to be discounted by the Clinton administration its last year or so. Listening to its spokesmen then, you would think this option had always been regarded as an abomination only the satanic would countenance.

Irony was to be compounded (or transmuted to flat contradiction) by a mood change discernible through the turn of the millennium in the pages of the *Aerospace Power Journal*, a quarterly known until 1999 as the *Air Power Journal*. It is recognized as seminal in regard to USAF doctrine; and, with military aerospace, the doctrinal domain is always important.

We were advised that, over the next two decades, the acquisition of space-borne war-fighting assets will transform the USAF "into an aerospace force that operationally employs both air and space platforms to achieve our national military objectives".[9] Another notion aired was that an arm thus integrated would be well placed to attack the ground support of an enemy's orbital satellites or the launch pads of any anti-satellite weapons he may have.[10] But if space-related surface facilities can thus be attacked, presumptively with high-energy laser, what about other surface assets? And what about adversary satellites in orbit and other potential targets aloft? And what of the reluctance of the Clinton administration to endorse the weaponization of space? One contributor pointedly recalled how President Eisenhower had been among those opposed to the weaponization of space but that the airmen of that generation "developed visions of space at odds with those of their political leaders".[11]

Fateful Contradictions

With George W. Bush in the White House, the quest for thin but inclusive BMD cover of the continental USA has been earnestly resumed as a top priority. Soon a total of 20 ground-based interceptors for use against InterContinental Ballistic Missiles (ICBMs) will have been installed operationally, 16 in Alaska and four in California.

But here is the rub. The warload lofted by an ICBM does not have to be unitary. Well before it comes within ground interceptor range, it could have split into several distinct Re-Entry Vehicles, each delivering a warhead to, say, a particular suburb within a large conurbation. Alternatively or additionally, the warload capsule may have released a lot of deception materials, mostly either metallic balloons for warhead simulation or else resonant metallic strips intended to cause radar to white-out: in other words, an enhanced version of the "chaff" or "window" distributed abundantly during strategic bombing raids in World War Two and afterwards.

Ever since a keynote article he wrote together with the late Hans Bethe in 1968,[12] Richard Garwin has been recognized as a considerable BMD authority. He has regularly stressed the efficacy of decoy balloons until their descent is appreciably retarded by air resistance in the lower atmosphere.[13] What the Pyongyangs of this world lack, however, is direct experience of how a threat cloud looks to defenders possessed of a diversity of orbital or ground-based sensors. Therefore they themselves would find it hard to contrive unaided a mix able to mislead such a panoply sufficiently. Accordingly they may incline towards the multiple warhead approach. Either way, one is talking about technologies which ought within a very few years to be attainable by any country already able to build ballistic missiles which can hit conurbations perhaps half-way round the world. Mixes of warheads and decoys are best contrived within one warload.

Tactically, the best way for the defence to negate these complex assemblages is to intercept every warload prior to their programmed dispersion. Shortly after rocket burn out is usually seen as the optimum time for this "debussing" to occur. So what is the altitude zone within which one should effect such a Boost Phase Interception (BPI)? The general answer (when countering an intercontinental strike) appears to be somewhere between 40 and 70 km up, depending on the designed burn-out height though also on warload composition. Decoys may not inflate satisfactorily where the air is relatively dense.

In principle, a warhead ascended to 50 km might be engaged from the surface across a slant range of up to 800 km. This could mean that ships or aircraft located in the Sea of Japan or in the Persian Gulf could cover for this purpose North Korea or much of the Gulf hinterland respectively. However, such a deployment would be beset by political and operational difficulties within the designated area. Nor could the scope for such a stratagem match up at all adequately with the global coverage the hard core of American support for strategic BMD has always sought. So through the turn of this decade, the great debate is going to hinge on the acceptability or otherwise of orbital interceptor platforms. The preferred means of interception from them would, by 2020 or thereabouts, be a laser.

Space-Based Lasers (SBLs)

More specifically, the laser beams would be directed from orbit to rupturing the thin hot casings of the rockets to which the ascending warloads would still be attached. Studies conducted to date have generally indicated that, located high above, first-generation SBLs may engage down to 50 km or thereabouts. But that prospect might be vitiated by making the casings shiny and/or rotational. Its application might in any case be limited by the adversary making his rockets burn a sight faster, this for a modest sacrifice of warload mass. On the other hand, lethal laser penetration down to sea level should in due course be on the cards. This might be effected by, say, the oxygen–iodine chemical laser which beams on a wavelength that lies inside a "window of penetration" within the atmosphere. It might alternatively be achieved by tunable Free Electron Lasers.

Writing in 1995, Henry Cooper, the last Director of BMDO, called *inter alia* for the deployment of SBLs by quite early this century. He believed SBLs would be able to intercept in boost phase and "throughout most of their flight trajectory" all ballistic missiles with horizontal ranges exceeding 300 miles. He even foresaw an SBL engaging any traveling 75 miles or more.[14] The broader implications are obvious. Depending on what wavelengths their beams were transmitted on, SBLs might operate to effect against a wide variety of surface or near surface targets.

This ramification could be further favoured by yet another SBL peculiarity, limited "dwell time" over target and the consequent absentee effect. In other words, an SBL platform orbiting relatively low above our rotating Earth (800 miles up is mentioned as quite likely[15]) would have a given target area within effective range a very low proportion of the time. Scenarios typically indicate dozens of absentees for every platform on target. Admittedly, the said platforms could be so orbited as not to reach any latitude higher than the highest in the territory being constrained. Thus one could present to North Korea without directly threatening Russian strategic forces. But this geometry would lend itself to swift adjustment. Space-basing is inherently for global coverage.

Advocacy of BPI has too often been marred by rather ghoulish celebration of the prospect of the detritus of encounter raining down on the launch country. Undeniably a possibility exists that any warhead intercepted could, if lacking design sophistication, explode on crashing back onto the Earth's surface; and that, even without explosion, it would cause significant contamination. However, an adversary is unlikely to fire strategic missiles from his palace gardens. He is much more likely to fire them from regions that are (a) borderland, (b) populated by ethnic minorities and (c) of low population density in any case. Studies of Iraq and North Korea indicate that the fatalities thus from a warhead of 20 kilotons (i.e. Hiroshima or "nominal" yield) would range from the low hundreds to the low thousands.[16] Such a toll may not sound a lot. But humanitarian considerations apart, it could be very damaging politically, if seen as yet another example of rain on the not unjust.

Backlash

What SBL therefore represents is another endeavour on the part of missile defence hard-liners to push their luck to the point where it all turns dangerously counterproductive. If Near Space does come unreservedly to be seen by the USA or the West as a whole as the "new high ground" from which forceful control can be exercised globally, a strong backlash could develop among the alienated of this planet.

Moreover, this could ultimately be expressed in a form that SBL would be no answer to – namely, biowarfare. Germ cultures may be best delivered to their target areas clandestinely rather than by ballistic missile. Moreover, the requisite biotechnology capability is spreading, for everyday civil purposes, intercontinentally. Witness the prowess South and East Asia already show in this field.[17]

Although informed world opinion is presently quiescent about the prospect even of global BMD, this could change in short order. Uncertainties about the operational efficacy of a Space-Based Laser screen could vitiate the arms control process. A worsening of debris accumulations at high altitude is to be apprehended as well. Any talk of warfare from space would compound apprehension.

Indeed, beams directed from on high could resonate with some disturbing mythology. The fact that "death rays" have been a staple of science fiction since H. G. Wells indicates how the mere mention of them being directed from on high evokes from deep within our psyche some graphic folklore: the "evil eye", the basilisk's stare, lightning shafts, and the firedrake. Even the prosaic *Anglo-Saxon Chronicle* tells how, among the "foreboding omens" that "wretchedly terrified" the people of Northumbria in the year 793, were "lightning storms and fiery dragons . . . flying in the sky". However phantasmagoric any latter-day revival of such concerns may appear to strategic analysts, they would do well not to dismiss it out of hand.

Over and against all of which may be set the precept of the counter-culture Left that "The heavens were made for wonder, not for war." To my mind, it has a more edifying ring about it than other slogans from that quarter – "Better Red than Dead", let us say.

Writing in 1985, Patricia Mische, the founder of Global Education Associates, saw SDI as a morbid diversion from the proper study of space, study which could help us to understand better our terrestrial situation. She concluded thus: "How each of us responds to the conflict in our own souls is what we will bequeath to the universe; this is what will become the soul of universe."[18] Lyrical for sure. But maybe the kind of lyricism our historical and prehistoric forebears would have cheerfully imbibed. During his first term, George W. Bush committed the USA to launch experimentally by 2008 missile-armed satellites, these as a preliminary stage in the development of fully layered missile defence. There is still time enough for a radical rethink. Those concerned can rest assured that any suggestion of the Earth's surface coming under the orbital domination of any state would engender a syndrome of resentment the world over, including not inconsiderably the United States itself. It would be a reaction akin in spirit with Lyndon Johnson's justifying Project Apollo on the grounds that he was not content to go to sleep under "the light of a Communist Moon".

BENIGN INFORMATION?

Meantime Near Space has become the old firmament writ new in terms of its providing the overarching framework for long-distance data exchange. However, this Information Firmament is liable to present major congestion problems, vehicular and electromagnetic. No doubt, too, there will be further backlashes against intrusive surveillance and "cultural pollution". As regards the latter, concern has burgeoned across the world lest imagery beamed from on high creates "cultural deserts" as traditional mores are eroded in favour of a glitzy cosmopolitanism with its accent on mindless titillation – pornography, violence, trivia, and a general dumbing down. In various religious traditions, of course, anything that smacks of a "craven image" is deeply suspect.

In 1986, Papua New Guinea deferred a decision about national television since "We do not want to become Americans or Australians", and because too many soap operas would exacerbate further the high incidence of violent crime, alias "rascalism". Within Europe, France has episodically assumed the lead in exploring the political scope for resistance.

The Authoritarians

Those who run police states often manifest a comprehensive concern to protect their national culture as they interpret it; and always have an obsessive desire to head off dissent. They therefore find the Information Firmament deeply disconcerting. In this regard, as in other attitudinal matters, the situation post-9/11 is still not easy to delineate. But in the preceding years, Albania, Iran and Libya were among the repressive regimes which either had failed or were failing to curb the stimulation of dissent by limiting the proliferation of satellite receiver dishes. Saudi Arabia and China were among those which doggedly continued to resist this tendency. The former still maintained an official ban on receiver dishes, though the privileged flouted it. Very much the same was being said of the collateral provision of Internet facilities.[19] Meanwhile, China had been striving to oblige all its urban dwellers to abandon private satellite dishes in favour of cable television from which every inkling of incisive comment had been excluded. Witness Rupert Murdoch's admission in June 1994 that he had felt obliged to have the BBC's round-the-clock news service removed from his Hong Kong-based Star satellite broadcasts, this in order to placate the Beijing authorities. But one of the contrary trends they and others like them have to reckon with is that towards smaller satellite dishes, ones easier to use clandestinely.

Technological Progress

Various exotic methodologies are being applied to satellite observation of the Earth's surface for civil or military ends. Infra-red scans of convective cloud tops coupled with microwave probes of the clouds interiors are being used to guestimate rainfall over tropical seas.[20] Likewise, surveillance from space can input predictive models of fish catches. This has served to resolve the paradox that the FAO global record of fish landed annually rose past 80 million tons in the 1990s though many particular fisheries conspicuously declined. The resolution lies, of course, in exposing implausible claims, made partly in anticipation of possible quota allocations in the future.

Meanwhile, NASA and German researchers are exploring the space-borne mapping of gravity fluctuations at the sea surface, these due to alterations in water temperature and salinity. Another technology which operates regardless of darkness is space-based radar. Wavelength choices give it a singular if limited ability to penetrate clouds, foliage, shifting sands and salt water. The utility thereof is considerable in the civilian context and can be invaluable to the military.

When the Superpowers began optical orbital surveillance in the early 1960s, the definition of imagery recovered was couched in terms of the diameter of the smallest white spot visible against a black background or *vice versa.* Quite soon the quotable figure shrank, in the American case, below 25cm.

In these days of integrated multisensing, comparison is more complicated. What is the mix of technologies and wavebands? What are light levels? What field of view is sought? How many transits are made? Currently "definition" is usually taken to mean the width of the smallest discernible feature. "One metre" is widely seen as a "sweet spot" for, let us say, distinguishing the individual vehicles in a road convoy. Commercial reconnaissance is now on that threshold while, in military programmes, below half a metre is achievable.

Integrated reception from several parts of the electromagnetic spectrum is the nub of advanced space surveillance. Its utility against camouflage, natural or contrived, is evident. Witness, on the civil side, the programme launched in 2003 to map the habitats of those mountain gorillas still surviving, however marginally, in the war-ravaged, poacher-ridden border zones between Uganda, Congo and Rwanda. Then in the military domain, there can always be room for debate about how to strike the balance between space and aerial surveillance in localized confrontations.[21] But for deep and wide coverage, continentally or whatever, the imperatives of space are obvious.

Additionally, considerable interest is being shown, certainly in the USA, in deploying for eavesdropping and communications swarms of quite or very small satellites. In this context, "quite" refers to nanosatellites weighing less than 10kg; and "very" to picasatellites weighing less than one. Taken along with the incipient emergence of space tourism, the prospect of constellations of these tiny platforms cannot but strengthen the case for an overriding international space authority. As of the present time, however, the world's political elites are evincing little interest in this possibility. Their current preference is to react *ad hoc* to immediate space developments.

Reaction to Surveillance

Nevertheless, it is none too soon to begin consideration of what the *modus operandi* of any such authority might be. Should a prerequisite be the complete demilitarization of Near Space? It is a fair question but one which would elicit a firm negative from the great majority of analysts. For one thing, they would see such militarization as has taken place to date as having been conducive to world peace. Communications satellites have aided crisis management. Reconnaissance of the Earth from space has assisted the West operationally but also reassured it strategically, partly through arms control verification and partly through revaluation of threat. A not untypical view in 1955 was that the Soviet army would be able to mobilize, given reserve recall, some 300 divisions within six weeks and 450 within six months:[22] a huge order of battle that would, in fact, have been extremely hard to commit operationally, even around the USSR's long borders. But in any case by 1965, the said order of battle was generally accepted to be 140 divisions, not more than half of them seen as at or near to combat readiness. This revaluation owed much to overhead transits of the Soviet Union by U-2 monoplanes between 1956 and May 1960; and then by orbital reconnaissance satellites from the autumn of 1961.

In any case, a thorough demilitarization of space would not be feasible in view of the large overlap already apparent between civil and military tasks. With meteorological reconnaissance, the concurrence of data base is almost complete. Likewise with navigational support, there is a pronounced commonality. Take the case of the US Global Positioning System (GPS) network of navigational satellites. An endeavour was made from the outset to preserve a distinction via the different signal modes. The Precision or P-code was intended for authorized military use; and can be encrypted. The Coarse Acquisition, or C/A mode, was for general civil use. However, deployed forces that are static or slow-moving might narrow the C/A resolution several-fold by multiple readings.

What was the world's first civil space-based Earth-imaging system, the US Landsat, has provided a 30-metre global resolution on the open market to anyone anywhere. Much of its information has been sold via licensing agreements and directly downlinked from the satellite to ground stations in a number of foreign countries. Yet Landsat data was also utilized by US forces during Desert Storm, the 1991 Coalition offensive against Iraq.

Nor can we yet assume that space reconnaissance will no longer engender resentment across the world at large. Only a decade ago, three prominent developing countries – Brazil, India and Pakistan – were evincing considerable hostility on this score. This was notwithstanding all three of them being already considerably involved industrially in aerospace high technology; and, in the case of Brazil and India, this included programmes for optical surveillance by civil satellite. In Brazil, there was protest against ecological surveillance data being processed outside the country. Meanwhile, India and Pakistan were specifically objecting to monitoring from space being provided for in a Comprehensive Test Ban.[23]

All of which served to confirm the sensitivity of world opinion about "Outer Space": a realm identified in the evocative terminology of the 1987 Brundtland report as one of the three "global commons", the other two being the Oceans and Antarctica. This keynote study spoke of "growing concerns about the management of orbital Space", not least as regards the threat of weaponization.[24] This then was and is the crux of the matter. "Demilitarization" would fail the first test, that of role distinction. "Non-weaponization" is a much more common-sense aim. Meanwhile, any Near Space regime planning must take due account of what already is a menacing and very intractable space debris problem. It must also take full account of the part the Near Space milieu is playing in the big advances being made, especially these last few years, in gathering data about the Heavens at large. The Hubble Space Telescope has been a splendid flagship in this regard.

PART FOUR

A Dissolving Heritage?

OVERVIEW

A Dissolving Heritage?

A premise underlying this study has been that a precipitate collapse of the great religions would create vacuo of belief liable to be filled by all sorts of unpleasant rubbish, political fanaticisms included. No less alarming, however, would be these obediences regaining momentum by turning on one another more. A constructive way through would instead be reconsiderations of doctrine in support of gradual interfaith conflation. In which connection, some of the most germane questions to ask about any particular belief system are as follows.

What does its customary teaching proffer as regards survival after death? Little or nothing? Reincarnation? Salvation in heaven, perhaps combined with hell as a diametric alternative? Merging into a cosmic life force? Then again, do polytheistic vestiges persist, perhaps via local saints? And what of spirituality and mysticism?

Otherwise, where is humankind placed in the cosmic order? How readily might it reach out to godliness? How personal or, at any rate, humanoid is any God? What part did one or more charismatic prophets play in the faith's origination? How do Creation legends, astronomy and maybe astrology relate to religious belief? How does Science in general? How inerrant are the original sacred texts deemed to be? Has there been much philosophic development across the intervening centuries? How much toleration is extended to contrary beliefs, within and without the said faith? What attitudes are struck by its hard core adherents? How well accommodated are contemporary challenges – e.g. shrinking priesthoods, population control, armed violence, homosexuality, women priests? How threatened is this belief system by fragmentation?

11

JUDAISM FULCRAL

IF ONE ASKS WHAT PEOPLE HOPE FOR FROM RELIGION, salvation or survival with redemption after death is most readily mentioned. Yet individuals may inwardly yearn as much for a sense of historical continuity, a sense of belonging to a process which far precedes and succeeds their own fleeting existence. No living faith proffers this more explicitly than Judaism. It does this most notably via the Hebrew Bible which was finally consolidated *c.* 100 BC and is basically, too, the Christian Old Testament.

The first five books, the Pentateuch (Genesis, Exodus, Leviticus, Numbers and Deuteronomy), constitute the Torah or sacred law. In the first chapter of Genesis, God is presented as without doubt the one and only. He creates everything uninterruptedly in seven days. Every day's work has been good. However, it is not quite Creation *ab nihilo*. Initially, everything was a "formless void" in customary translation or "wild and waste" in a recent alternative.[1] Either way, it all happened on the face of the deepest of oceans.

The Pentateuch concludes with the death of Moses (at the ripe age of 120) shortly after the patriarch, under God's guidance, has looked down over the Canaan promised by the Almighty to Abraham and his descendants for "their perpetual holding" (Genesis 17.8). The last word in Deuteronomy is "Israel". But this betokens a contradiction still not resolved. Is this the God of all Creation or the YAHWEH of a chosen people?

What then of Abraham, the patriarch who more than any other human is seen as a bond between Judaism, Christianity and Islam? Quite likely he, along with Isaac and Jacob (identified in the Bible as his son and grandson), were historical figures around the middle of the second millennium BC. Abraham could well have led his tribe from Ur to Canaan via the great Euphrates bend.

Today, Abraham's tomb in Hebron is venerated by all three faiths. Even so, the persona who, at the age of 99, was promised Canaan for his people is not quite the same image-wise as the Christian elder statesman envisioned by John the Baptist and St Paul. Surely, too, each is different from the Islamic Abraham, the father of Ishmael, with whom he worked to make the Kaaba in Mecca a place of pilgrimage for the worship of Allah. All the same, the three creeds can be seen as "religions of faith" – in contrast to those of mystical union in India or the wisdom religions of China.[2]

For the Jewish people, the most weighty moral to be derived from their history, biblical and otherwise, appears to be that one can repeatedly survive what look *a priori* like terminal societal crises. The supreme example has to be the Exodus across the Suez isthmus of those Jews who had been working in Egypt as impressed labour. If the biblical account be broadly valid, this land bridge was likely crossed via an ancient miners' routeway along its intricate southern coast,[3] a sector *tsunami* susceptible. The Book of Exodus (14.21–31) rather gives the impression that the Egyptian pursuit force was hit by a *tsunami* return phase.

This episode *c.* 1230 BC could be said to mark the onset of a prolonged regional phase of ecological instability identified in a pioneering study of climate in history by the late Rhys Carpenter, a classicist at Bryn Mawr. By 1200 BC, he said, "Mediterranean man has begun to suffer the most severe cultural recession which history records or archaeology can determine". Kingdoms like the Hittites in Anatolia collapse "without apparent adequate reason". Refugees surge onto the Mediterranean's eastern shores. The fortified palaces of Mycenae and Crete are burned by persons unknown. Throughout Greece, communities revert to primitive subsistence. Egypt sinks "into helpless apathy" for a full 400 years, after the Exodus and the "nine plagues".[4] Volcanic eruptions and earthquakes figure prominently.[5]

Then in the eighth century BC occurred – suddenly, it seems – a pronounced renaissance, led by Greece. It gave rise to Homer then Hesiod. Soon a new culture spread round the Mediterranean as the Greeks founded colonies. A central theme was the further development of phonetic scripts, rudiments of which from the fourth millennium BC are discernible in the Fertile Crescent.[6] Hebrew was poised to benefit. A text held in the British Library dates from the early tenth century.[7]

But firstly one must address the Exodus epic in its Sinai phase. Nobody sees it as having been a route march to the Promised Land, not at a pace of several miles a year. Rather it was an extended interlude of nation building against a most demanding background. Moses was the master builder for most, if not all, the escapees; and his keystone was the worship solely of the

God he felt face-to-face with, high on Mount Sinai. To Küng,[8] Moses epitomizes the prophetic archetype Friedrich Heiler vividly sketched, "Born of a tenacious will to life, immovable confidence, reliance and trust firm as a rock, bold adventurous hope breaks through at last . . . The prophet is a fighter who ever struggles upwards from doubt to assurance . . . "[9]

The invasion of the Promised Land by Joshua, the successor to Moses, was also beset by difficulties. Iron weaponry was then proliferating regionally. But intrusive and penurious migrants may not have accessed much. At all events, the strategy at this stage was to by-pass strong fortifications. However, ancient Jericho could not readily be avoided since it overlooked a key ford across the Jordan. Moreover, it was surrounded by two highish walls, the inner twelve feet thick and the outer six. The opportune earthquake gained the day (see Chapter 1).

The annexation thus begun was brought to its prosperous and expansive zenith under King David (*c.* 1012–*c.* 972 BC) who *inter alia* moved his capital from Hebron to Jerusalem. There his son, King Solomon (d. *c.* 932 BC), built the first Temple. Partly thanks to onerous taxes, however, the kingdom split after his death. The tribes of the north formed Israel; and those of the south, the smaller but closer knit Judaea. During the next two centuries these two kingdoms were regularly under pressure from predatory empires. In 721 BC Samaria, capital of Israel, fell to the Assyrians; and, says the Bible, there were major deportations, the elusive Lost Tribes. In 586 BC, the Babylonians overran Judaea. The Temple was destroyed and many Judaeans exiled to Babylon. But many who so opted were allowed to return from 538 BC, the year Babylon fell to the Persian-Midian King Cyrus. The rebuilding of the Temple was completed in 516.

In the meantime, however, History had observed a distinctively Jewish contribution to the renaissance of the eighth through the seventh centuries – namely, the writing prophets. They ranged from Amos and Hosea in Israel to Isaiah, Jeremiah and Ezekiel in Judaea and, ultimately (*c.* 460 BC), to Malachi. They deplored sundry current tendencies. These included discarding a "history of salvation" in favour of reliance on national power. Included, too, were adulation of the priesthood and alleged legalistic chicanery. Religious syncretism and geopolitical manoevrings were likewise abhorred.

Nevertheless, YAHWEH would, they hoped, forge a lasting peace between the nations. The late eighth-century prophet who contributed the first half and more of the Book of Isaiah (Chapters 1 to 39) was especially eloquent about this. But the rub here was that entry into Canaan had rekindled the internal threat from polytheism. Now this came principally from the extended influence of the Phoenicians whose every town had its quasi-private local god, Ba'al if male, or Ba'alat if female. Adonis, the tragically youthful Greek god was also impacting.[10]

Not until the sixth century had the severe and jealous YAHWEH (alias Jehovah) gained an ascendancy among the Israelites sufficiently categoric to allow of His being projected as the only God everywhere[11] (see Jeremiah 16.19). The prophets, though few in number, had played a big part, this by virtue of essaying a direct approach to Jehovah as well by extolling social justice as a check on the allure of exotic or erotic local deities.

Closely related was an upwelling of interest in a personal afterlife, hopefully as contributory to a glorified existence for Israel. Customary Hebrew belief was that a very, very few (Elijah and Elisha, for example) made it to Heaven, the abode of the gods or God. The rest of us were more likely destined for a shadowy, lonely existence in subterranean sheol. In any case, the overriding concern had to be the survival of the Hebrew people.

As the seventh century drew to a close, Josiah, King of Judaea (reigned 640–609 BC), assumed the leadership of a reform movement which also embraced the kingdom of Israel. Subordinate religious centres were downgraded or closed in favour of Jerusalem. Likewise, new legal taboos about corpse disposal were intended in part to nullify any vestiges of local ancestor worship. Then came the Babylonian exile, searchingly stressful yet also a source of new religious insights. Prominent among these was Zoroastrianism with its belief in titanic struggle between Good or Evil as well as in paradise or perdition for individual souls, depending on which side they had taken on Earth. A later entry in the Book of Isaiah (i.e. one written after the deportation) tells how "for Jerusalem's sake I will not rest; until her vindication goes forth as brightness; and her salvation as a burning torch" (62.1). In other words, restoration of the Holy City will be part of an enveloping, transcendental experience. Ezekiel, a prophet of the Exile, deftly combines his dislike of the Zoroastrian/Babylonian practice of leaving "dry bones" unburied with a promise of personal resurrection, "Thus says the Lord God to these bones: Behold, I will cause breath to enter you and you shall live."[12]

The next Israelite transformation began in the fourth century BC. Soon prophecy would die out. Instead there was a swing towards social and religious conformity, to a regime led by the High Priest, subject to Persian suzerainty. This adaptation was not easily effected. A ban on non-Jewish wives aroused particularly fierce opposition. Increased recourse to the Hebrew scriptures came at a time when Aramaic was well established as the popular *lingua franca* in and around Palestine. Then in the second century, hellenization triggered an armed uprising under Judas Maccabaeus. The independence won lasted until Herod gained the throne in 37 BC as, in effect, a Roman puppet. Thus the scene was set for the philosophic and social tension which precipitated the rebellion of AD 66. In the interim, two major schools of religious thought emerged, the Sadducees and the Pharisees. The latter was the more populist, its militant wing having Maccabean connections. The Pharisees stressed the Torah as the scribes interpreted it, though also the Oral Law. Suffused with apocalyptic anticipation, they professed belief in a life hereafter, a coming Messiah and Judgement Day.

Over against them the Sadducees were a movement extant by 200 BC with the philhellene urban elite at its core. Pretty much accepting the literal truth of the Hebrew Bible, they had reservations about Oral Law as well as about Temple cultishness. They did not believe in immortality nor in a Messiah. The AD 66 to 70 insurgency war all but eliminated them. Beforehand they had controlled the Sanhedrin, the council which helped the High Priest run things.

Millenarian tendencies were also abroad, ones extolling forceful direct action. The Qumran caves, where the Dead Sea scrolls were discovered in 1947, revealed a community which (drawing on Babylonian creation mythology) was committed to serving the Prince of Light in his end-time struggle against the Prince of Darkness. It was probably Essene, an ascetic sect believing in an afterlife spent in either paradise or perdition. Alternatively, Qumran may have been linked to the Zealots, a group persuaded that resisting alien rule would bring forth the Messiah. As the Roman army finally stormed their Masada stronghold AD 72, all bar seven out of a holding force of 960 died by their own hands. Those seven were either women or children.

Characteristic of the apocalyptic writers of that generation was a preparedness to put their dire predictions into the mouths of great forefathers – e.g. Noah or Moses. The gist usually was that an evil power was tightening its grip on all mundane things and its arrogance was increasing. The one hope was divine intervention to establish God's kingdom on Earth or (particularly in later versions) in Heaven. Only then would the beast indescribable give way to the Son of Man.[13]

The Dispersion

The year AD 70 saw Jerusalem sacked and its temple destroyed. A major dispersion began, largely to pre-existing Jewish colonies in the Mediterranean or Near East. With Emperor Hadrian's crushing of another revolt in AD 132, the Jewish presence in the Promised Land was drastically diminished. Nevertheless, several factors assisted identity survival elsewhere. Since the Babylonian exile or soon thereafter, the synagogue had emerged as an alternative house of God and community. The Babylonian experience had also stimulated a Torah piety which was little affected by rabbinical endorsement of the whole Hebrew Bible *c.* AD 90. Also important, in due course, was the Talmud – further oral traditions and sacred commentaries from Babylon and Palestine, finally compiled in the sixth century. It served to interlink the Jewish communities forming up across Christendom (especially the Ashkenazim in Central and Eastern Europe), along with the Sephardim within Islam. The Ashkenazim came to speak Yiddish, basically a middle German dialect written in Hebrew letters.

Overall, Islam in the High Middle Ages was considerably more tolerant towards Judaism than Christianity was. With the latter, the Crusades were a big aggravation. Especially troublesome were the *tafurs* latched onto the First Crusade by Peter the Hermit, charismatic fanatic and compulsive coward (see Chapter 3). That his involvement was unwelcome in both Rome and Constantinople goes without saying. What cannot be denied, however, is that anti-Jewishness was a recurrent theme within the cultural renaissance then under way in Europe. In the next century, it would be endorsed by the likes of Peter the Venerable, Abbot of the pacemaking Cluny Abbey; and even, albeit more temperately, by Peter Abelard.[14] The proscriptions imposed on Jewish people by the 1215 Lateran Council would confirm discrimination as received practice.

Anti-Jewishness also figured in European response to the Black Death (bubonic plague?) of 1347–51. Granted, many people assumed a kind of positive denial, keep things a-moving as if nothing untoward had happened. But savage reactions did occur. From 1348, there were violent attacks on minorities presumed to have wilfully spread the malady – lepers, pilgrims, Moors, Jews . . . At least 200 Jewish communities are believed to have perished.[15]

With access to agrarian pursuits impeded in various ways, Jews settled in built-up areas emphasizing handicrafts but especially financial services. This latter was further encouraged by the reluctance of Christian clerics to see their faithful engaged in "usury". The medieval Jews thereby became subject to the kind of opprobrium widely felt nowadays towards "market-dominant" minorities (Chinese, Indians, Lebanese, Ibos, Croats, Jews . . .).[16] In this medieval situation, resentment could always be supercharged by casting the Jews as Christ's killers.

Reasonable debate on that particular score is still precluded by ignorance of the legal-cum-political circumstances of the Crucifixion. If we knew enough to hold helpful opinions, many might simply say that Jewish inputs and Roman ones are hopelessly entangled. In any case, there could never be grounds for blaming all Jews eternally. Besides which, Christian theology has seen the summary execution of Jesus as the key to his redemptive victory. In short, the said blame game could serve as an argument against having truck with religion at all.

As almost always in such matters, however, it is wrong to paint pictures of unremitting reciprocal hostility. Roughly from the first Christian turn of millennium, the tendency across much of Christendom was for Jewish people to concentrate in a particular part of a given city – i.e. in a ghetto. Sometimes this trend owed most to concern on the part of those involved for their physical security and cultural identity. At other times, it became, or was throughout, a means of external social control. Often there was an enclosing wall. Within what became Ottoman Islam, the comparable arrangements were the *mellahs*. Here, too, the spectra of possibilities ranged between constriction and consolidation, though often with more emphasis on mutuality. Thus in the cities of Egypt (*c.* AD 1125–1275), Islamic and Judaic quarters interrelated closely.

Maimonides (1135–1204)

Enough security, physical and psychological, was afforded one way or another for the Judaic philosophic tradition to survive. Quite the most important contributor, for his times and today, was to be Moses Maimonides, alias Rambam. Driven in 1148 from his birthplace, Cordoba, by Islamic extremists, he and his family eventually settled in Egypt.

Perhaps as early as 1160 he codified Thirteen Principles of Faith still much acclaimed.[17] They can be condensed thus. The non-corporeal Creator (exclusive, unitary and eternal) is unlike any other entity. He knows our every deed and thought; and dispenses rewards and retribution. The prophets proffer the fullest available insight into His exalted nature. The Torah received on Mount Sinai by Moses, father of the prophets, is changeless. The Messiah may come any time, soon or remote. There is personal resurrection.

The cosmos Maimonides saw as probably originating *ab nihilo* out of a single Creation event. But he did not believe the alternative understanding that everything had been and would forever be much as now necessarily excluded a transcendent Creator. Textual references to miracles he played down; and he allegorized angels and demons. His influence extended in many directions far beyond the confines of Judaism.

Mysticism and the Book

Late medieval Christian mysticism had a Judaic counterpart in the Kabbalist movement prominent between the late thirteenth and seventeenth centuries. Emergent mainly in Spain and Provence, it stood outside the philosophic tradition Maimonides belonged to. Yet it was rich in its own way. Accepting that the godly presence – *En sof* – spread through endless realms higher than our own, it looked to the esoteric mysticism of loving yet ascetic individuals to help keep things in harmony or restore them thereto. The YAHWEH of the Hebrew bible was never far from its envisionings. But it found concordance with *sūfī* mysticism as well as with Christian.

It was something of a haven for astrology, too. Perusing the Hebrew Bible, one gets little sense of astrology or, indeed, astronomy being influential among the Israelites. What research in train may show, however, is that this owes something to editorial concern to marginalize any challenges to what was bound still to be a very insecure monotheism.[18] The anti-astrology disposition was part and parcel.

The Kabbalist tradition was also to influence strongly the Hasidist movement. Revived in the Carpathians by Baal Shem Tov (1700–1760), this had by 1830 or so waxed strong among the Jews of Poland and the Ukraine. Believing the Divine Being to be present in all Creation, it found everyone to contain goodness. It stressed how meaningfully an individual can relate to God on a workaday basis. Joyous, egalitarian and spontaneous, it scorned (especially early on) much rabbinical and scriptural authority. In carefree manner, it fostered local charismatic leaders; and thereby became fissiparous. It is reckoned to have had by 1914 around 30 quasi-dynasties with thousands of followers apiece.[19]

A question sometimes raised is why was there no Judaic Reformation. One answer could be that there more or less was. From the early Middle Ages, there had been (in Islamic Persia, Byzantium, Catholic Christendom . . .) a Karaite dissentient perspective whereby the Hebrew bible was deemed the sole source of Judaic revelation. However, the geography of the Dispersion hardly favoured the development of a mass movement along these lines. Instead, the Karaites turned inwards to ascetic rigour. In 1970 in Israel, an estimated 7000 still survived.[20]

The Renaissance Continuum

The contradictions of the Renaissance, as it is commonly defined, are exemplified by the Jewish experience. With the completion in 1492 of the *reconquista* of Moorish Spain, 100,000 Jews were constrained to depart, leaving still more behind to adapt to obligatory conversion. Then a generation later, the Reformation would be causing across the Catholic realm tensions acute enough to undermine tolerance in whatever direction. Regarding Judaism, one could expect Calvinist Protestants to be less antipathetic than Lutheran since theologically they were the people of "the Book", especially the Old Testament. To an extent, this did apply. Neither Calvin or Zwingli sought to match *On the Jews and their Lives*, a frenetic diatribe an elderly Luther promulgated in 1543. Even so, it is hard to discern in them much magnanimity. Meanwhile, the Counter-Reformation peak years, 1555–1590, saw several anti-semitic popes.

A more emphatic advance towards mutual acceptance came out of the eighteenth-century Enlightenment. By 1750, Jewish people had won more formal equality across much of Europe. Meantime, convergence from their side was led by Moses Mendelssohn (1729–86), a North German philosopher and mathematician. He advocated assimilation to the fullest extent consistent with retention of faith. For instance, speak High German not Yiddish (see below). He extolled, too, plural pathways to God, averring one could commune with a Confucius or Solon without feeling obliged to proselytize.

However, there were limits to Enlightenment accommodation. Judaism was, after all, a well-defined minority faith, not an easy phenomenon for Voltairean rationalists to address, even had their rationalism been more finessed than it was. As Emperor, Napoleon Bonaparte squared this circle with more resolve, aplomb and liberality than anybody, despite his facing the distraction of upwelling anti-semitism in rural Lorraine and Alsace.

The Emergence of Zionism

But as the nineteenth century progressed, outright anti-semitism revived within the context of resurgent state nationalisms exacerbated by (a) the long world recession post-1873, and then (b) that sense of frontiers closing round a shrinking Eurasian peninsula possessed of too much energy for its narrow confines (see Chapter 4). Depending on how "energy" may here have been visualized, these considerations bear on Jewish vulnerability. *L'Affaire Dreyfus* broke within the French military in 1894. A decade earlier, after savage pogroms and anti-Jewish laws in Russia together with Poland, Jewish emigration therefrom had begun *en masse*. In 1820, there had been in the USA perhaps 8000 Jews of Eastern European origin. By 1908, there were 1,400,000. By the 1880s, too, there had been a significant increase in the emigration of European Jews to Palestine. Zionism was in the air.

However, debate continued as to its prescriptive agenda. In due course, the viewpoint of the charismatic Viennese journalist, Theodor Herzl (1860–1904), effectively prevailed. Shocked by the framing for treason of Captain Alfred Dreyfus, Herzl became convinced that Jews must have their own nation state, a modern secular polity committed to justice and toleration. After much equivocation, he accepted that this should be in Palestine, not Argentina or Uganda. Save that statehood was not mentioned for fear of Ottoman reaction, this thinking found expression in the Basle programme adopted by the first World Zionist Congress held, with Herzl as president, in 1897 at Basle.

Twenty years later, against the background of the war to end all wars, the famous Declaration was sent to Professor Chaim Weizmann (1874–1952), the wartime head of the Admiralty munitions laboratories, by Arthur Balfour, Britain's Foreign Secretary. It said His Majesty's Government would back a Jewish "national home" being established in Palestine provided nothing "be done which may prejudice the civil and religious rights of existing non-Jewish communities in Palestine or the rights and political status enjoyed by Jews in any other country". Already London and Paris had covertly agreed Britain should manage the post-war Palestine mandate.

The Mandate Years

One cannot recapitulate again the conflict between Zionism and emergent Palestinian nationalism. But neither can one escape the fact that, three decades after the Balfour Declaration, the dye was cast for the first of what have been five wars between Israel and neighbouring states and much other armed violence besides. Suppose the Arab League and the Palestinians had resolved to accept the UN Partition Plan of November 1947; and that the Irgun (see below) had been constrained from wrecking it. Then a confederation of Israel-cum-Palestine might now have been a Holy Land beacon for all mankind. For one thing, the plan's geography was so interlocking as to preclude separate development.

By then, however, tensions were too acute. Nor could the Palestinians be overly blamed, in any case, for a rejectionism which would have been evinced in such circumstances by countless other societies, the modern Israelis certainly included. If the British were looking for something more edifying, they should have gone for it from Versailles onwards. After all, the Palestinians were arising from Ottoman torpor. Local chiefs, imams and so on ran competing fiefdoms. There were urban elites not too antipathetic to modernization. Otherwise what passed for national ideology had a fascistic streak in it. Yet Winston Churchill as Colonial Secretary (1921–22) simply opined that, under the rule of law, every inhabitant of Palestine stood automatically to gain from Zionist enterprise.[21] Correspondingly, he expressed satisfaction with how Palestine was covered at the keynote Cairo Conference of March 1921. Yet a signal feature thereof was each side waxing furious at any talk of basic guarantees for the other.[22]

Soon intense debate developed across the Jewish diaspora about the merits of Zionism in whatever mode. Two Leftist perspectives might now be considered, those of Albert Einstein (1879–1955) and Hannah Arendt (1906–75).

Einstein was a committed Zionist within the social democratic mainstream that led Israel to statehood in 1948. Regarding the Palestinians, he looked by 1940 towards forging "an advantageous partnership which shall satisfy the needs of both nations".[23] Then in 1944 he warned that partnership will succeed only if both peoples conquer "that childhood complaint of a narrow-minded nationalism imported from Europe and aggravated by professional politicians".[24]

Had he been a shade younger, Albert Einstein could have served Israel admirably as its second President. One does have to acknowledge, however, that his generosity could slide into naïvety. Chaim Weizmann criticized his weak sense of realpolitik.[25] Then again, there was something rather callow about Einstein's extolling Gandhi as *the* greatest man of the twentieth century, notwithstanding his having studiedly to remind the latter that "non-violence" would avail little against Hitler.

Einstein was also too sanguine imagining that, as a rule, the volunteer elites labouring to create settlements in the desert would be the people with the "humane and worthy" approach needed to "establish healthy relations with the Arabs which is the most important political task of Zionism".[26] Surely, the adaptive talents required are too contrasting.

Hannah Arendt comes across as having been an estimable person who combined courage with sensibility, nay spirituality. She has much to say to the present day. For one thing, she stood out early on against the Manichean disposition to make "terror" a corral word which encompasses all insurgency action but nothing anybody else does. She reminded us that an authoritarian regime may operate a "terror" so ubiquitous that "often its victims are innocent even from the point of view of the prosecutor".[27] Though here Arendt was particularly referring to Soviet Communism, her point can be more widely taken. Then again, her stringent but scrupulous coverage of the trial of Adolf Eichmann in Jerusalem drew upon her a lot of opprobrium. But it can serve as a model for resolving the hatred engendered by "the banality of evil" as well as stand as a warning about the ease with which our species can talk itself into the ugliest conduct. This moral may be especially pertinent in an age in which we are becoming, as never before, globally interconnected yet disconnected.

She had also been her own woman when, in the sombre victory autumn of 1945, she had warned that "If Zionists persevere in retaining their sectarian ideology and continue with their short-sighted realism, they will have forfeited even the small chances that small peoples still have in this none-too-beautiful world of ours."[28] The admonition was pertinent in relation to one of the structural factors impeding the search for peace in the Holy Land. This was the split within Zionism between the social democratic mainstream by then led by David Ben Gurion (1886–1973) and the Revisionists under Vladimir Jabotinsky (1880–1940). The latter was a multilinguist from Odessa, a gifted orator. Careless of Arab rights, he and his colleagues aspired to create a Jewish state extended to biblical borders beyond the Jordan. Ben-Gurion, a Pole, had come up through trade unionism.

In 1931, a Revisionist guerrilla group, the Irgun Tzvai Leumi, broke with a Haganah by then established as the main Jewish defence force. The Irgun sought to shatter British as well as Palestinian cohesion through escalatory retribution. Its worst action was the massacre of 250 peaceable Arab villagers at Deir Yassin early in 1948, just as Britain was completing withdrawal. This terror was condemned by social democratic Zionism as well as the world at large. Moreover, the Revisionists were not to come to power in Israel until Menachem Begin assumed the premiership in 1977. Even so, Deir Yassin long epitomized, for many Palestinians, Israeli ruthlessness and international indifference. Working as a journalist in September 1967, I met on the Jordan's East Bank a number of the 90,000 refugees who had fled across the river due to the recent June War. Some told of advancing Israeli troops pressurizing them to "Go to King Hussein". But overwhelmingly, the chief reason given for exiting was the folk memory of Deir Yassin. Here was the mirror image of Israelis surveying the Palestinian scene and seeing "terrorism" as the overriding *leit motif.* In each direction in this respect, the sword was proving mightier than the pen.

Holocaust and Response

Looming over the formal creation of Israel as war became full-scale in May 1948 was the systematic Nazi annihilation of two-thirds of European Jewry. May this erstwhile early teenager of 1945 recall how shocking the year of victory was. A welter of information about human brutishness pulsated through the air waves. Two images particularly stay in the mind. One is Hiroshima after "the bomb". The other is the Belsen concentration camp after its liberation by the British Army, including a sergeant my folks knew well. Of these revelations, the latter shocked one most.

Yet I have never seen the Holocaust as History's worst atrocity in scale and character. That satanic accolade goes to the African slave trade, Christian and Islamic. But the Holocaust has to be prominent in the hall of infamy. The atrociousness commenced with inflictions on a German Jewish community which had considerably assimilated, the standard progressive prescription for resolving the Jewish question.

Since those camps were liberated, little has changed to alter intrinsically the basic record. But the episode can now be read contextually. Six million Jews died there but a million others did as well, many within further groupings (gypsies, Polish intellectuals, homosexuals . . .) targeted for elimination. Also the total number of Soviet people who died because of the Nazi invasion is today officially put at 27 million, under a third of them killed on the battlefield. With gratuitous cruelty all around Eurasia, World War Two (1937–45) rates as the nastiest span in history.

Nevertheless, the Holocaust was for international Jewry a trauma any people would find hard to cope with without sinking into negation. Particularly afflicted was Eastern Europe, the heartland of a rich Jewish culture. Despite spurious "wonder-workings" early on by certain *zaddikim* (i.e. holy men), Hasidism had and has survived. Then again, by 1850 *haskalah* – the "Jewish enlightenment" – was particularly entrenched in Lithuania. Come 1931, however, some 80 per cent of Polish Jews gave Yiddish as their mother tongue.[29] This blending of medieval German dialects with a vocabulary built up multi-lingually was very much a preserve of East European Jewry.

Resources for Recovery

Still, the reality that none of these traditions could be wiped out afforded a basis for recovery. So, too, did the dramatic progress Jewish people were making professionally throughout the free world. Music and the arts, commerce, the media, political philosophy, and natural science come readily to mind. Nobel Prizes are as searching a test as any. Between 1901 and 1955, Jews won 20 per cent of all awards in Physics and 29 per cent of those in Medicine, the Queen of the Applied Sciences.[30] Not bad for, on a recent estimate, 1.2 per cent of the world's population.

Also contributory to resurgence was the Zionist experiment and the sense it gave of Jewish renascence, Holocaust or no. Until 1973 at any rate, war winning was seen as a cause for celebration. But many Jews and others enthused, too, about the social innovations, above all the *kibbutzim*. Since the 1970s, this sector has wilted somewhat as the Israeli climate of opinion has turned more conservative, consumerist, econometric and footloose.[31] But beforehand it was widely seen by Jews and Gentiles alike as an exciting communitarian departure. Not inconceivably, this pristine vision will one day be regained.

New Beginnings

A subject of remark still is the extent to which Communism was Jewish in formulation and leadership. It is or was a sensitive matter because a standard fascist allegation used to be that Soviet Bolshevism (colluding with Wall Street plutocracy) was a Jewish plot for world domination. It is something the study Hilary Rubinstein has led tackles head on. During the 1920s, Soviet Communism turned savagely anti-Jewish and was to remain so quite markedly. Nevertheless, perhaps a third of the senior leadership of the initial Soviet regime was Jewish. Besides which, we all know Marx and Engels were Jewish though, in Marx's case for sure, this circumstance made him a desperate inverted snob.[32]

If there is within Marxist Communism a Jewish disposition, it will be as follows. A routine criticism of classical Marxism has been its failure to discuss how things may proceed after the revolutionary overthrow. The assumption was that, after one had surmounted that well-defined crisis, everything would be fine. This, after all, was consistent with the Jewish legend, mythic as well as historical: Eden, Noachic flood, Israel in Egypt, Exodus, Jericho, Dispersion, Reconquista . . .

Obviously the Holocaust was an extreme case. But Israel's wars of survival also fitted the pattern: 1948, 1956, 1967, 1973 . . . In 1956, the Israeli pre-emptive armoured thrust across the Sinai was completed inside five days, thanks considerably to the immobilization of Egypt's air force by British air strikes. Had advance been slower, it might have stopped due to a breaking down of the armour's improvised supply train.[33] Yet perhaps Zionism's narrowest scrape was one that, in those terms, goes entirely unremarked. In the winter of 1940–1, only a brilliant pre-emptive strategy by the British imperial coalition prevented the whole Middle East from being engulfed by vastly preponderant Italian forces.

Towards Partnership

The war which was easiest for the Israelis to win, 1967, was the one which left the most disastrous legacy, their seizure of much populated Arab land. Within it, Jewish settlements arose almost from the outset despite the urgent advice of *inter alia* David Ben-Gurion. As realization dawned of the demographic contradiction, intimations were heard from the Israeli Right about either ethnic cleansing or apartheid, perhaps while casting Jordan in the role of Palestine. This whole question would soon become quite the biggest barrier to conflict resolution. Access to the biblical heartland of Judaea and Samaria will, of course, remain a legitimate Jewish and Christian interest, as will accessing Jerusalem to all faiths.

Each of the peoples in this conflict has proved incredibly resilient in the face of continual pressure. Yet each has continued to suppose the other has no resilience; and can therefore be curbed by force. On each side, too, children are regularly raised (at home, in school and at worship) in the belief that their side has never done anything wrongful or stupid except be too restrained. Meanwhile, both are ridden with insecurity. Evidently there is a reciprocal fear of land denial. But each harbours more existential anxiety. The Israelis are conscious of how, across the diaspora, the Jewish identity is tending to contract or dilute. For the Palestinians, the culture shock of adaptation to the modern world has been sharpened by the Zionist dynamic. Comparisons can instructively be drawn between them and the Tamils in Sri Lanka, the other people famous early on for recourse to suicide bombings. The Hindu Tamils rose up because they saw their Sinhalese compatriots embarked on an especially zealous neo-classical Buddhist revival.[34]

Of late, it has looked as if the Holy Land could, with American pressure, be on course to a "road map" two-state solution. What this has to connote to be constructive and, indeed, viable is not separation but partnership. That is needed for a raft of reasons relating to the local scene. But perhaps the most important is one consistently overlooked,[35] to whit, climate change. Admittedly, the complicated geography of the Eastern Mediterranean/Near East means that local tendencies often buck hemispheric or global trends. The long slow process of medieval warming affords examples.[36] But on present showing, the polewards shift of climatic belts concurrent with global warming is bound within decades to cause or aggravate desertification in Palestine and Israel.

Besides, collaboration between Palestinians and Israelis in the Holy Land could provide a solid nub for a gradual convergence worldwide between the major obediences. It is something the Jewish people could attune to well. It is consonant with their historical experience, inter-faith scholarship included. Perhaps helpful, too, for this purpose is the sparing attention most customarily accord an afterlife. This could be conducive to flexibility all round. Collaboration with Palestine could be their finest chapter yet.

12

CHRIST AND THE HUMANISTS

GREAT RELIGIONS TEND TO APPORTION THEMSELVES a heartland region in which they come to fullish maturity. In the case of Christianity, this has been Europe, a continental area extraordinary in terms of location and configuration. A French map of 1768 innovatively presented the world as two hemispheres. The one, centred a few hundred miles north-east of New Zealand's North Island, is very predominantly sea. The other, centred in western France, comprises appreciably more land than sea. Europe thereby enjoys a pivotal advantage in global seafaring.

Its configurative idiosyncracy was succinctly extolled by David Hume (1711–76), the lead philosophic sceptic of the Scottish Enlightenment. He noted that it was that "of all parts of the Earth, Europe is the most broken by seas, rivers and mountains . . . and most naturally divided into several distinct governments". Especially remarkable is how deeply the continental bounds are penetrated by two well articulated arms of the sea – the Mediterranean and the Baltic.

Unfortunately, however, it is hard to relate much of Europe's history the last two millennia or, indeed, two centuries to what we understand to be Enlightenment values. The fractured geography Hume celebrated has sometimes favoured local freedom but has otherwise called forth tyranny to impose cohesion. The disparate has become the atrocious, not infrequently via religious conflict.

Alongside which must be set another consequence of geography. The three great religions which originated on the Arabian desert fringe have generated sacred texts which are each seen as core historical narratives informed by the word of God, directly or indirectly communicated. Such Hindu sacred narratives as the Bhagavad Gita, on the other hand, unfold in settings more detached from concrete experience. This could make them more able to accommodate new interpretations. The USA's National Association of Evangelists defines the Bible as already "the inspired, the only infallible, authoritative word of God".[1] The claims made within Islam for the Koran are, of course, very similar.

Since the late Middle Ages, the deepest cleavage within Christendom has been between Protestants always disposed to fall back on the Bible and Catholics more reliant on priestly authority. This dialectic is not concluded yet. But it is overlain by those with (a) other obediences, (b) secularism in sundry manifestations and (c) Natural Science.

Love Your Enemies

It does appear that Christianity was predestined by geography to assume exceptionally many forms. Yet suffusing them all has been a general tendency for the creed to ask a very great deal of its adherents as a more holistic morality is superimposed on tight-knit Mosaic law. Quite extraordinary in this connection is the injunction in the Sermon on the Mount to "Love your enemies and pray for your persecutors", this following hard upon specific advice that if "someone slaps you on the right cheek, turn and offer him your left" or "if a man wants to sue you for your shirt, let him have your coat as well".[2] Admittedly, the effect of this and of further enjoining to "Pass no judgement and you will not be judged" is weakened somewhat by abrasive allusions in the same discourse to Pharisees, doctors of law, hypocrites, heathen, pigs and pearl, and false prophets. All the same, precepts so explicitly counter-intuitive applied so open-endedly to conflict situations can be making demands so onerous that failure to meet them could be condoned. One gets the feeling that, within Christian witness, a huge gap too readily opens up between what is said and what is likely done. Never mind the rewards of Heaven nor the fires of Hell.

Nor is this assessment at all novel. In 1896, Charles Sheldon published his novel *In His Steps*. In it a clergyman asks his congregation to undertake nothing for a whole year without first asking "What would Jesus do?". The results were arresting. The minister sent a poor family on holiday in place of himself. The local editor declined to report a prize fight; eliminated whisky and tobacco advertisements; and stopped the Sunday edition. He was saved from bankruptcy by a slumming millionaire. And so on . . . By 1948 this book had run to 20 million copies and was still being widely read.[3]

The Jesus Persona

Still, the Sermon on the Mount made a big impact then and has ever since by dint of impressively diverse attributes – workaday imagery, homespun wisdom, plain speaking and the easy assumption of peerless authority. If it was a unitary statement delivered by Jesus without props on a Galilean hillside, it will have been awesome yet never intimidatory. The same can in due measure be said of many other discourses He engaged in during His three-year ministry. Likewise, quite a proportion of the miracles claimed by the gospel writers could well have been effected, this through force of personality coupled with depth of concern.

As a small-town carpenter, Jesus belonged to that genre of skilled artisans which has perennially been to the fore pressing for social change. Archetypically He felt righteous anger towards the Jerusalem temple establishment as well as to the Pharisees, acceptable though their movement will have been to most people from His background. He depicted riches as inherently corruptive without explaining why this must always be: "it is easier for a camel to pass through the eye of a needle than for a rich man to enter the kingdom of God" (Matthew 19.24–25). Likewise, He defiantly empathized with and conversed with those of either gender fallen by the wayside. Yet He retained a strong sense of Evil as a proactive force in being, Satan and his fallen angel band. Paul also projects such a vision, though perhaps more abstractedly.

By the time He embarked on His ministry, Jesus of Nazareth evidently saw God the Creator as the loving Father of Mankind, someone with whom He Himself had a special affinity. Moreover, His link with "My Father" would be confirmed when He returned Himself as Messiah, this within the lifetimes of some around him (St. Matthew 16.28). The Kingdom of God/Kingdom of Heaven would then be something more than a state of mind or aspiration. It would have become an immanent reality. The Palm Sunday entry into Jerusalem with His apostles betokened for some this blessed day drawing near.

Jesus combined a glowing universal inclusiveness with deep roots in the culture of Jewish nationhood. But He showed no familiarity with Graeco-Roman thought. Given His overriding sense of end-time urgency, He largely left doctrine and practice for others to evolve. This they did once it became clear enough that a Second Coming ushering in the Kingdom of God would not be that soon. However, to Edward Gibbon, the era AD 96–180 (pre-Antonine and Antonine) was without question when the human condition was at its "most happy and prosperous".[4] Even allowing for Gibbonesque exuberance, it was not going to be a time ripe for apocalyptic transformation.

Pauline Linkage

Saul the tentmaker from Tarsus, alias St Paul (*c.* AD 3 to 65), was a crucial link between the Ministry of Jesus and the halcyon era just delineated. A Hellenized Jew, he may have been under Stoic influence yet seems to have owed little to Plato or Aristotle nor to the specifics of Greek cosmology. He inspired adherents of this new Christian religion with a strong belief in the Resurrected Christ as the divine Son of God. He consolidated the instinctual universalism of Jesus Himself by not requiring Gentile converts first to become Jews through adherence to the Mosaic Law as interpreted through the synagogues, including the rite of circumcision. At the same time, he sought to make Christianity a matrix for social peace as its communities proliferated into Asia Minor and beyond.

Yet for much of the last century, Paul got a bad press from liberal commentators exercised by his perceived disposition to keep women in subjection. Lately, however, he has gained more favour. Karen Armstrong, a former Roman Catholic nun and strong campaigner for the admission of women to the Catholic priesthood, has evidently been a trail blazer. She avers that, while both Paul and Jesus were committed to gender equality in the new Christian order and while both felt those who spread the gospel should best be celibate like themselves, they differed on an important particular. This was that Paul did not believe that somebody who already had a family should *ipso facto* desert them in order to spread the Word.[5] It would all depend.

But there is a major sticking point as Armstrong acknowledges. It is the eleventh chapter of Paul's First Epistle to the Corinthians. In it, women are harshly advised to wear a veil when at prayer, under pain of having their hair shaved or cut off. "A man has no need to cover his head because man is the image of God, and the mirror of his glory whereas woman reflects the glory of man. For man did not originally spring from woman but woman was made out of man" (1 Corinthians 11.6–9). The rest of the passage in question was more equable. But the damage had and has been done.

So Karen Armstrong concedes that what she calls Paul's "Jewish chauvinism does burst through" on this occasion. But she seeks to place this within the context of Christian cohesion in Corinth being put at risk by the fast and loose life style of a dissolute port city.[6] Even so, this outburst must surely be seen as a Pauline backlash erupting through his usually tight web of emotional control. Jesus would never have spoken thus regardless. On the other hand, only Paul could have galvanized the way he did a proliferating archipelago of Christian communities under the shadow of a suspicious imperial Rome.

Original Sin

This doctrine basically says that all humans inherit a disposition towards wrongdoing which closely correlates with a woman-led tendency towards sexual intemperance first apparent in the Garden of Eden. Contrary to common assumption, its full formulation did not owe much to St Paul. In the fifth chapter of his Letter to the Romans he does say that Adam burdened mankind with sin, a situation not redeemed until Christ's death on the cross. However, he does not mention Eve in this context. Nor did he ever ask how sin was passed down the generations. Nor, indeed, did he ever discuss Eve's part in the Fall nor how Original Sin related to sexual activity.

Development of the concept within Latin Christendom was undertaken largely by Tertullian (*c.* 160–230), Jerome (*c.* 347–420) and – above all – Augustine of Hippo who made it a point at issue in his struggle against the Pelagians. All three were what William James dubbed "twice born". In other words, each one had arrived at their convictions via a psychological crisis.

In the Eastern Orthodox Church, less emphasis was placed on interpreting the Garden of Eden story in terms of original sin, manifested sexually. For one thing, scholars were at home with sacred texts written in Greek. For another, the Eastern Roman Empire had stayed more stable internally and less exposed to the external barbarian threat.

In Western Europe a sense of sinfulness and doom resurged during the crisis-ridden fourteenth century. Correspondingly, Martin Luther and John Calvin restored original sin to centre stage during the Reformation. Then in the early nineteenth century, it received an evangelical fillip. By now there was less focus on *femme fatale* and more on children, the latter being considered in special need of redemption through correction. Nowadays, sin functions for most Christians merely as a perspective on the human condition, not as a sufficient explanation of why evil continues.[7] Rendered thus, it has some resonance with Arthur Koestler's notion of our "dangerous altruism", our willingness to do combat on behalf of our close-knit tribal group, perhaps a thousand strong. That social category is now shrinking remorselessly into oblivion. It is an entity we need robust substitutes for.

Hell and the Humanists

In the popular understanding, Original Sin and Hell are but different takes on the same theme, the paying of due attention to the downside of human nature. However, the derivations are different. The Hell stereotype was a lurid follow-through of the gloomy Mediterranean underworld. It was one driven by (a) ecclesiastical concern to avert social collapse and (b) the fascination of artists with the savage or even the grotesque.

As society became somewhat more solid and confident, however, perspectives altered, in part through monastic influence but also lay education. Seventy-five years ago, Charles Haskins introduced the notion of a twelfth-century Renaissance, this to accommodate the humanistic spirit abroad in Latin Europe by 1070. Witness the specific "germinal" advances mentioned in Chapter 4. More fundamentally, a change was taking place in how humankind's place in the order of things was seen. Granted, Man was still a fallen being who had lost immediate knowledge of God and finds reason and instinct often to be in conflict within himself. Yet now he was the noblest of God's creatures, even in his fallen state. Furthermore, the structuring this indicated meant that Nature as a whole was no longer mere chaos. It was seen as possessing order.

Not that this cultural sea change was conclusive. Nor was it entirely to the good. On the contrary, Sir Steven Runciman opined that "in the perspective of history, the whole Crusading movement was a fiasco";[8] and there was, of course, the associated anti-Semitism. More generally, progress brought its usual contradictions, some personified by Frederick II (1194–1250), Holy Roman Emperor and King of Sicily and a profound religious sceptic. At one time, his Moslem–Byzantine–Catholic–pagan court of all the talents at Palermo matched any fount of extrovert humanism in the greater Mediterranean Renaissance so imprinted on our consciousness. He himself was actively involved in art, music, medicine, astronomy, astrology and – not least – ornithology. Yet with a badly abused harem guarded by eunuchs, and with his opportunistic bellicosity, endless machinations and boundless conceit, this self-styled *stupor mundi* ("wonder of the world") elevated himself to a spiritual isolation as barren as any a virtuoso prince of the sixteenth century might experience.

Haskins himself concluded that, by 1250, the work of this medieval Renaissance "was largely done".[9] Nevertheless, Richard Southern, another very distinguished medievalist, adjudged "the period from about 1100 to about 1320 to have been one of the great ages of humanism in the history of Europe: perhaps the greatest of all".[10] Undoubtedly, the first three crusades to the Holy Land afforded some stimuli. But from *c.* 1275 a big constraint on further development must have been the hemispheric climatic reversion to cooler, wetter, more erratic conditions.[11] As noted earlier, this by no means excluded important societal advances in the fourteenth. But it recurrently meant a reversion to a dismal view of the human condition as famine, pestilence or war took control. By then, the Catholic Church had firmed up on its doctrine of purgatory, an intermediate stage between Hell and Heaven for those possessed of Grace though not yet perfect enough to face God. However, it is not clear to me what impression this prospect made in circumstances so stressful.

The representation of Hell in painting and literature continued to figure strongly, especially in Italy. *The Divine Comedy* by Dante Alighieri of Florence (1265–1321) was a 14,000–line rendering of hell, purgatory and heaven with Virgil cast as a guide. In Hell, Mohammed is tormented endlessly while Islamic and Christian scholars discourse above. Written in the vernacular, it made a deep impression on Italy's High Renaissance and later upon William Blake (1757–1827) as well as on *The Gate of Hell* by Auguste Rodin (1840–1917). But Blake and Rodin were concerned with hell on this Earth while Dante himself had raised such issues as predestination and the gradation of human sin. However, what got painted on the walls of parish churches was Hell unreconstituted, Hell a long way from the Sermon on the Mount.

By the time of the High Renaissance (fifteenth to sixteenth centuries), humanism was interwoven deeply with classical revivalism. On this basis, it might have established a kind of sceptical sub-culture. But the whole situation would soon be much complicated by the Protestant Reformation and then by Copernicus. Suffice to observe that mainstream Calvinism, with its rigid predestination as between the elect and the damned helped stoke the hellfires. Many parish churches in Eastern England, for instance, were portraying these by 1650. Also in vogue were Leveller doctrines of straitlaced utopian socialism. So, too, was chiliastic prophecy, "I was born in this setting of time". Mercifully, the persecution of witches was already receding in Western Europe.

A century later and Hell was widely coming under that most insidious of all forms of offensive action, ridicule. The French philosophes were well to the fore, Voltaire and Diderot in particular. The Evangelicals, Franco's Spain and Southern Baptists notwithstanding, the Hell concept was destined markedly to recede across the next two centuries. The word itself continued to be freely used but more and more to connote real life situations which might fall well short of ultimate horror but were taxing or, at the very least, a damnable nuisance. To the radical Romantic poet, Percy

Shelley (1792–1822), "Hell is a city much like London . . . populous and smoky . . . ". Yet Charles Darwin (1809–1882) still felt constrained finally to abandon Anglicanism primarily because of its continued concern with eternal damnation.

A Vacuous Enlightenment

Conceptually, Hell had probably been the most tangible blocking factor as between those who called themselves Christians and those not thus persuaded. Therefore, its undermining paved the way for more freely ranging dialogue all round, if not forthwith then eventually. What this has not yet led to is a new philosophy up to ensuring world amity.

A particular altercation serves to illuminate the reasons. In the middle 1750s, those arch-rivals Rousseau and Voltaire were each residing regularly in Calvinist Geneva. This common fate pleased Rousseau little and Voltaire less; and will continually have reminded each of their mutual antipathy. Then when the big Lisbon earthquake struck in 1755, Voltaire penned his famous essay averring that this was a random occurrence, hardly the work of a caring or even a punitive Providence. The response this elicited from Rousseau was that Voltaire was displaying his characteristic *hauteur* towards the world of the underprivileged.[12] He further remarked that far fewer people would have been killed had they all been living in villages as he himself favoured, both socially and in terms of democratic participation.

Rousseau will certainly have hit a raw nerve in pinpointing a contradiction between Voltaire's ultra-liberal principles and his disdain for the masses. But he himself was open to the charge that his impetuous personality sought early and decisive change, whereas the key proposals he had for this (village communities and small statehoods) might take generations to implement in full. In the meantime, any projection of the "general will" of mass society would be very prone to distortion, accidental or intended (see Chapter 4). The yearnings of millions of Frenchmen for more responsive and modern government led in 1789 to a revolution with a problematic end-game. By November 1793, the Reign of Terror was in full swing. As part and parcel, churches in Paris and elsewhere (starting that month with Notre Dame) were being transformed into "Temples of Reason" dedicated to rationalism, anti-clerical fervour, well spun Science and, above all, ruthless zeal. As a political religion, "deChristianization" had its own rituals, hymns and saints.[13] There was also savage suppression of a Catholic backlash outside Paris. Thus in the Vendée region of Poitou, some 250,000 were killed, 1793–5.

By early 1795, the revolutionary regime itself was reacting against all this but proffering no alternative except a nebulous "liberty of cults". Inexorably a counter-current set in. In 1801, Napoleon as emperor signed, with wide popular approval, a papal concordat. The next thirteen years were consumed with yet another European geopolitical struggle, one waged more continentally than ever before. Hardly a quintessential Enlightenment consummation.

Then through the mid-nineteenth century, the Manchester School of British Liberalism sought lasting peace via universal free trade. This time round, the denouement was to be the First World War. Again, philosophic inadequacy had made its fell contribution. In a nutshell, nationalism was not enough but neither was ill-managed internationalism.

Enlightenment Redeemed?

Arguably, the advanced countries our ancestors knew as Christendom face today a spread of self-imposed problems far more daunting than in the eras of unhappy outcomes just cited. Are they equipped to address them? Can they bring social values, religious affiliations, natural science and metaphysics sufficiently into harmony to ensure a happy outcome? Focusing awhile on Christendom is warranted because (along with a complementary Judaism) it still affords much of the material and intellectual resource that drives our world in a perilous direction.

By insisting that God's Only Begotten Son lately lived as one of us, Christianity elevates our species even more than other faiths do. But can this belief be squared with the existence of a hundred billion billion stars? If not, can Christians accept a less predetermined role within the cosmic life presence? Must not taking this step involve relinquishing our exclusive claim to eternal life? But might this enable the churches to receive on board, in one way or another, many non-believers who reject the theological formalism they have known yet look earnestly for metaphysical guidelines?

Might all concerned thus draw inboard the cultish melange variously identified as New Age, nature cults, druidic paganism . . . ? And might this help to save 'wild nature' as a concomitant to redeeming ourselves? Could there ultimately be scope for the generation of a world order which proffers a better setting than consumerism *à l'outrance* does for sensibility, for spiritual values? Would not the loosening up all round this consummation would require be assisted by empathetic interaction between Christianity and other major faiths plus, necessarily, Science?

Divisive Tendencies

The transatlantic gap in religious commitment has been long remarked. Of late, two things have transpired. The one is that the gap has widened. The other is that it is Europe which is now deemed to be different. As the Bishop of Oxford has put it: "In the 1960s what sociologists couldn't understand was America – the most modern country in the world and the most religious. Now they are coming to a kind of consensus that . . . the more modern a country is or the more its going towards modernisation . . . the more important the role occupied by religion."[14]

One could ask what meanings may attach to "modern" or "going towards modernization". Moreover, one can reflect on the dynamic of Christian witness observed in different locales. No matter. For a clutch of professional opinion surveys the last decade do confirm the divide here identified. Nearly 80 per cent of Americans profess belief in Heaven as against 50 per cent of British; and over 70 per cent claim to believe in Hell as against 28 per cent of British. Nearly 45 in 100 Americans attended church more than once a week. The percentages for Britain were 13, France 10 and Iceland two.[15]

More critical, however, are clerical doubts. In 2002, Christian Research in Britain published a survey of 2000 Anglican priests. A third expressed doubts about or outright disbelief in the physical resurrection of Jesus. Only half are convinced His was a virgin birth; and only a half see faith in Christ as the sole route to salvation. Still, a good 75 per cent accept that He died to take away the sins of the world.[16]

Traditionally, the Catholic church in France has been integral to her distinctive nationhood. Yet the number of priests has fallen by half the past 40 years; and a half of the present incumbents are liable to die by 2012. Only 150 ordinands completed training in 2004. Sometimes one priest will manage a score or more parishes; and many rites of passage are conducted by laity. The precipitate decline of the Church as a mass movement has matched that of its erstwhile rival, the Communist Party.

The American Scene

The last 40 years, the balance of American opinion has moved Right in both the secular and religious domains. Whether this trend continues and whether it presages long-term Republican rule has been a matter of intense debate, especially since the Bush re-election in 2004. Within the said shift, there has been an evangelical revival, a resurgence of Protestant piety within Southern Baptism and – increasingly – other denominations. It is a major driving force behind belief in the inerrance of much or all of Scripture and in the forthcoming, maybe near time, second advent.

Episcopalian-turned-Methodist, President Bush has deserved more credit than he has had for moderating such influences in important respects. His utter colour blindness is much at variance with southern Protestant traditions. He has lately stressed that faith-based social initiatives must only be an ancillary part of "compassionate conservatism". His outreach explicitly extends to all faiths. Assertions that he has no liberal international vision akin to those of 1919 and 1945 are belied by his, maybe too visionary, drive for world democracy.

Yet there have been acute shortcomings. The President's view of the world as a "moral design" (meaning, apparently, a Creation singularly crafted by a very personal God) can easily prove too discordant with his declared commitment to deep Space exploration to allow of coherent interpretation. Also, his weak academic background leaves free rein for a very Manichean streak. Hence, for instance, his "intellectual love affair" with Natan Sharansky, the Likud Minister who had a sterling record as a Jewish *refusnik* in the USSR but has a thoroughly perverse attitude to Palestinian rights.[17] Worse, Bush has lately revealed on the campaign trail a presumption that your median American voter is utterly uninterested in working with allies and international institutions. A survey by the Chicago Council on Foreign Relations suggests this is not the case.[18]

Consequences have followed. An asymmetrical definition of "terrorism" has led *inter alia* to Guantanamo. Policy towards the Holy Land has been unsteady. The liberation of Iraq has been flawed in certain critical respects. A trend towards missile defence from Space is ominous. Nor is the domestic political landscape free of clutter. A lead journal of science report and commentary ought not to feel constrained to consume precious pages countering Creationist banalities.[19] Nor should televangelists get away with the thesis that 9/11 was divine retribution for the liberal consensus on homosexuality and education.

The general trend will likely continue awhile. Likely, too, religious liberals will remain riven. But this does not mean the USA is headed for Rightist theocracy. The number of Americans who admitted to no religion doubled to 29 million (i.e. an eighth of the population) between 1990 and 2001. This compares with the quarter of the electorate aligned with the Born Again/Evangelicals.

The Born Again set store by authentic experience rather than historical re-enactment. Nowadays, too, a strong Pentecostal influence favours emotion and spirit as opposed to puritanical proscriptions. A prayer emphasis can likewise be self-directed. Figuring, too, is a home church movement, small groups interacting spontaneously. Organized religion counts less than does simple personal faith seeking oneness with the divine.[20] Not that this precludes dangerous lurches in foreign policy. Witness those Christian Zionists who declaim that there will be no peace in the Middle East until Christ sits on David's throne in Jerusalem. Even so, lifestyles prioritized along the lines just indicated are not the stuff out of which are made upheavals, social or geopolitical.

America's Roman Catholics (and recent Hispanic arrivals in particular) tend to be Leftist on social welfare but Rightist on personal behaviour issues and sometimes, too, foreign policy. They are unlikely to form a militant bloc for reasons that go beyond current demoralization over priestly sex abuse and falling recruitment to the priesthood and orders.

Further Afield

The sixties can be looked back to as a decade in which, guided by Cardinal Léon-Joseph Suenens of Belgium (1905–96), Rome nearly seized a great opportunity for reform. The Second Vatican Council (1962–5) endorsed the principle of collegial governance. A few years later, contraception was well up the agenda. High profile, too, was the anti-capitalist dependency analysis formulated by the new Liberation Theology in Latin America. It envisaged will and energy lifting social consciousness to a higher level.

However, Pope Paul VI overrode commission advice about liberalizing birth control policy. By doing so unilaterally, he compromised the collegial principle. Cardinal Suenens left the Vatican. All but 40 years on, both these issues have assumed a new urgency. So has the question of married and female priests.

Thanks to the French, Belgian, Portuguese and Spanish, the Catholic church is strongly represented in Black Africa – a region lately termed "the centre of the Christian world", taking as the criterion the proportions of active Christians in the respective populations.[21] Now Africa is at the top of the international development agenda, this largely because of the way social mores leave many of its people vulnerable to the spread of AIDS through improperly protected sex. It presents the Vatican with the severest test yet of its birth control policy. Can and will it impose on its Church worldwide a family planning strategy that is (a) suitably humanist and (b) not liable to cause the Church to break up? What would Jesus do?

Still, the international church most immediately in danger of breaking up is the Anglican. It has been so since 2003 when the American branch ordained an openly gay bishop; and an Anglican diocese in Canada approved a rite of blessing for same-sex couples. The devolved character of this association of autonomous churches ought to leave scope for variety. But Anglicanism's "Global South" has a powerful capacity for anger on such matters.

A particular growth area for Christianity in the developing world may be China. Granted, more indigenous philosophies – Buddhism, Taoism, Confucianism – may also be gaining ground but Christianity is the most closely associated with modernization via capitalism and science. The current guestimates for church membership range between 2 and 7 per cent of China's population. It does seem clear that Christianity is (a) drawing in more young people, (b) spreading among the urban intelligentsia, and (c) attracting attention on the campuses, formally and informally. The attitude of the government is equivocal thus far.[22]

Eastern Orthodoxy

In 2001, the Patriarch of Moscow and All Russia reported that his register of parishes had risen in twelve years from 6800 to 14,000; and that the Russian Orthodox Church had 15,000 priests and deacons plus 570 monasteries and convents. True, this priesthood was less than a quarter of the pre-1917 strength. However, there has since been a resurgence of Russian patriotic pride in relation to the Soviet era. The huge sacrifices of the Great Patriotic War (1941–5) are being recognized along with the great military achievements. There has also been some restoration of pride in the USSR's pioneering contribution to Space exploration, its demonstration that even Marxist atheists can reach out to Heaven.

If this mood is sustained, it could afford more scope for Russian Orthodoxy and maybe, by extension, for sister churches within the Orthodox obedience. For through the Byzantine era, the interaction between state and church became strong. The Emperor had religious as well as political responsibilities, with society itself being viewed very holistically. It was a perception complementary to received theology about the transcendent spirit of God. The sacramental importance attaching to icons was part and parcel. So was a eucharistic Divine Liturgy more elaborated than a Latin Mass. A mystical tradition likewise reinforced transcendence.

Though there were no monastic orders as such, individual monasteries waxed strong in the wake of mystic hermits entering lonely places to fight demons and resanctify the Earth. But they were also a counterbalance to statist authority. Therefore, those in the USSR were to be severely persecuted by Stalin.

With the fall of Constantinople to the Turks in 1453, Moscow effectively became the chief node of the Christian East, the "third Rome". But this served to emphasize separateness. If one allows that the Renaissance came to Russia at all, one has to say it did so low profile and belated. Printing arrived only in the second half of the sixteenth century.

The closest Russian approximation to the Reformation, the Great Schism, was triggered by a Russian Orthodox Council decision, taken in 1667, to accept guidance from the Greek Orthodox and even the Catholic persuasions about corrections to the Bible and reforms of the liturgy. In this the council had Tsarist backing but was fiercely opposed by a loose alliance of middle-ranking priests, peasants, Cossacks and merchants – the "old believers". Instinctually, they were gripped by an overwhelming sense of "the Other". They would always react against any whiff of spiritual pollution from Western Europe.

It is doubtful whether a spirit of outright exclusiveness counts for much in modern Russia. But a sense of exceptionalism clearly does. All else apart, it has been reinforced by those literary titans – Fyodor Dostoevsky and then Alexander Solzhenitsyn. It is a circumstance which makes a general dialogue with the Russian Federation both more difficult and more needful. In the religious domain, this is something Pope John Paul was very appraised of. His sending back to Moscow in 2004 of the miracle-linked icon, Our Lady of Kazan, betokened a desire to visit, once the thorny question of "poaching" converts had been resolved. A review of papal relations with the Christian East through the turn of the millennium is therefore salutary.

The Pope and the Patriarchs

In 1999, Pope John Paul went to Romania, the first pontiff in over nine centuries to visit an Orthodox country. He was made welcome at all levels but later that year was received more coolly in Georgia. In 2000, the Vatican's Congregation for the Doctrine of the Faith published *Dominus Jesus*, a document which many saw as retrogressive in that it perceived fullness of faith as exclusive to Roman Catholicism. The Russian Orthodox Church indirectly responded by declaring itself to be the "one holy, catholic and apostolic Church", the keeper of the sacraments throughout the world.

In May 2001, Pope John Paul visited Greece following an invitation from President Stephanopoulos. Their going ahead in the face of much local opposition, priestly as well as popular, was noteworthy in itself. Yet more so was the Pope's using the occasion to apologize outright for the sacking of Constantinople by the Fourth Crusade in 1204. His overture was all the more creditable since (following their prior sacking of Zara, a peaceable Catholic seaport) these crusaders had arrived in the great metropolis already excommunicated.

Nevertheless, the very next month Pope John Paul was criticized by the Russian Patriarch for conducting an open air mass in the Ukrainian city of Lvov. His visit was said "to exonerate the harassment of the Orthodox which takes place in that region". Traditionally, Lvov has been a staunchly Catholic enclave within an Orthodox countryside.

Throughout 2002 an apprehensive Russian Orthodox Church continued to protest against Rome's "proselytism", an especially contentious issue being the upgrading to diocesan status of four apostolic administrations within Russia, previously seen as temporary. Then in November a new law came into force in Belarus. Closely resembling legislation passed in Russia in 1997, it encompassed a strategy for elevating the Russian Orthodox Church to a lead role in "shaping the spiritual, cultural and state traditions of the Belarussian people". This departure showed that the deep insecurity of Eastern Orthodoxy was undiminished by the new international setting. Yet the way through has surely to be extended dialogue. One can glean a modicum of hope from the battle lines not having been too tritely drawn in recent warlike confrontations. Russia supported Moslem Adzharia against Orthodox Georgia while Romania backed NATO against the Serbs.

The Outlook

Arguably, the future of religion hinges on the rediscovery of spirituality in everyday life. But it is by no means certain that in the West the lead role in this regard will be assumed by the United States. For one thing, Western European sensibilities may be awakened by a Russian Orthodox Church suffused anew by a mystical sense of the "Russian Soul". Anyone who doubts the latent strength of Russian mysticism should study contemporary landscape paintings of the new-found Siberian *taiga*, 1750–1850.

The lack of a single overriding authority must hamper systematic doctrinal development within Eastern Orthodoxy. But its various national churches can, in communion with one another, be adaptive in other ways. Moreover, one can anticipate that the need for certain spiritual values will be impressed on the European mind by the pressure of population on Nature; and that, throughout Europe, religious and secular spirituality may gradually blend.

A question of mounting interest is whether interfaith dialogue between the monotheisms is compromised by the visualizing of Jesus as sitting at God's right hand, thereby erecting a barrier between Him and both Moses and Mohammed. Is not the biblical rationale for this understanding very much a matter of contention? And does not this assignation also inhibit a three-dimensional appreciation of His striking personality? His words and deeds are reported at great length in the gospels; and, duly allowing for evident embellishments, He comes across, not as somebody overly concerned with the heavens, but as a practising humanist who combines a firm sense of purpose and moral courage with admirable sensibility and homespun friendliness. He inspires confidence and loyalty. He also appears remarkably free from the psychological complications which so often beset aspirant prophets. To understand this fully, it is better to envisage Him as reaching up to the Cosmic God rather than descending from that realm.

Coming in from a Quaker background, Rex Ambler of Birmingham University explored in 1990 the relevance of Jesus to the comprehensive world crisis he presciently saw as continuing through the end of the Cold War. Ambler noted how, following on from John the Baptist, Jesus saw scope for the Kingdom of God arising non-violently out of the crisis of His time: scope, in other words, for turning the other cheek to effect.[23]

Today we do face a much bigger and more complex world situation. Nevertheless, one can still insist on non-violence being a dominant precept within a holistic approach to a cultural or spiritual revolution in the real operational world. As Colin Powell has well said, force should only be a last resort. But, as he went on to say, it has to be a last resort.

13

COUNTERVALENT ISLAM

Western commentators have often noted how Islam originated six centuries after Christianity, the suggestion being that its collective persona remains more youthful, therefore more forceful.[1] Yet on several counts, this argument may be misleading. An organic model with maturation built-in hardly befits a religious movement. And who can make blanket judgments about philosophic maturity? And what follows either way?

Arabian Origins

A more instructive line to pursue is that, in the Arabian birthplace of Islam, humankind stood more exposed to a pitiless Nature than it did almost anywhere else. It was so mainly on account of rainfall vagaries. These were largely due to climate dynamics intrinsic to South-West Asia. But in 540, something of global import happened by Sumatra and Java. Krakatoa exploded with almost unparalleled force (see Chapter 18). In his pioneering study of this event, David Keys linked to it *prima facie* a marked rise in climatic erraticism, through the rest of the sixth century, in Arabia Felix – i.e. the salubrious south-west of that wide peninsula.[2] This region had long been a nodal point in Indian Ocean trade as well as a lead producer of perfumes/medicines from the frankincense and myrrh group of tree resins.

A kingpin of this economy internally was the stone-faced dam (600 metres wide) close by Marib, the capital of Saba, alias Sheba. During a flood in the 540s, this dam was breached for the first time since 450. A decade later the same happened, the silting then being especially severe. After a third breach in 590, the dam was finally abandoned. A wan allusion to its demise can be found in the Qur'an, the inference being that beforehand the locals had turned away from Allah.[3] Meantime, the spread of Christianity elsewhere was reducing the demand for oils and perfumes for embalming.

A considerable emigration from Arabia Felix therefore began. Not least was this to Yathrib, an oasis and the second city of the central Hejaz. However, it seems that this whole region was already in the throes of a droughty trend extending over centuries.[4] Yet discomfited by ambient climate change or not, the Hejaz and especially Mecca had since the third century AD experienced something of a cultural uplift led by the locally dominant Quraysh tribe. Classical Arabic was further refined and enriched. Towards the end of the sixth century, an Arabic script had been set down.

Furthermore, the Quraysh appear to have been already trending away from single-tribe gods and towards a broader monotheism. In this connection, one can surmise that the black meteorite encased in the Kaaba at Mecca was by then an object of pagan veneration sufficiently compelling to connote a single universal deity; and could draw added stature from the removal of the scores of idols dotted around the site. Under early Islam, in fact, the meteorite was identified with Abraham; those idols were got rid of; and the site as a whole was treated as an especially sacred locale.[5]

To see how directly ecological trends conditioned the initial emergence of Islam (620–632), we should return to the two Hejaz cities. Yathrib had languished, due to the refugee influx and to tensions between Arab and Jewish merchants. But the business leaders of Mecca seemed to be making headway despite the regional malaise. To others, however, the spiritual cost appeared too high in terms of economic polarization, self-indulgence and, yet again, primitive idolatry.

Into this milieu was Mohammed born *c.* 570. His history and personality are a shade obscure. However, the gist of his early years is as follows. Born into a none-too-prosperous Quraysh household, he was orphaned in infancy. Yet these and other familial traumas notwithstanding, he grew to well-rounded manhood. At 25, he embarked upon what proved a happy monogamous marriage with Khadijah, a wealthy Damascan widow. Said then to be aged 40, she was still able to bear him six children.

In due course, however, he took to cave meditations; and *c.* 610 these gave rise to revelation visions he found extremely distractive. As to any linkage with deep mental disorder, this comment by an ordained Christian priest (and one time Professor of Arabic at London and Princeton universities) is pertinent. Those who question his "stability do so only by ignoring the overwhelming evidence of his shrewd appraisal of others and of the significance of what was going on in the world of his time; and his persistence in the face of constant opposition until he united his people in the religion of Islam".[6] Important as that judgement is, however, it cannot exclude the influence of transient stress factors in an exacting desertic ambience.

Some three years after that first revelation, Mohammed publicly emerged as the messenger of Allah, the one true God. In due course, his witness would cover every aspect of life. But emphasis was placed throughout on *zakat*, charitable expenditure on a regular basis "For your kin; For orphans; For the Needy; For the Wayfarer; For those who ask; and for the ransom of slaves" (sūra II v. 177). It is appropriate thus to cite the Qur'an by way of confirmation of the main priorities as Mohammed comprehended them. It is declared to be, after all, a compilation of his inspired oral pronouncements. A simple structure existed within 30 years of the Prophet's death; and the full text was generated over the next two centuries.[7] To devout Moslems, it has customarily been inerrantly authentic.

A millenarian climax is foretold. It will usher in a final Day of Judgements which totally transcends the Lesser Judgement effected at an individual's death; and, of course, the awakening of an Inner Light their soul may experience at any moment. At this fell time, "the sky is cleft asunder . . . the stars are scattered . . . the oceans are suffered to burst forth . . . the graves are turned upside down" (sūra LXXVII v. 1–4). All souls are then weighed on the "Scales of Justice" for their every thought and action on Earth; and are duly consigned to either Heaven or Hell. The former is foreseen as untrammeled relaxation within a lush Garden of Eternity. The latter is portrayed as a ferociously fiery ambience – furious flames, bubbling pitch, boiling water . . . Those thence consigned "will dwell therein forever. Except as God willeth" (sūra VI v. 158). It is the starkest of divine messages relayed by one utterly convinced of the authenticity of its transmission to himself via the Archangel Gabriel.

To say the Prophet's message went down ill with the Mecca establishment would be to understate. His insistence on voluntary redistribution of income hit many pockets. So, too, did his condemnation of the idols which pilgrims paid onerously to engage with. Undaunted, he seems to have been highly successful recruiting and consolidating a cadre perhaps a few dozen strong. Beyond that, however, progress was and is hard to register.

The opportunity for break-out came from Yathrib, by then torn by intertribal and anti-Jewish violence. A body of locals, much impressed by Mohammed's personality and prophesy, invited him to come and build a stable consensus. With his close colleague Abu Bakr, he covertly left Mecca in 622, preceded by a small band of followers. Once in his new post, he built social peace via Islam. Yathrib, now renamed Medina, thus became a base for warlike operations against Mecca, starting with a sustained campaign of caravan raiding.

Come 630, he was able to return home in almost peaceable triumph. It was to a Mecca that, the past year or two, had been plagued by a drought so bad that (according to Arab literary sources) the people were turning to cannibalism. No doubt the warmth of his welcome (and then the explosive spread of Islam to embrace half of Arabia in what proved to be the remaining two years of the Prophet's life) owed something to a concurrent easement of the droughtiness.[8]

Out of Arabia

To those at the receiving end, the expansion out of Arabia will have come across as a cadre one, extending a new religion and language but not an entire people. However, the Hejaz is unlikely then to have carried a population of more than 250,000 which probably connotes a warrior roll a tenth that altogether. Expeditionary forces five to ten thousand strong figure in accounts of Islam's early decades. In fact, the first wave to gallop into mighty Egypt may have been a mere three thousand strong. They would have had to function very much as overlords. The population of Egypt early in the seventh century has been conservatively estimated at 2,600,000.[9]

Alexandria fell in AD 642. Damascus and Jerusalem had done in 635 and 637. The phase thus entered upon did involve the brain draining from Arabia of elites led by the Ummayads, the most powerful Quraysh clan. Expansion acquired its own momentum as the incoming Muslims fused, however factiously, new-found courtly lifestyles and traditional Arabian mores.

The advance down the North African coast to Tripoli took but two more years. From Gibraltar to Toulouse took under ten. An immediate Islamic conquest of Europe was finally precluded by the Frankish victory in the Battle of Tours on the River Loire in 732, just a century after the Prophet's death. In 759, the Muslims withdrew for good behind the Pyrenees.

To which one must add that on the eastern side Basra fell in 656, Kabul in 664 and Bukhara in 710. The involvement of Islam in India began in the seventh century (see Chapter 13). Then under its fifth caliph, Hārun ar-Rashīd (ruled 786–809), Baghdad became "the richest city in the world . . . Arab merchants did business in China, Indonesia, India and East Africa. Their ships were by far the largest and best appointed in Chinese waters or in the Indian Ocean. Under their highly developed banking system, an Arab businessman could cash a cheque in Canton on his bank account in Baghdad. In Baghdad, everything was plastered with gold".[10] Muslim traders were settling in Indonesia (short-term or otherwise) by the fourteenth century; and by 1600, Islam was the dominant religion within the archipelago. By 1100, the West African kingdom of Ghana had been taken over by an Islamic confederation.

Explanations

Still, it is the advance up the approaches to continental Europe which has attracted most attention from Western historians. The weakness of the opposition everywhere was a follow through of a late Roman/early Byzantine urban crisis. The decay once again of Carthage "to a shadow of its former self... appears to be typical of cities large and small, all over the Mediterranean"[11] in the century or two before Mohammed. It is at one with how every walled city of the Fertile Crescent fell inside four years (635–39) to those smallish contingents of Arab irregular light horse with no siege equipment. Invoking his expertise on Byzantium, Cyril Mango further remarked how "the absence of any urban resurgence after the enemy had withdrawn shows... military hostilities were just the last shock that brought down a tottering edifice".[12]

What seems clear is that secular droughtiness was again a factor, making some of the target regimes all the more supine. Widely across the Fertile Crescent from Palestine round to Upper Mesopotamia, there is evidence (i.e. in church location; olive planting; and agricultural terracing) of a rainfall trend peaking out in the sixth century. In the Negev much the same applied. For the Nile valley, the relevant data base is rather patchwork but points towards a similar conclusion. Likewise, Cyrenaica was manifestly drier than in Graeco-Roman times. What cannot yet be demonstrated, it seems, is that the turnover decade, 632–642, was especially characterized by climate adversity.[13]

Islamic Civilization

It is tempting for Westerners to see Islam as a bedouin philosophy evolved under the desert stars. However, neither the Holy Qur'an nor modern scholars encourage this perception. Quranic entries on this score vary. But not unrepresentative is one which says the "Arabs of the desert are the worst in unbelief and hypocrisy, and most fitted to be in ignorance" (sūra IX v. 97). Correspondingly, one Islamicist has spoken of the faith as able to "multiply cities at will".[14] Another stresses how the location of the central Hejaz plus the way the winter rains usually ensured enough groundwater the year round created possibilities for trade and hence a measure of urban growth.[15] Nor should one exaggerate the part nomadic fighters played in spreading the faith. Horsemen in the numbers cited above could not have seized and held big cities without local influences working in their favour. A further reason for the passivity of those being occupied may have been that the transition could never have been too brutal. The incomers had not these inclinations, let alone that kind of strength.

Something that plainly emerges from Mohammed's witness as recorded in the Qur'an is that local idols are anathema. Local *sūfī* saints of Islam (see below) came also to be viewed with much scepticism, interposed as they were between the laity and Allah. Nevertheless, a marked proliferation of sainthoods did occur as Islam extended and consolidated. Generic accounts of them appear from the late ninth century onwards. The saint is the *wali*, the friend of God, and duly enjoys his special protection. His miracles are termed *karāmāt*, "charismata". However, they rank lower than those performed by prophets.

What one here observes is, to an extent, a spontaneous human inclination to extol an individual who is a profound local influence for the good, to acknowledge him as a *qutb* – the axis of time around which all else revolves. Operative, too, is the influence of the *Sūfis* (see below) and of their precept that "If someone has no Shaikh, Satan is his master".

Also to reckon with, as in emergent Christendom, is a disposition to take over more ancient deities residing in charismatic locations – wells, springs, mighty trees . . . This practice was very prevalent in North Africa with its deep pre-Islamic culture. The same came to apply to India where ancient Hindu sanctuaries were often transformed into Moslem shrines.[16]

The Qur'an

Some of the tendencies just noted also indicate a predilection for gaining converts, this through persuasion and accommodation. A sterner message goes out from the Holy Qur'an. Should Unbelievers "from unbelief" desist, "their past would be forgiven them. But if they persist, "the punishment of those before them is already a warning" (sūra VIII v. 38). Nor should one discount the risks inherent in collaboration. Suppose "they were to get the better of you. They would behave to you as enemies" (sūra LX v. 2).

What needs be recognized, however, is the context within which the Qur'anic message was originally conveyed. It was a time of strife to gain a Mecca held by people who were unbelievers not by dint of their moral philosophy or metaphysical understanding but simply through brutal self-interest.

As with other texts sacred to major obediences, it ought to be possible for all sides in the on-going debate about Islam to agree on the following. It is too formalistic to see the Qur'an's social philosophy as in detail a blueprint for life in the modern world, a blueprint to be either accepted *in toto* or rejected outright. The question to ask is how progressive its message was in that day and age. The bounds of "enlightenment" for this particular purpose cover the affairs of this life, not transcendence to ultimate bliss.

In these terms, the Qur'an appears enlightened in various cardinal respects. The injunction "Kill not your children for fear of want" (sūra XVII v. 31) took courage to enunciate at a time of acute aridity. The same goes for liberal rules for slave manumission; and the proscription on forcing female slaves into prostitution (sūra XXIV v. 34). The status accorded women in general (sūra IV, sections 1 to 6) may still seem retrogressive by modern standards. But to quite an extent their pre-existing status was thereby consolidated or enhanced, institutionally as well as in terms of codes of male conduct.

To my mind, however, a most edifying aspect of the Qur'an concerns the references made to animals. Though few and brief, they explicitly accord every member of the animal kingdom an independent place in the scheme of things: "the birds of the air with wings outspread. Each one knows its own mode of prayer and praise. And God knows well all that they do" (sūra XXIV v. 41). Indeed, all animals on Earth live in communities just as we do and "Shall be gathered to their Lord in the end" (sūra VI v. 38).

These are beautiful sentiments, beautifully expressed. Would they have gained some currency in the non-Islamic world, rather than so much being made of Quranic allusions to war-waging. Not that these are actually at all extreme. The text recognizes a contingent need to "Fight in the cause of God. And for those who, being weak, are ill-treated . . . " Admittedly, the Qur'an does not always spell criteria out. But the general understanding appears to be that (a) war should normally be just for self-defence and the restoration of the *status quo ante*, (b) women, children, the elderly and infirm should not be molested, (c) trees, crops and water sources should not be destroyed and (d) peace should not be withheld once the enemy is prepared to come to terms.[17]

Often remarked are those materials in the Qur'an which bespeak a common heritage with the other two great monotheisms that developed on the western fringes of the Arabian desert. In the Qur'an, Abraham is depicted very much as an archetypal desert father, an inspiration to all within that region who are committed to the one true God. It is in this sense that one should interpret the designation of the Mecca Ka'aba as the "Station of Abraham", supposedly built by him and his son, Ishmael (sūra II v. 125–7).

Meanwhile, Adam is presented as the truly universal spirit, an edified one to boot. In the Bible, he has been given charge of the Garden of Eden. But through the Qur'an, Allah makes him "vice-regent on Earth" (sūra II v. 30), someone the Angels must pay obeisance to. This broad designation lessens the significance of the expulsion from Eden, an imposition in any case described as temporary (sūra VII v. 24 and sūra XX v. 129). Absent is the Biblical disposition to cast starkly the eviction, making it a pointer to Original Sin, sexually expressed. Instead Adam is God's "most important creation".[18]

A critical test of the prospects for concordance with Christianity must be the Quranic coverage of Jesus and his circumstances. Mary is told by God of His ability to will a virgin birth (sūra III v. 47). Jesus Himself is accepted as "a righteous prophet (sūra VI v. 85) whom God raised up "unto Himself" (sūra IV v. 158). But there a line is drawn. He was never crucified by the Jews (sūra IV v. 157); and so, presumptively, experienced no bodily Resurrection.

Nor can He be considered part of a three-in-one headship. Addressing Christians as the "People of the Book", the Qur'an advises them to "Say not 'Trinity'. Desist. It will be better for you. For God is one God, Glory be to Him" (sūra IV v. 171). After all, God could not have had partners in Creation; and so would have had none since (sūra X section 4). Yet notwithstanding such aberrations, God has instilled "compassion and mercy" in the hearts of those who follow Jesus (sūra LVII v. 27).

Still, the main import of the Holy Qur'an concerns the nature of Allah. A cardinal axiom is that "No vision can grasp Him but His grasp is over all vision" (sūra VI v. 103). It locates Him much higher above us than the loving Father of Christianity or than YAHWEH, God of Israel. He may be adjudged nearer to Man "than His jugular vein" (sūra L v. 16) but this merely follows from His comprehensive knowledge of the whole of Creation. Humankind may be comparatively close to Him but only in so far as He has set us over all Earthly life. Allah does hear our prayers but can respond only on a proportional basis. To ask for more would be to "transgress beyond all bounds through the Earth" (sūra XLII v. 27)

Therefore as Allah metes out justice, He is eminently fair, totally informed and always prepared to reconsider, should the culprit repent and also make amends. Yet in the final analysis, Allah is stringent, contingently prepared to destroy a whole people "when its members practice iniquity" (sūra XXVII v. 59). This austere objectivity is further ensured by His detachment from familial concerns. He has no consort. He has progeny only in the ultimate sense that every living being is within His scheme of things (sūra VI v. 101). He is a supreme authority who looks less hard to reconcile with cosmological perspectives than do most other supreme deities humankind has identified. To its believers, Islam basically means submission to the will of Allah. What Moslems reject, however, are "most malicious" suggestions that this connotes fatalism, determinism or resignation. We are advised that were "man not to be free, religion would have no meaning".[19]

Islamic Spain (*al-Andalus* alias Andalusia) is where an adapt-and-adopt strategy found most conspicuous expression. Roman irrigation was restored and extended, a renovation assisted by technology transfer from Egypt and Mesopotamia. Around Córdoba alone, some 5,000 waterwheels were installed.

Over 100 varieties of fruit and vegetable were cultivated, rice and sugar being among those featuring in Iberia for the first time. Woodland cover reached perhaps its greatest Iberian extent these last two millennia.[20] Emphasis was also placed on grapes for wine, even though the Holy Qur'an brackets wine-drinking with gambling as activities in which "the sin is greater than the profit" (sūra II v. 219). *Al-Andalus* Christians may well have remained in a majority through AD 1000, at least in the countryside.[21] Inevitably they influenced the Moslems in their midst. In this respect as in others, the conquerors were drawn into the Mediterranean "oneness" An obvious contradiction was kept in check by wine taxation.

Heavenly Horizons

By retaining Mecca as its paramount Holy City, Islam set roots deep within Arabia's existing religious culture. The impact is hard to gauge though there are accounts early on of dialectical interaction. There are *hadith*, non-Quranic statements attributable to the Prophet. Pre-Islamic poetry is extant, too. Meanwhile oral traditions lasting into modern times assist assessment corroboratively.

The Islamic lunar calendar, primarily used to fix religious festivals, was derived from a pre-Islamic one into which was inserted triennially an intercalany month (*nasi*), this to align it better with the Earth's seasons. Intercalation, reputedly Judaic in origin, was to be proscribed by the Qur'an (sūra IX v. 37). Even beforehand, however, there was much recourse to the stars to track the seasons. Here, as so much elsewhere, the Pleiades received especial attention.[22] Their dominion during the *wasm* ten weeks of winter rains was being celebrated by Sinai and Negev bedouin in the sixth century AD.[23]

Mohammed discerned little spiritual depth among the bedouin he encountered during his decade of decisive strife, AD 622–632 (AH 0–10).[24] Arabists in the modern West have tended to credit those they observe with much. Several explanations might be adduced. An enduring "noble savage" strand in the literary romanticism of the West has to be one. Another is that the cultural revolution within the Quraysh remained urban elitist in character. Islam shared these urban roots, though it did prove able, soon and ever subsequently, to appeal profoundly to desert dwellers. Another consideration could be that, in ecologically less stressful times, even the pre-Islamic bedouin would likely have had higher morale and deeper spirituality. In his 1997 study, Dr Nasr spoke of desert dwellers' "great love for Nature . . . a deep intuition of the presence of the Invisible in the visible".[25] On the face of it, this suggests a God more immanent than Allah. Still, absorbed into Islam early on, this perception will have prepared the ground for the precept of indivisible understanding. All learning is thus open to religious infusion, Divine Unity.

From whence arose the *madrasah*, a "place of learning" usually located within a mosque complex. Just over a millennium ago, this concept was further developed with centers of higher learning teaching subject mixes ranging from religious law to astronomy. A hospital might well be attached.[26] This pattern was to serve very explicitly as a model, from the twelfth century, for university development in Catholic Europe. By then, however, the religious and legal establishment within Islamic academe was turning defensive, fearful lest doctrinal ructions furthered the general disintegration of *dār al-Islām*, the Islamic orbit.[27] Intrusions by Crusaders then Mongols compounded its apprehensions. Accordingly, it proved unable to accommodate the accelerating growth in knowledge diversity and volume or to endorse the single-subject specialisms this apparently necessitated. The upshot was intellectualism being constrained all round. The *madrasahs*, in particular, turned inwards unduly.

Yet before that, Islam's territorial expansion had allowed of interaction with major founts of scientific enquiry. Not least did this apply to schools of cosmology: Nestorian Christian in Persia; Hindu–Greek in India; and Alexandrian Greek. Alexandria had had its fabled library destroyed by Christian zealots in 389 but would long remain quite the most important Egyptian city. In due course, Hindu numeration became the norm within Islam; and would soon be received into Christendom as "Arabic". The incorporation of a "zero" symbol was, of course, the key to its highly economic decimal notation.

Aramaic translations of classical Greek authors (surviving via the Nestorians) were translated into Arabic. From 786 to 833, Greek texts were purposively collected in Baghdad by Hārun ar-Rashīd and his successor as Caliph. Ptolemy's *Almagest* was in translation by 827. Aristotle, Plato and Plotinus were also influences on the cosmogonic front.

By 800, routine astronomic observation was under way within *dār al-Islām*, it often being more refined than Antiquity had managed. Granted, most of the participating teams were overly concerned just to consolidate existing tabulations. Still, this involved the development of astrolabes in the eighth century AD and celestial globes by the ninth. Moreover, until the seventh AH/thirteenth AD, highly adventurous spirits periodically emerged in the realm of basic speculation, cosmological and philosophic. Bukhara's Ibn Sina (alias Avicenna) lived AD 980–1037. Famous as a physician, he also kept astronomical records and edited a compendium on Ptolemy. One of the respects in which he bucked the mainstream view within Islam was in seeing Allah's present power as limited by virtue of His having endowed Nature with intrinsic laws for preserving equilibrium and harmony and for progress towards goodness and perfection.[28]

Avicenna always insisted he was a committed Moslem scholar. However, he and a few near contemporaries were strongly appraised of a rational obligation to seek progress through doubt,[29] more so than anyone in Christendom appeared to be at that time. Similarly distinctive was his notion of love pervading the cosmos. He saw the human soul as having a special role here in that its transcendent love for Allah affords linkage between the mundane world around us and His supreme intellect. Meanwhile, Avicenna adjudged alchemy unable ever to effect "any true change of species" on the evidence to date. Likewise, though astrology worked from sound cosmological foundations, it was unable to predict the future. For one thing, the effects of all the bodies in the Milky Way could not be evaluated. This, in fact, well matched the standard theological position which was that "although astrology, in placing intermediary causes between man and God, is opposed to a certain aspect of the Moslem perspective, its contemplative and symbolic side confirms closely to the basic spirit of Islam, which is to realize that all multiplicity comes from Unity".[30]

Likewise, by the eleventh century AD speculative theologians within a Sunni school of thought derivative from al-Asharī (873–975) were respectively advancing several propositions at variance with received understanding. One was that Allah had ordained Creation out of nothing.

Another, not unconnected, was that Creation comprised minute and irreducible elementary particles. The mainstream objectors alleged each tenet flouted the idea that everything was created out of formless uniformity; and any other postulate would involve Allah's own state fundamentally altering during Creation. Yet He is changeless.[31]

What can veritably be described as the apotheosis of this Islamic intellectual tradition is afforded us by the life of Ibn Rushd, alias Averroës (AD 1126–1198). Born into an eminent legal family in Córdoba, his career epitomizes the contradictory tendencies already evident within *al-Andalus* as a peripheral and vulnerable frontier society, closer to Western Europe in some respects than to the rest of *Dār al-Islām.* His influence on Islamic thinking was to remain minimal until quite recently. From the outset, however, he had "a far more successful afterlife among the Jewish communities in the medieval world, and a widespread effect upon the Christian world".[32] His commentaries on Aristotle were one of several classic works consulted by Copernicus prior to the formulation of his heliocentric hypothesis.[33]

Widely cast though his own interests were, Aristotle was easily the first love of Averroës's intellectual life. He revered the Macedonian philosopher for his logic, reasonableness, empiricism, comprehensiveness and clarity. In a discussion in 1169 or thereabouts, the then Almohad Caliph encouraged Averroës to explore Aristotle's belief that the world we know has always existed. There had never been Creation or even fundamental change. About the same time, he received an official commission to rework and clarify (via the Commentaries) the Arabic translations of Aristotle.[34] Despite this endorsement, however, Averroës (along with close disciples) was expelled for a year near the end of his life for being too distractive in fractious Córdoba.

Still, the late medieval Islamic thinker best known across the world nowadays is probably Ibn Khaldun (AD 1332–1406). He was less unequivocally close to the Qur'an than were the neo-classical philosophers like Averroës. His big interests were strictly terrestrial. His most famous work, *Muqaddimah,* deals with human environments; and is, in fact, quite strong on climate determinism. Ibn Khaldun manifestly had a big posthumous influence on the late Arnold Toynbee just as he had previously had on Montesquieu.[35]

By his day, however, the great tradition of Islamic cosmology was indeed in decline operationally. In the fourteenth century AD, the Damascan astronomer Ibn al-Shair produced mathematical models for solar, lunar and planetary motions very similar to those Copernicus was to use.[36] But Persian mathematics had reached its zenith with the founding of the

Maragha observatory in AH 637/AD 1259, accuracies of one part in ten million being achieved in its table of tangents. Maragha has been described as the first observatory "in the history of science as such institutions are understood".[37] Its successful foundation owed much to 1259–60 also having been the time when events elsewhere induced the Mongol conquerors of Persia to turn decisively away from being part of a universal "tents of felt" revolution in favour of assimilation downwards. Later on, in Persia as elsewhere, the cosmological debates "had lost their urgency and had been transformed into mere topics of scholarly tradition".[38] Another fine observatory was to open in Istanbul in 1575 but soon shut down, ostensibly because the Caliph disliked its astrological findings.

Islam the Disparate

The decline of Islamic cosmology could be seen as a simple failure of philosophic accommodation. But it also relates to how *dār al-Islām* encompassed no territorial heartland sufficiently solid and central to mobilize properly the whole zone. According to hallowed principle, the one caliph should be the religious and political head of all Islam. But the interposed deserts and seas precluded this. Between AD 780 and 1171 there were always two or three caliphates. At the turn of the millennium, their respective locations were Córdoba, Cairo and Baghdad.

These individual statehoods found it hard to engender allegiance. Witness the impressment of infidel captives into military service. Nor was the Ottoman Empire to prove a helpful departure. All else apart, its *leitmotif* was Turkish not Arabic. Besides which, its capital from 1453 – Istanbul alias Constantinople – became a study in morbid obscurantism, a collective mind set typified by the rambling Topkapi (the Sultan's main palace) and, above all, by its harem. In the latter, hundreds of women and eunuchs lived lives of tedium relieved only by fear: "a minor breach of etiquette . . . resulted in instant death. In the seventeenth century, Sultan Ibrahim had 280 of his women sewn into weighted sacks and thrown into the Bosporus".[39] The outlook that allowed of such atrocities had also long engendered a deeply negative attitude to the adoption of gunpowder, watermills and windmills, and printing. So different from *Al-Andalus* in its heyday.

The Shi'a

The most significant sacerdotal and political cleavage within Islam remains that between Sunni and Shi'a, a cleavage originating in early clashes between the family of 'Alī (cousin and son-in-law of the Prophet) and the 'Umayyad dynasty. The key Shi'ite claim has been that 'Alī and next his grandson, Husayn should have succeeded the Prophet as spiritual leader. Instead in AD 680, Husayn along with his immediate followers were lured to Karbala to be slain. Their martyrdom is honoured each year at a lunar-calendar Karbala festival. Scarification of one's own body is practiced by not a few at this event. Shi'ites remain in a minority in virtually every polity save Lebanon, Iraq and Iran.

In early Islamic times, Shi'ites established Abbasid caliphates in Baghdad (witness Hārun al-Rashīd) and Córdoba. More generally, however, the mainstream has regarded the caliphate concept as too corrupted by 'Alī's exclusion from the succession. So their leadership, religious and secular, is regularly invested in twelve Imams – i.e. prayer leaders. However, one of them, Muhammad, disappeared in childhood long ago; and this "hidden Imam" is expected eventually to make a messianic return. Suffice now to add that the Alawī and Druze in the western Fertile Crescent, as well as the Isma'ilis (who are cast more widely) are Shi'ite derivatives.

Sūfīsm

The Qur'an recounts how the Prophet Mohammed received the word of Allah through mystic experience (sūra LIII v. 1–18). Therefore one need not be surprised that a Sūfi ascetic-cum-mystical tendency is abroad by the eighth century AD. During the tenth and eleventh, it proliferated rapidly, not least among the famed poets of Shi'ite Persia. Outstanding among these was Omar Khayyam, now a poet of world renown but then a peerless mathematician and court astronomer as well.

A sūfi practitioner is required to embark on an inner "spiritual itinerary" through a succession of "stations", each of which is associated with a cardinal virtue – abstinence, gratitude, trust . . . Again, the ultimate aim is to draw nearer Allah through sheer love of Him, reducing the void caused by our inability to comprehend Him through reason or describe Him prosaically. To quote the female pietist, Rābia al-Adawīya, "that purest love which is Thy due . . . the veils which hide Thee fall". Entrancing music figures prominently in the drawing nearer.

The influence of sūfism has ramified. It has spearheaded interaction with other creeds, inputs from the New Testament and from Buddhism being especially evident. Sometimes it is forging links within the Community of the Faithful, the Umma. At other times, it is urging social change. In the modern context, it may be more open than other Islamic traditions to religious syncretism.

Old versus New

By capturing Istanbul/Constantinople in 1453, Islam gained its biggest victory ever against Christendom. But 1492 was to be a milestone year in a much broader offensive by Christianity. The surrender of Granada marked the end of *Al-Andalus*. Meantime, Columbus set Renaissance Europe on the way to establishing a network of sea routes that would, among other things, circumvent Islam.

The resultant clash between this intrusive culture and the respective indigenous ones progressively became ubiquitous. It was eventually to turn acute in the Islamic swathe of territory extending from the Horn of Africa, Arabia and the Gulf to central Asia. There the visible impact of the West had long remained but tangential and episodic (see Chapter 17). Even so, from 1763 Muhammad ibn Abd al-Wahhāb (*c.* AD 1703–1791), a former sūfī adept, waged war to impress on Arabia and hopefully Indian Islam the view that the vanquishing of Satan required a return to the Quranic doctrinal purity of the first three centuries AH as confirmed by *ijmā* – i.e. consensus, be it popular or scholarly or priestly. No matter that Sharīa law, a very important aspect heavily dependent on a formalistic interpretation of the Holy Qur'an and of *hadith*, was not to be stabilized until the fourth or fifth centuries AH.

The Wahhābī leadership regularly portrayed the moral degeneration it thus campaigned against as brought on by sūfism, idolatrous shrines, and other local factors. One positive departure, in the nineteenth century, was the establishment (in well-watered locales) of agricultural colonies.

Wahhābīsm survived and thrived sufficiently to put Ibn Saūd on the throne of what thereby became the Kingdom of Saudi Arabia in 1932. By which time, Western imperialism was inducing culture shocks in a region which still remained the spiritual heartland of Islam.

However, it may be that Islam (along with Judaism, Christianity and other major faiths) will become more polycentric in the course of this century. There are currently indications of this in, for instance, South-East Asia. Sūfism shows signs of some regional revival; and, in Malaysia, the new Islam Hadhari movement is stressing the ethical worth of being both "cosmopolitan and Islamic". Much may depend on whether Indonesia can achieve lasting political stability in spite of the persistence of an acute demographic imbalance between overcrowded Java and the other islands. Much may depend, too, on a re-engagement with scholarship. The broadly-cast Amman conference of July 2005 (involving 170 Islamic scholars from 35 countries) may have been ground-breaking in this respect.

14

INDIAN PLURALISM

Of the great centres of dawning civilization, the most elusive is India. Across much of the subcontinent many early buildings, public and domestic, were made of wood. Moreover, much of the earlier writing was not on papyrus or clay tablets nor on stone or metal surfaces but on leaves or bark, materials perishable by water or air and heat. Nor, for many centuries more, did the Indians adopt the Chinese invention of paper which dates from *c.* AD 100. The oldest major text extant, the *Rig Veda*, was composed mainly between 1500 and 1000 BC.

What can confidently be said, none the less, is that a brilliant civilization flourished in the Indus Valley between 2500 and 1900 BC, its sudden demise likely being largely due to climate change capped (*c.* 1500 BC?) by Aryan military intrusion. At 35,000 people apiece, its two biggest cities never matched in size what Ch'ang-An, Rome or Constantinople were to achieve. Even so, at every level from rectilineal master plans to the provision of flush lavatories these Indus new towns were virtuosity itself. They could have been created only by people adept at measurement and calculation.

What the Indians did not do then, nor long after, is develop mathematical theorems the way the Greeks were to. Indeed, they seem not to have matched the Babylonians in this respect. By the time of King Hammurabi (*c.* 1750 BC), the latter were "solving linear and quadratic equations with two variables; and even problems involving cubic and biquadratic equations".[1] In contrast, the pristine skills of India were arithmetic rather than algebraic. The adaptive system of notation we know as Arabic numerals would be passed by the Indians to the Islamic Arabs around AD 700. By then, it afforded a symbol for zero, thus allowing for decimal notation[2]. This provision seems to us an obvious must. But the Greeks and Romans declined to acknowledge the requirement. So, apparently, had the Babylonians. Interestingly, the Mayas did not.

Polyculturalism

Those who trace the evolution of Sanskrit literature see the centuries following the Aryan invasion as characterized by the Vedas, the classical sacred texts. This era is deemed to phase, through 800 BC, into that of the *Brāhmanas*, a time noted for the production of several hundred *upanishads* or informal sacerdotal commentaries. By then, too, the application of numeracy to astronomy was making headway. One commendably accurate estimate was that both the Sun and the Moon were averagely 108 times as far from the Earth as the width of their respective diameters. There is even report of one arbitrary reckoning yielding a remarkably good guestimate of the speed of light.[3] Next one should consider how far fundamental science was nurtured and moulded by the prevalent religious culture.

In the sixth century BC, Jainism emerged as a subsidiary alternative to Buddhism in the articulation of a burgeoning resentment at such negative aspects of Hinduism as undue recourse to the Vedas, excessive ritual, and the promotion of the caste system. A scientific precept it has often been credited with, for either originating or giving impetus to, is "atomism": the notion that the cosmos is comprised of *anu*, particles which are minute, underived, and irreducible.

In the classical Jainist rendering, the cosmos is composed of two basic ingredients, life-monads and quanta of non-living matter. The former are the basis of the soul. They are ubiquitous, infinite in number, and suffused with bliss. However, their attributes are largely submerged by the *anu* attaching to them in life. The balance thus struck between good and evil in our current existence will determine the soul's starting point in the next of what are liable to be interminable cycles of incarnation.[4] By the same token, "life monads expand or contract to fit the bodies which they respectively animate . . . ".[5] Fritjof Capra felt justified in saying that "the notion of atoms is prominent in the Jaina system".[6] All the same, the early Jainists did not live in an atomic world in any sense a Bohr or Einstein might comprehend.

Granted, a form of atomism closer to that of modern physics was to be fairly widely endorsed within India within several centuries of the emergence of Jainism. However, the Jainist input may never have been decisive on this score. The Greek philosopher Democritus (*c.* 460–*c.* 370 BC) had brought a mechanistic perspective to bear to formulate a theory of atoms of varying sizes but all minute, indivisible and everlastingly in motion. Nor should one underrate Greek influence within India around this time. A legacy of Alexander the Great's invasion (327–325 BC) was the establishment of a Greek state in the North-West. This had been vanquished by the turn of the century but cultural linkages survived in the empire which Asoka (d. 232 BC) was soon to extend over virtually the entire subcontinent. As much is most evident in Buddhist art. The halo is the device Alexander had learned of during a visit to the Kharga oasis in Egypt's Western Desert. More generally, indeed, art historians discern Graecian idealization in Buddha representations from the North-West.

Germane, too, is the flow from Greece to India, between the first and seventh centuries AD, of astronomical concepts and techniques. The surviving materials owe a lot to Babylonia or to Hipparchus. Some are Aristotelian. There is, for whatever reason, a stark absence of Ptolemaic material.[7] A general reduction of the flow from *c.* AD 400 closely coincided with a secular contraction in intra-Eurasian long-distance trade.[8]

Cosmography

Meanwhile, as India's Buddhists strove to retain due prominence, their scholars articulated other cosmographic wisdom long-received on the subcontinent. This was that, beyond our world system, lay an infinity embracing many others, billions some intimated. Granted, certain modern commentators have read these entities as having been separate galaxies rather than other universes. But this distinction would have been lost on all concerned originally.

More to the point, the whole issue must have been complicated by another early tenet, one which said the centrepiece of our world system (i.e. Mother Earth itself) was flat.[9] So what might a flat Earth rest on? And could other flat Earths be congruent? Only, it would seem, by being universes in quite separate dimensions.

Another Hindu notion which classical Buddhism endorsed was that of each and every cosmos pulsating. But in this instance, it has been the Hindus who have since pursued the theme the more overtly. Most notably have they done so in the *Bhagavad Gita*, the "Song of the Lord" – an episode within an epic maybe dating from the second century BC. The central figure, Arjuna, is sorely troubled by the imminent prospect of participation in a great battle. His distress is compounded by several of his relatives being in the opposing army. Disguised initially as one of Arjuna's own charioteers, Krishna (an incarnation of Vishnu, see below) offers solace and spiritual guidance.

Within its text is the legend of *lila*, the divine rhythm. Brahman, the ubiquitous holy spirit, starts out configured as humanoid male. However, He then transforms Himself into a world system that pulsates eternally. Every contraction to nothingness culminates in re-inflation. As the *Gita* puts it, "At the end of the night of time, all things return to my nature; and when the new day of time begins, I bring them again into light."

Mathematical or *Siddhantic* astronomy develops, with Greek stimulation, from *c.* AD 500, towards the end of the great Gupta empire. A big preoccupation in the next millennium was the calculation of geocentric planetary orbits, including the development of the algorithms this required. But sufficiently systematic and sophisticated observation got under way only from *c.* 1370, its emergence being stimulated by contact with the Near East, notably via Islam.[10]

Hindu cosmism finds strong material expression in the Khmer Empire, the best known of several Hindu–Buddhist temple states that emerge in South-East Asia in early medieval times. Centred originally on Angkor Wat but later on Angkor Thom, it was at its height between the tenth and fourteenth centuries AD. From the eighth to the tenth, as state institutions consolidated, accumulating surplus wealth had been considerably directed towards "theocratic hydraulics". The Angkor cities (and, less completely, the rural hamlets) were thus turned into replicas of Heaven as per Indian cosmology. Remarkably, no instance is known to the specialist here cited "where a temple pond was equipped with a distribution system to water the fields".[11] The two mighty networks were kept strictly apart.

Later the great temple to Vishnu at Angkor Wat (built AD 1113–1150) incorporated extensive built-in provision for lunar and, most particularly, solar observation. To the Khmers, astronomy was "the second science".[12] Not that this edified stance inhibited hideous maltreatment of delinquents and of the many slaves.

Classical Hindu Theology

A notable aspect of the post-Vedic sacred writings – the Vedanta – is the depiction of the Hindu pantheon. At its apex are the three gods, Vishnu and Shiva plus Brahman, the creator. Specialist commentators discourage the drawing of parallels with the Christian Trinity. For one thing, Hinduism never aspires to be as synoptically coherent nor as definable in detail as do most branches of Christianity. For another, there is *brahman*, the pure consciousness behind our illusory world. Evidently, its connotations are more cosmic than are those of the Holy Ghost.

There are indications that Shiva may partially derive from a supreme god of the Indus Valley civilization who was at once "lord of beasts and the great ascetic", a force of life and yet of release therefrom. Thus this god of yoga is also Lord of the Dance. In his exuberance, he may exhibit a fearsome ferocity. Vishnu, though still awesome, presents a kindlier aspect.

The most prominent of his Earthly manifestations are Krishna and Rama. The overall effect is that what otherwise rates descriptively as a rich polytheism is given strong integrative connotations.

The *upanishad* informal sacred writings were a follow through of the Vedic tradition. The term derives from "sitting close to" a teacher, here meaning getting into a tightly focused discourse about a defined issue. Out of those *upanishads* with this stamp about them came doctrinal advances. One was the full development of the *brahman* concept. Another was probably the elucidation of *atman*, a word etymologically related to the German *atmen* – "to breathe" – as well as to the Greek *atmos*, a breath. What the Sanskrit word means is the individual's inner self, an ultimate which can survive *samsāra* – the endless cycle of growth, death and rebirth. A collateral concept is *karma* or overall moral worth.

As in Jainism and Buddhism, it is believed that one's moral worth in this existence determines where one starts from in the next. The *samsāra* alternation can only be transcended once and for all by achieving *turiya*, a higher state in which the individual *atman* blends into the Absolute Self of Brahman. Attaining it, one blissfully loses all destructiveness, cognition, susceptibility and responsiveness. There were and are many variations on this and related themes. There were, after all, a number of different schools of classical Hinduism.

The said doctrinal evolution broadly coincided with the extension and consolidation of Aryan hegemony across and around the Gangetic plain. An operational link between these two tendencies was the emergence of a very hard-and-fast social caste system. This was cited above as one of the factors that aroused organized religious protest, Buddhism and Jainism being its primary manifestations. This radicalizing propensity was further enhanced by the priesthood – the Brahmins – being positioned at the very top of the caste hierarchy.

Jainism Further Considered

In certain cardinal respects, this creed may merit more attention than its recognized following would indicate. Its classic philosophers were within the South Asian mainstream in so far as they believed Time was eternal in each direction and Space infinite in all directions. However, they saw our world-system as the only one within these timeless and limitless voids. Moreover, their understanding of its morphology was curiously idiosyncratic; and manifestly owed too much to the backward extrapolation of concern for the human condition. The whole of creation actually assumed a humanoid form; and, as such, could be said to be (in modern parlance) 70,000 light years tall.

A horizontal transection at waist level presents the world we inhabit. Below it are several million purgatories where evil-doers are incarcerated pending arduous reincarnations. Above it are a series of heavens within which gods, along with life monads (i.e. souls, more or less), are accommodated hierarchically. As accretions to Jainism from Hindu and Buddhist philosophy, the gods confuse the issue rather. In all three traditions, any soul may sometimes be reborn as a god.

Progress towards transcendent bliss is made chiefly through asceticism, this involving self-mortification and ideally some monastic seclusion. Advance or retreat apropos this consummation is registered by the ascent of one's life-monad up through the heavenly zone to the very summit of the humanoid cosmos. So what happens to the process as and when the proportion of monads making it has increased from very low to extremely high? The totality of life on Earth is subject to immensely long cycles of decay and regeneration. At the start of a generative process, people are titanic in every way; and they draw on boundless resources. At present we are well into a decay phase which ostensibly explains why the number of committed Jainists may barely top a million.

One can easily scorn Jainism: its chronic failure to make popular headway within the subcontinent, let alone beyond; its doctrines frozen in time; its contrived cosmography . . . None the less, it has made a philosophic contribution. Closely related to its asceticism is an absolute insistence on *ahimsa*, no violence (be it calculated or casual) towards living beings. All life is somewhere on a scale compounded of sensing and cognition. Gods and humans have five sense organs plus a mind. A plant has a sense of touch but little cognition. Microbic organisms have no sense organs *per se* but may cluster together for collective action. This looks like a more cogent attempt to elucidate Consciousness than any made by the other faiths here being specifically considered.

Their taboo against harming life has long obliged practising Jains to eschew many occupations. So they have applied themselves with all the more dedication and integrity to those "white-collar" and craft professions they have felt free to enter. What ought to be acknowledged as well is that Jainist philosophers have long sought also to address fundamental questions akin to those Einstein wracked his mind over: ones concerned with the interaction between space, time and motion. This they have done partly to exclude the possibility that energetic motions within the humanoid world-system would lead to materials exiting into boundless Space forever.

Their view has been that matter can only move through empty Space when the latter is filled, however elusively, with a medium of motion. This postulate has alluring analogies with the pervasive "aether" proposed by nineteenth-century Newtonian physicists to bear throughout the cosmos the radiant waves of electromagnetic energy. Unfortunately, the existence of this aether was called into question by how the very accurate Michelson–Morley experiment of 1887 failed to show the Earth's motion supposedly through it to have any effect on the recorded velocity of light. Then in due course Quantum Theory was to make the aether concept entirely superfluous by deducing that electromagnetic waves were comprised of discrete photons. Therefore one might conclude, too, that Jainist cosmography really has nothing to contribute. However, the religion's particularly strong insistence on the special value attaching to all life, however humble it be in our instinctual estimation, may yet prove invaluable. Very similar things might be said of certain other customary faiths. Zoroastrianism and Quakerism come readily to mind.

Indian Buddhism

The birth of Siddhartha Gautama (alias Gautama Buddha[13]) was probably 10 to 15 decades earlier than the 476 BC customarily cited. At that time, the middle to lower Ganges and its eastern surrounds were divided into a number of statelets, each one extending outwards from a city of some size. Tolerably good relations between them were underpinned by economic progress; and, of course, the converse held true as well. However, this benign situation favoured flux in other respects. Brahmin authority was declining; and the whole caste system weakening.

Siddartha was the eldest son of the king of the Sakya clan. According to historical legend, he left the royal palaces to seek through self-denial the truth about human existence. Six years later, he experienced profound enlightenment while meditating under the Bo tree. Duly he walked all over central India imparting the revelatory message. Various alluring elaborations surround his life story, a number of them strikingly similar to some attaching to Jesus.[14] What reads like a very mythic account of his death amidst his disciples at the age of 80 has been handed down. Reportedly his demise occurred not long after his Sakya home territory had been ravaged by hostile incursion.

How far classic Buddhist doctrine originated with Gautama is unlikely ever to be resolved. At all events, the core precepts are encapsulated in the Four Noble Truths and the Eightfold Path. The former are (a) life is perceived by us as essentially a matter of suffering, (b) this perception is down to our obsession with stretching out for or clinging onto all the mundane aspects of our transient existence, (c) this disposition can be curbed by purging ourselves of lust and acquisitiveness, and (d) the approach to adopt is the Eightfold Path. The eight aims in question involve the pursuit of rectitude in three prescribed spheres. Trust must be gained through rightful attitudes and sound beliefs. Ethical conduct is imperative in respect of speech, deportment and general life style. So, too, is thorough self-training in regard to application, self-awareness and meditation. The ultimate goal is the attainment of buddhahood via the melding of one's ego into *nirvana*, a permanent state of blissful quiescence. One is bound to say that such a consummation looks quite incompatible with the retention of individuality.

However, this is almost bound to require quite a number of human generations; and must therefore involve the repeated reincarnation of a specific soul, of *atman* as the Hindus would say. Yet this runs contrary to the Buddhist emphasis on the impermanence of everything. So does it to the faith's denial of any divide between the material and spiritual domains. It does likewise to Buddhist overt acceptance that every sentient being is, along with all else, a product of random chance.

A legendary endeavour to square this contradiction took the form of a dialogue between a Buddhist seer, Nagasena, and Menander (alias Milinda), the Greek king who ruled in North-West India in the wake of Alexander the Great. Nagasena tells how only a kind of silhouette is reborn, nothing more. Even so, he does aver that the regenerative parameters are determined by the balance between good and bad in the last life. So must not a tangible identity thereby be continued?

Nor was the inherent contradiction resolved at all convincingly by a modern rendering by a sympathetic Western observer, the late Ninian Smart. He acknowledged that it was not open to Buddhists to posit "the migration of a permanent self from one psychophysical state to another". Therefore in the likely event of an individual failing this time round to attain *nirvana* on death, "a fresh sequence of psychophysical events is set up".[15] But if the worthiness or otherwise of the previous existence figures, a new departure cannot be entirely "fresh". Otherwise the notion of generational progression loses its meaning. One must further ask how past merit is arbitrated in the godless cosmos of classical Buddhism.

The Development of Buddhism

The early evolution of Buddhism is not easy to unravel. Correspondingly, its marked progress within a competitive religious environment is hard to explain. Gautama may, in principle, have prescribed a "middle way" between mortifying asceticism and hedonistic indulgence. But, in practice, the lifestyle laid down for a Buddhist monk was harsh and considerably reclusive. It could not have related easily to what was an urban-led economic development under way in that middle to lower Ganges region, the Galilee of Buddhism. But here the key was a monastic progression away from the wandering mendicant and towards settled religious communities, towards monasteries.[16] Being free from arcane brahminic rituals as well as from caste, these could assume a salient yet corrective role in the progress of a market economy.

What must more specifically be said is that Buddhism eventually received a big fillip from Asoka, the Maurya Emperor (reigned *c.* 273 to *c.* 232 BC). He brought under his sway nearly all the subcontinent plus Afghanistan. Then hard upon a bloody conquest of Kalinga (*c.* 261 BC), he turned remorsefully to *ahimsa*, non-violence. While showing tolerance towards every faith, he especially favoured Buddhism in various ways. Witness his underwriting of Buddhist missions to Sri Lanka, Syria and Greece.

Meanwhile, Indian Buddhism was polarizing more deeply between the Theravadin (i.e. conservatives) and those plumping for radical change. Both camps, though particularly the latter, were further prone to split into sectarian groups, these usually possessed of substantial patronage and dogmatic aloofness. In due course the more radical of them came collectively to be entitled *hīnayāna* (small wheel). The proclivities just identified related in their case to a belief that the Buddha always existed on a much higher spiritual plane than they could ever attain themselves. Their only hope was to struggle hard to achieve *arhat*, ordinary sainthood. An anguished sense of obligation was further evidenced by a formalistic preoccupation with the exigesis of the classically sacred texts.

With the end of the Maurya dynasty (*c.* 180 BC), invaders poured into India via the Khyber Pass, effecting geopolitical instability but also extensive cultural intercourse. In this milieu, *mahāyāna* (big wheel) Buddhism spread through the population readily. Its basic understanding was and is that there were and are countless buddhas, actual or prospective, over and beyond the historical one. Most importantly, there have always been *bodhisattvas* as well, individuals who have reached the ninth in ascending order of the ten levels of enlightenment but who have deferred their transition to *nirvana* in the interests of their fellow earthlings. The *mahāyāna* accent is thus on the spiritual advancement of humanity as a whole.

As regards Buddhism abroad, *hīnayāna* zeal ensured its becoming the primary received faith across South-East Asia from Sri Lanka to Cambodia. But across East Asia from Nepal to Japan, *mahāyāna* made the running. All but three of the buddhist schools of thought established in China from the first century AD were of *mahāyāna* persuasion. However, the cultural adaptiveness that favoured expansion in other lands led in India itself to convergence between Buddhism and Hinduism; and this largely meant the latter assimilating the former. By the high middle ages, Indian Buddhism had all but disappeared; and, one must add, caste was again a major force to reckon with.

Islam

In the near elimination of Buddhist influence in India, Islam was to be instrumental. In the seventh century AD, the creed had established a presence in Sind from whence Moslem trading communities distributed themselves further afield. Eventually Islam was to gain geopolitical ascendancy, this after Rajput military power had been broken in campaigns waged (999–1026) by the Emperor of Afghanistan, Mahmud of Ghazni. Suffused with pious disgust at the use of craven imagery, advancing Moslem soldiery attacked many religious sites – Hindu, Jain and Buddhist. The monasteries were prime targets.

Subsequently, there was relatively little syncretic relationship between Hinduism and Islam. All else apart, no major obedience (unless it be Judaism) is as disinclined to compromise with polytheism as Islam has been. Moreover, *sūfi* preachers put the brahminic establishment under considerable pressure because their mysticism was usually interwoven with a forthright condemnation of caste.

However, three caveats should be entered. One is that Indian Islamic society did and does incorporate caste to an extent. Another concerns Sikhism, a Punjabi offshoot of Hinduism engendered by the prophet Nanak (*c.* 1469–1539): an offshoot that was and considerably remains of consequence, especially in and around its local region. Overtly seeking eventual concordance between Hinduism and Islam, Nanak taught monotheism, the fundamental commonality of all religions, and the realization of God through meditation and religious exercises. He opposed idolatry, excessive ritual and caste. To which one should add that, by 1700, military prowess was a core element in the Sikh ethos. It has remained so though very generally to laudable ends.

The third caveat concerns governance on the subcontinental plane. Akbar, the Mogul Emperor from 1556 to 1605, twice married Rajput princesses. Correspondingly this "Guardian of Mankind" surrounded himself not just with Muslim divines but also with Hindu Brahmins and Jesuit priests. He promulgated in 1582 *Din-I-ilahi*: a faith eclectically derived from Islam but also Hinduism, Zoroastrianism and Christianity. It was a monotheism, tolerant in its approach and with Akbar himself as prophet. Alas, this bold departure gave rise to violent rebellions by Muslim extremists; failed to take root more generally; and declined into oblivion soon after Akbar's death.

By the nineteenth century, India was experiencing a burgeoning debate about how to preserve what was best in tradition while achieving balanced modernization. *Suttee* was outlawed in 1829, it being the practice (then still fairly prevalent among Hindus) whereby a widow would immolate herself on her husband's funeral pyre. Precious few people wax nostalgic today about its virtually complete disappearance. But those same philosophic liberals within Britain's East India Company who promoted the *suttee* ban also presided over a headlong rush to imperial and international free trade which was to cause much structural unemployment and social distress among India's handicraft workers, cottage cloth spinners and the like. We should therefore turn next to the most prominent rebel against modernization under the British Raj; someone whose own persona was dichotomized as between the high medieval and the post-modern.

Mahatma Mohandas Gandhi (1869–1948) [17]

More than any other individual, Mahatma Gandhi epitomizes Indian syncreticism as made manifest in the turbulent first half of the last century. He was brought up on the Kathiawar peninsula, almost all of which today lies within the state of Gujarat. In these opening years of the twenty-first century, Gujarat has been sullied by violence between Hindus and Moslems, especially the former against the latter. But through the 1870s, the young Mohandas found Islam accessible as, indeed, was Jainism. Meantime, his mother actively practised Bhakti, a Hindu devotionalism that studiedly peers through the priesthood and the pantheon to seek a personal rapport with the supremely divine.

In 1893, not long after completing his legal training in London, Gandhi went to practise law in South Africa; and soon gained prominence in the struggle to reduce discrimination against the large Indian minority. His political activities included, from 1907, *satyagraha* or civil disobedience campaigns. He returned to India triumphally in 1915 and joined the indigenous independence struggle. From 1919, he masterminded its new *satyagraha* strategy and, in due course, complemented it with his personal "fasts unto death". In 1930, in protest against the government's salt tax, he led the famous 200–mile march to the coast to distill sea salt. For this he was imprisoned.

However, he was released in 1931 to attend the London Round Table conference on India's future, this as the sole representative of the Indian National Congress. Yet his relationship with Congress was always ambivalent, most basically because of his oft-stated desire to revive the rural cottage industries, though how was never made clear. There may also have been some unease within Congress about how categoric Gandhi was waxing in his opposition to caste "untouchability". Nevertheless, he did strike up a working relationship with Pandit Nehru, a man twenty years his junior. In the absence of immediate and tangible progress towards Indian independence, both Nehru and Gandhi declined to lend broad support to the British war effort, 1939–45. Instead they were instrumental in launching the Quit India movement – this in 1942, the menacing high watermark of fascist aggression worldwide. In particular, Imperial Japan arrived at India's eastern border.

Following release from further imprisonment in 1944, Gandhi opposed vehemently any partitioning of the subcontinent, portraying it as contrary to his wider vision of the unity of mankind under one divinity. But when independence came in August 1947, it was on a partition basis. Throughout the terrible communal violence that so marred this transition, Gandhi repeatedly walked into the thick of the fray, continually reiterating the need for pacification. In January 1948, he was fatally wounded by a Hindu fanatic.

Gandhi is little discussed these days, even in India. Yet up to the climacteric State of Emergency of 1975 (imposed hard upon a succession of internal crises), his personality and philosophy were continually being evaluated. In 1952, Einstein saluted him as "the greatest political genius of our time". In 1935, however, he had advised the Mahatma in writing that, while a strategy of non-violence might be appropriate against the British, it would be no good against the Nazis. Einstein also adjudged Gandhi "mistaken in trying to eliminate or minimize machine production in modern civilization. It is here to stay".[18] Later Arthur Koestler was among those who scorned his ostensible rejection of Western medicine and his commitment to a revival of the cottage spinning wheel. That revival never materialized.[19]

However, he was to be criticized far more caustically by V. S. Naipaul, the Nobel prize winner born in Trinidad in 1932 of Indian immigrant extraction. Naipaul felt India had paid a heavy price for Gandhi's shortcomings, especially his unpredictability and his inability to manage the social pressures he had himself engendered. These had caused her independence to be delayed a quarter of a century; and then to be achieved only on the basis of territorial partition. Naipaul's anger was compounded by his being influenced by a then widespread tendency to exaggerate the bloodiness of the ethnic cleansing engendered by partition. Actually well under a million died. Communal massacre must still be seen as the ugly heart of a dreadful displacement process. But it was not "as great a holocaust as that caused by Nazi Germany".[20]

Naipaul minced no words about Gandhi's persona, in its psychological and iconic aspects. As he grew older, we are told, the Mahatma's obsessive asceticism (not least as regards sexuality) took his public self over: "Knowledge of the man as a man was lost; mahatmahood submerged all the ambiguities and the political creativity of his early years, the modernity (in India) of so much of his thought. He was claimed in the end by old India, that very India whose political deficiencies he had seen so clearly with his South African eye."[21]

To George Orwell, Mohandas Gandhi was not exactly lovable but was entirely admirable, bearing in mind how consistently he eschewed corruption, vulgar ambition, malice and fear. But with characteristic concision, he dismissed the Mahatma's reversionary political economy as "obviously not viable in a backward, starving, overpopulated country". He also describes Gandhi's asceticism in terms that, especially when the sexual domain is addressed, border on benign farce.

However, Orwell's main concern was with *satyagraha*, the doctrine of non-violent resistance to foreign occupation. Oddly, he finds it separable "to some extent" from Gandhi's other precepts, a separation the Mahatma himself would never have thought of. At any rate, Orwell finds this pacifism more persuasive than some if only because its progenitor was, so he felt, prepared calmly to address objections.[22]

A peculiarly insightful interpretation of one side of Gandhi came in 1992 from what some would deem an improbable source – namely, Daisaku Ikeda. He was lecturing in the National Museum, New Delhi, in his capacity as President of Sōka Gakkai International, the contemporary revisionist rendering of Japan's Nicheren Buddhism (see Chapter 15). He compared the Mahatma with Josei Toda, Sōka Gakkai's second President. Imprisoned by the imperial authorities in World War Two, Toda similarly upheld anti-militarism and social justice. With Gandhi, this approach was considerably directed against modern fragmentation: against "the lines of separation that have been drawn between the human being and the universe, between humankind and nature . . . between means and ends, between the sacred and the secular . . . ".[23]

This assessment is persuasive as far as it goes. Undeniably, there was in Gandhi a spirit of outreach which he himself identified with a universal inclusiveness inherent in his native Hinduism. But the zeal of the reconverted drove his personal faith, the reconversion having largely taken place in South Africa. About his prior sojourn in Europe, Orwell comments (no doubt with some accentuation) that "he wore a top hat, took dancing lessons, studied French and Latin, went up the Eiffel Tower and even tried to learn the violin – all this with the idea of assimilating European civilization as thoroughly as possible".[24] Then during his reconversion, he incorporated a lot of religious traditionalism into what thus became a decidedly disorganized world view.

Nor was he above giving tradition a nasty twist. A selection of his writings confirms this all too amply.[25] Thus the *brahmacharya* doctrine he espoused indicated that "men and women should refrain from carnal knowledge of each other . . . they should not touch each other with a carnal thought, they should not even think of it in their dreams" (p. 159). Likewise the linchpin of his vision of "Hindu–Mohammedan unity" in a united subcontinent was social distancing: "If we make ourselves believe that Hindus and Mohammedans cannot be one unless they interdine or intermarry, we would be creating an artificial barrier it would be impossible to remove" (p 166). Let us consider, too, his averration that caste "has saved Hinduism from disintegration. I consider the four divisions alone to be fundamental and essential. The innumerable sub-castes are sometimes a convenience, often a hindrance" (p. 171). Was not such reasoning discordant, logically and psychologically, with his outright condemnation of untouchability? After all, sub-caste diversity may be one way to make inequality less stark.

Then again, one may sympathize with Gandhi's feeling that "the rise of the cities like Calcutta and Bombay is a reason for sorrow rather than congratulation" (p. 332). Yet this hardly commits one to his further opinion that "India's salvation lies in unlearning what she learnt during the past fifty years. The railways, telegraphs, hospitals, lawyers, doctors and such like will have to go".[26] One is told that a literal translation of *satyagraha* is "incessant quest for truth". Yet the fact surely was that, while Gandhi could be overtly objective in his pursuit of truth up to a certain point, there were higher levels to which he pursued neither truth nor much else. Hence the philosophic incongruences.

Arguably, his most fundamental shortcoming in this regard was a lack of metaphysics, not least such as he might have set within a cosmic framework. His failure to repair this omission will have been due in part to the demands made on his time and energy over many years as mobilizer of the Hindu masses. All the same, he seems not to have had the instinctual rapport with the starry heavens Pandit Nehru evinced. On the other hand, he did comment in 1927 that "Churches, mosques and temples, which cover so much hypocrisy and humbug and shut the poorest out, seem but a mockery of God and His worship under the vast blue canopy inviting every one of us to real worship . . . ".[27]

Eternal India?

An imperative for this century is confluence between the great religious obediences but also between them and those many of us who subscribe to no established faith but who would lay claim to a broad ethical concern and spiritual aspiration. In his angry frustration, V. S. Naipaul portrays his ancestral home as part of the problem, not of the solution: "India will at the end be face-to-face with its own emptiness, the inadequacy of an old civilization because it is all men have but which no longer answers their needs."[28] Maybe the remedy rests in part with outside observers. Thus far, however, a synoptic overview is rarely achieved by them. One moment one is told how spiritual the Republic of India still is.[29] The next one reads of what may now be its great material prospects.[30] Apprehension of an AIDS epidemic also figures. Little attempt is ever made to conflate these several lines of enquiry. Is it not time?

15

JAPANESE TOGETHERNESS

NEAR THE OLD CAPITAL CITY OF XI'AN IN WEST CHINA is the Neolithic village of Banpo. Skilfully excavated and presented, it proclaims the continuity across the millennia of "we Chinese". This heritage is similarly betokened by the imperial geophysical records, mainly meteorological but also astronomic. Nowhere has Halley's comet been recorded as consistently for so long, the first surviving record being from 613 BC. Moreover, by 1100 BC, a lunar-solar calendar is known to have been in use. From 104 BC, imperial calendars, lunar and solar, were regularly promulgated. Extant, too, in the Han (202 BC – AD 220) was the concept of an infinite universe, with the stars floating in empty Space.[1] The constant though irregular efflux of sub-nuclear particles we know as the 'solar wind' was picked up in AD 365, mainly through how cometary tails slew away from the Sun.[2] A gravitational perturbation during the AD 837 Halley fly-past is discernible from the surviving record.[3]

The interchange of Buddhist scholars between India and China was by then well established. Several Indian astronomers and mathematicians were appointed to high positions in the Astronomical Bureau during the resplendent Tang dynasty (AD 618–907). One became President of the Board of Astronomy and hence masterminded a famous astronomy compendium. Links with Islamic central Asia were likewise important.

Astronomical science evolved more slowly in Japan; and did so very much as a derivative from China and Korea. The first astronomy text known to have come from the mainland arrived in 602. Thenceforward, Chinese astronomy and astrology overlaid previous received wisdom. All else apart, traditional Japanese cosmology had never proffered a commanding overview, metaphysical or aesthetic. Granted, the principal female Shintō deity, Amaterasu had been linked with the Sun within a complex creation myth. Granted, too, Buddhist influence would soon reinforce strongly a disposition to reverential Moon-viewing at festival times. Moreover, the depiction of steadily falling snow can be traced back to eighth-century chroniclers. Since the "Little Ice Age", which troughed out *c.* 1700 in Japan as elsewhere, it has figured prominently in paintings and prints.

In the decades after 1549 (the year they began so successfully to spread the Christian faith), Jesuit priests in Japan also propounded to effect a Ptolemaic view of the cosmos.[4] None the less, an awareness of the new Copernican astronomy would be slowly spreading towards the end of what had proved to be two centuries of national seclusion, *sakoku*, embarked upon in 1639 hard upon an uprising of Christian converts. Never mind that Japanese contact with the West almost throughout *sakoku* was solely through the Dutch factory on Deshima, a consolidated refuse tip out in Nagasaki Bay.

The Chinese community, too, was confined to a Nagasaki enclave. However, attitudes towards it were rather less constrictive, a circumstance which facilitated the introduction from China from 1654 of the Obaku sect of Zen. Over the next century, Obaku established 1000 temples. Afterwards it went into decline through assimilation.[5]

In general, the criteria for deciding exactly what *rangaku* (Dutch learning) might be admitted were narrowly utilitarian, a basic imperial concern throughout being not to undermine the national self-image with foreign influences, especially Christianity. Even so, astronomy eventually came more to the fore than a narrow definition of utility would have required. Tokugawa Yoshimune (ruled 1716–51) was an ambitious pragmatist. He made the Edo shogunate more serious-minded and authoritative. Part and parcel of this thrust was his lifting, in 1720, the ban on importing foreign books on astronomy as well as ones on medicine. Then in 1744, the country's first astronomical observatory was founded.[6]

Over the next half century, first Copernican cosmology and then Newtonian mechanics were introduced. No doubt their subsequent progress was assisted by the indigenous ethos being "neither anthropocentric nor geocentric".[7] The local gods had been too strong to allow of either. They were so because of localized natural threats. Take earthquakes. At present over 1000 seismographic stations report to the Japan Meteorological Agency; and identify 80,000 earthquake foci a year. Volcanoes, typhoons and monsoonal floods also impact acutely yet erratically.

A year after Commodore Perry's "black ships" arrived in Edo bay (alias Tokyo bay) in 1853, *sakoku* was formally renounced, a *volte face* consolidated with the Meiji restoration in 1868. The adoption in 1859 of the Sun and in 1868 of a Sun-like chrysanthemum, as national and imperial motifs respectively, betokened a concern to preserve identity within modernity. For over 2000 years, Japan had recurrently adapted from a rather agnostic Chinese civilization the precepts of religious eclecticism and pluralism. Periodically this had engendered outright conflict though more often it had led to constructive synthesis.

Shintō

The most authentically indigenous of Japan's salient religious traditions, Shintō emerged out of an ambience of *kami* (nature gods) effected by the confluence of two shamanic tendencies, the one from Altaic Siberia and the other from the central Pacific. The emergence occurred, from the third century BC, in reaction to immigrant waves from a troubled continent.

Oral folk tales of ancient derivation are recorded in *Kojiki*, the earliest known chronicle of the Japanese language. Its original compilation is dated now at 712.[8] It proffers *inter alia* a version of Creation built around Amaterasu. The initial stages are recounted thus: "Now when chaos had begun to condense but force and form were not yet manifest; and there was nought named, nought done who could know its shape? Nevertheless Heaven and Earth first parted and the Three Deities performed the commencement of Creation; the Passive and Active Essences then developed; and the two Spirits then became the ancestors of all things." The dualism here expressed was evident, too, in the notion of an afterlife, one maybe involving a heaven populated by deities and an underworld where the dead reside. This religious consolidation took place at a time when the balance between population and resources was turning much more favourable, thanks to the introduction from the continent of rice for wetland cultivation. Shintō duly assumed a distinctly positive mien.

There are in modern Japan over 100,000 listed Shintō shrines. About a third are dedicated to *Inari-sama*, the *kami* of the rice harvest. Many of the rest have non-human identities: a mountain, volcano, waterfall, mammal, bird or whatever. At festival time, a *kami* may descend from heaven or (in still more ancient tradition) advance from over the horizon. In fact, Shintō is suffused with a sense of the ubiquity of Nature. And though the Japanese people have constantly been exposed to its destructive aspects, the great majority of their *kami* have been benign.

Shintō says nothing about early fathers having revelation experiences. Nor does it about received dogmas or sacred texts. Its broad appeal has derived not from systematic dogma but from ceremonial rites. An authoritative commentary of AD 833 insists that "all gods in the Heavens and on the Earth shall be revered according to the traditional rituals".[9] Prominent among the Shintō rites are those concerned with personal purification. Often these relate to starfield immutability.

Confucianism

Confucius (*c.* 551–479 BC) is the first Chinese luminary individually identifiable as an influence on Japanese philosophic development. His overriding ambition for China, mainly expressed in Socratic discourse with his disciples, was to see the restoration of a unified imperial regime, one sufficiently talented and assured to promote a culture not unlike Renaissance humanism. Throughout he laid great stress on the psychological factor. Much the same applies to Sun Tzu, a close contemporary of Confucius and the author of *The Art of War*, a text still accepted as being the first known scholarly analysis of military strategy. Witness its averration that "to take intact a regiment or company or squad is better than to destroy it".[10]

Since 1945, Sun Tzu has undoubtedly been cited more than any other classic strategist, Clausewitz probably excepted. Likewise since 1980 or thereabouts, Confucius has been typecast as a primary source of the "Asian values" that have well sustained economic progress. However, that half-received wisdom is sharply at variance with the judgement of Max Weber (1864–1920). This pioneering political sociologist blamed China's continuing material backwardness on the rigidities engendered by the integration of the Confucian literati into the imperial court, the past two thousand years and more.[11]

During the Han dynasty, Confucianism became firmly established as a primary element in China's philosophic outlook. As part and parcel, this credo was invested with a metaphysical dimension rooted in Taoism – an indigenous philosophy strongly emergent from the fourth century BC. This was of course, *yinyangdao,* the interpretation ascribing all flux to the continual alternation of *yin* and *yang*. Essences emanating from *taiji* – the innermost cosmic reality – control the five cardinal elements: fire, earth, gold, wood and water. They also mediate the afterlife, a declared belief in which had waxed strong (with imperial encouragement) since the Qin, the unifying dynasty preceeding the Han. The entombed terracotta army (see below) is a legacy thereof.

Tradition firmly has it that Confucianism arrived in Japan via Korea in AD 404. Alongside Chinese legalism and Buddhism, it contributed to the

Seventeen Article Constitution promulgated by Prince Shōtoku in 604. The Confucian input is especially evident in moral exhortation, a quintessential Confucian propensity.

During the Sung dynasty in China (966–1279), a neo-Confucian reaction took place against the alien credo of Buddhism. None the less, the precept of a cosmological setting for everything continued to be endorsed. Within the neo-Confucian school, which was led by Zhu Xi, a cosmic dualism was propagated: the *li* (Japanese ri) or conceptual bases versus the *qi* (Japanese *ki*), the fundamental contents out of which the material world originally condensed. Applied to the human condition, *li* was usually seen as positive and *qi* as negative.

Within the Zen community in Japan (see below), there was a certain sense of affinity with neo-Confucianism, not least apropos a commitment to law and order. During the Kamakura period (1185–1333), some 100 Zen monks either visited China or had hailed from there to start with. During the Muromachi (1533–1568), Zhu Xi was extensively studied by the Zen monks of the Gozan temples.

He was probably less agnostic than Confucius had been, more ready to reject outright the idea of a personal deity. He openly scorned talk of sinking one's individuality in the universe as opposed to seeking its full development. After his death in 1200, disputes broke out among Zhu Xi's followers (conspicuously in China but also in Japan) as to the nature of *li*. Regarding Japan, the view has been advanced that an on-going debate on this score engendered a scientific spirit which, come the eighteenth century, favoured recourse to *rangaku*.[12]

The Character of Japanese Buddhism

An especially striking aspect of Japanese Buddhism has always been the extent to which it achieves accommodation with other faiths but, above all, with Shintō. Witness the millions who get married in Shintō shrines but will eventually be buried by Buddhist temple priests. Likewise, combined services may occur in either temples or shrines. No matter that individual Buddhist sects have exclusivist inclinations, emphasizing as they do master–disciple lineage.

So whereas continental (and especially South Asian) Buddhism has tended to be considerably reclusive, Japanese has been less so. Its disposition has been to get involved in the everyday. This in turn has often led to broad acceptance of the *status quo* coupled with a certain disregard for analytic enquiry.

Primary Buddhist Sects

Buddhism first entered Japan via Korea in the mid-sixth century. In the eighth, it officially became the imperial religion. By the thirteenth, the five major sects active today were established in-country. The two that interactively assumed prominence in pre-medieval times were Tendai and Shingon. The initial introduction from China of the former was effected through a working alliance between a restless monk, Saicho (767–822), and Japan's Emperor Kanmu (reigned 781–806), a common adversary being the temple establishment in Nara. As part of an official mission, Saicho visited China in 804–5. There the Tendai school impressed him most. On his return, he founded what would eventually grow to be a community of 3000 Tendai temples. The first was established at Enryakuji on the slopes of Mount Hiei, north-east of Kyoto. The site was auspicious in that reputedly it protected the city a bit from predatory devils, not least those borne in on wintertime's bitter north-easterlies.

Tendai attached singular importance to the *Lotus Sutra,* the most widely revered of the Buddhist sacred texts. Furthermore, as moulded in Japan by Saicho, it had affiliations with Esoteric Buddhism (see below). But in some respects, too, Saicho and his disciples were committed to a more general outreach. For one thing, they were insistent that any human being could attain buddhahood. Also there was a place in their scheme of things for all sacred texts, Buddhist and otherwise. Above all, there was a philosophic emphasis on the holistic unity of the universe, a reality more profound than anything our workaday experience reveals. As a celebrated maxim put it, "One thought is the 3000 spheres (i.e. *the whole universe*); and the 3000 spheres are but one thought." To look at this rendering another way, one can say that inner truth is absolute. Everything derivative from it is phenomenal and transient.

The monks of Enryakuji were an unruly lot. Indeed, they could be disposed to extract favours by pretty literally throwing their weight around. Even so, this Tendai movement did something to bring Buddhism to the common people. More particularly, it paved the way for the rise of Shingon. The founder thereof, an itinerant monk called Kūkai (774–835), went to China in 804 on the same trip as Saicho. He found a remarkably universalist atmosphere prevailing in the capital city, Xi'an, as the broad-minded Tang dynasty encouraged not a few of the 10,000 non-Chinese residents to take up positions of influence.[13]

He stayed on an extra year, becoming the while much involved in Esoteric Buddhism. On Kūkai's return, Saichō humbly received from him *kanjō*, the esoteric ceremony for initiation in advanced mysteries. Yet despite this and other signal overtures, Saicho died without resolving what role Esoteric Buddhism might have within Tendai.

After his return, Kūkai had established the Shingon sect. Being set within the esoteric tradition, it developed very determinedly the sense of a Cosmic Buddha, *vairocana* (or, in Japanese, *dainichi* – "the great one"). The supreme spiritual experience is defined as "breathing" in harmony with the universe, with *vairocana.* This state of grace might be approached via the concurrent application of three mysteries: ritual finger signs, sacramental recitation and deep contemplation. *Vairocana* could be said to epitomize an Oriental sense of the sublime aspect of nothingness. For it is numerically zero yet contains all things that exist in the "universe; and to unify oneself, just as one is, with that zero is to fulfil the ultimate aim of Esoteric Buddhism, the attainment of Enlightenment in this very existence".[14] In other words, buddhahood can thus be realized within an individual lifetime. Since Kūkai's demise, Shingon has experienced little further development philosophically. More generally, however, an important departure in medieval times was to be the acceptance of *Amaterasu Ōmikami* (the Shintō Sun Goddess referred to above) as an aspect of the Cosmic Buddha.

Kūkai exhibited much energy and virtuosity. He compiled what now rates as Japan's oldest extant dictionary as well as over fifty religious works. He was sculptor, painter and calligrapher. He dammed lakes. He could be a severely ascetic pilgrim. His Esoteric Buddhism with its yogic discipline, rich ritual and complex iconography appealed to the Heian court and to many beyond. Aided by some myth-making, he became a legend in his lifetime and has considerably remained so. He has gone down in history as "*the* monk who went to China".

Some 35 metres high, the Daibutsu at Kamakura rates as the second largest Buddhist statue in Japan. It was completed *c.* 1250 in honour of Amida, the mythic Buddha of Light who presides over the Western Paradise, the Pure Land. Though present in Japan by AD 600, Amidaism effectively stayed low profile until the Heian period (794–1185) when the nobility waxed keen to anticipate Pure Land bliss through calligraphy and painting. One priest, Kūya (903–72), danced through the streets, singing Amida's praises. He introduced, not least to ordinary people, the concept of *nenbutsu.* What this comprised most basically was an invocation, "I take my refuge in the Buddha Amida." Soon another scholar monk, Genshin (942–1017), was comparing this with the, to him, ineluctable alternative to the Pure Land – namely, the torments of Hell.

Eventually Hōnen (1133–1212), a one-time novitiate at Enryakuji, became persuaded that, in the current age of *mappō* (i.e. "the end of law" or spiritual and behavioural decline) formalized meditation and every other accepted route to buddhahood should be abandoned in favour of *nenbutsu*. Unfortunately, his declamations to this effect incurred the forceful opposition of the buddhist establishment. None the less, he launched the Jōdo sect which then spearheaded a considerable extension of Pure Land influence.

So was it not needful, or at any rate helpful, regularly to reaffirm the *nenbutsu* commitment? Eventually Hōnen's disciple, Shinran (1173–1262), sought to resolve matters by averring that Amida proffered salvation to all who had absolved themselves thereby, even if only the once. Over many years, he prominently promoted this precept, so easy to act on, to a diverse underclass (ranging from small peasants to prostitutes) from which establishment Buddhism had all but disengaged. Come the turn of the century, perhaps two million Japanese had avowed *nenbutsu*. Today Jōdo and its breakaway, Jōdo Shinshū remain the largest Buddhist sects in Japan.

Through the late thirteenth century, Japan's crises of faith were probably compounded by the global climatic shift towards weather more erratic and, in general, less clement. The *kamikaze* ("wind of God") storms of 1274 and 1281, which shattered two Mongol-led invasion armadas in turn, were seen by contemporaries as part and parcel of this worsening. That of 1281 came well after the normal typhoon season. By reputedly conjuring up these storms, the Shintō gods revered at Ise Shrine on the southern coast of Honshu had evinced a consummate capacity to turn climate decline to advantage.[15]

Against this background, the Nichiren sect (founded in 1253) by the Tendai monk, Nichiren (1222–82), put down its roots. By 1253, Nichiren himself had lost patience with routinized religion. Instead he was gripped by an acute sense of *mappo*, of impending apocalyptic crisis – religious, social, political . . . Duly he had seen the Mongols as harbingers of divine retribution. Therefore their double débâcle left him severely disoriented. Quite likely it contributed to the rapidity with which the ill-health which had dogged him since 1278 turned terminal in the course of 1282.

His discomfiture can hardly have been eased by credit for the *kamikaze* being taken by those Shintō gods residing at beautiful Ise. Not that triumphalism survived long anywhere. For a good twenty years to come, the Japanese would be preoccupied with a possible Mongol return.

The philosophic base for Nichiren and his followers was simply the *Lotus Sutra*, especially its enunciation of the doctrine of universal eligibility for buddhahood. From that vantage point, Nichiren himself declaimed with fervour (and, alas, venom) against all and sundry. Kūkai had been "the greatest liar in Japan" and the Shingon entourage "traitors". The Hōjō ruling family was condemned for endorsing Pure Land pluralism as was the

Pure Land itself for its preoccupation with the afterlife. Zen was purportedly "a doctrine of fiends and devils", misguided in the stress it laid on paths to enlightenment outside the scriptures (see below). Ahead of the 1281 invasion scare, Nichiren urged the various sects be crushed in the interest of national solidarity. He deemed his own exempt by virtue of his being a reincarnation of the Buddha.

However, many people would agree that quite the most significant of the sects here identified was and is Zen. This is on account of its pervasive influence on Japan's religious landscape; and on her national culture viewed in the round. Yet so is it by dint of the wider world's interest in Zen philosophy as it is understood. For many outsiders, indeed, "Zen" is a portentous manifestation of Buddhism in general. In 1966, doing a tour of South-East Asia as a defence correspondent, I interviewed Lee Kuan Yew, then seven years into his long stint as Singapore's dynamic Prime Minister. One remark he made was that the Americans must understand that in Vietnam they were "up against Zen Buddhism". He was well placed to say as much. For all his Cambridge virtuosity, "Harry Lee", as the British dubbed him, was and is quintessentially Straits Chinese.

The Essence of Zen

But if instead one does define Zen strictly as a sectarian variation within the Buddhist mainstream, one still finds the school to be articulate in Japan, more so than virtually anywhere else. Yet like so many other "things Japanese", it is a cultural import. Most decisive in effecting this transfer was Eisai (1141–1215), an Enryakuji monk who had become disillusioned by the laxity of that community. He visited China in 1168 and again from 1187 till 1191. He is now considered the main progenitor in Japan of the Rinzai school of Zen, the one that works towards the eventual achievement of a sudden revelation experience. His disciple Dogen (1200–1253) is seen as the mainspring of the alternative Sōdō school which seeks gradual revelationary transcendence. This doctrinal difference seems subtle but in China had already been bitterly divisive.

What the Japanese call "Zen" but the Chinese "Chan" is thought to have been introduced in China by the Indian monk, Bodhidharma (d. *c.* 532). However, it only gained a firm lodgement under Huineng (638–713). Quasi-mythic though this surviving persona is, he was undoubtedly charismatic. The forests screening Chan monasteries were said to emit a "thunderous roar" expressive of a "mystical density" which drew in seekers of "the Way" from all over.

Chan advanced within China through interaction with Taoism. There may be a tendency not to appreciate how much the Tao contributed to the syncretic outcome. Taoism bespoke rejection of Confucian notions of progress. Its political and social ideal was "a small state from which the cocks and dogs of a nearby state could be heard but whose people were so content that nobody ever bothered to visit this neighbouring village".[16] Any higher level of material development was felt to result in detachment from Nature, social fissures and strife.

By the same token, Confucian scholarship was adjudged a snare and a delusion. Take Zhuang Zi, a Taoist classic writer who lived *c.* 369–286 BC. Portraying the Tao spirit of Creation as a formless being which makes no distinctions – big or small, life or death . . . – he avers that "The most extensive knowledge does not necessarily know it; reasoning will not make a man wise in it." This sounds very Chan. Witness, too, the terracotta army famously created at Xi'an, *c.* 200 BC. There is surely something Chan-like about the sublime commitment expressed on every face.

With its translation to the Japanese homeland as Zen, Chan's emphasis on meditation was to be strongly reaffirmed. Nor was Eisai the first to play things thus. The eminent monk, Dōshō (629–700), had taught Zen meditation – *zazen* – when back from a sojourn in China. A century or so later, Saichō, too, was to expound several meditative modes. Among them was *jōza sanmai*, sitting upright and cross-legged in the "full lotus" posture for many days on end. This came in due course to be seen as the epitome of *zazen*.

More generally, Eisai was able considerably to claim he was simply restoring lapsed practice, this in the interests of a stable and secure society. He did so in his treatise *Kozen Gokoku Ron* ("the propagation of Zen for the protection of the country"). But this hardly commended him to sectarians preoccupied with ritual and/or recitation. Nevertheless, shogunal and samurai support enabled Zen to thrive. So perhaps did a gradual burgeoning of contacts with China.

The aim of *zazen* is the achievement of a meditative bliss, *satori*, which is free from conceptual or factual lumber and devoid of the delusions of language and ego consciousness. *En route*, the ultimate superfluity of such accoutrements is demonstrated by *kōan*, Rinzai riddles to which there can be no answer in terms of everyday logic. An implicit corollary is that little importance attaches to routinized worship. Celebratory interaction with the world during the approach to *satori* is best maintained by workaday pursuits: "How wondrous this; how mysterious; I carry fuel; I draw water."

Those who assuredly experience *satori* describe a sense of merging with the whole universe; and cosmic themes likewise figure in Zen hymns and poetry as well as many *kōan*. Even so, Zen is not much guided by metaphysics and cosmogony. The wholeness it really comprehends is that of everything immediately around. A most important aspect of this comprehension philosophically is the fundamental unity of what language presents as opposites: action and passivity; attack and defence; subjective and objective . . . Duality implies interaction. Due recognition of this is prospectively of great importance, not least in relation to human conflict situations.

Hakuin

Zen is something of an exception to a tendency for Japanese Buddhist sects not to develop much philosophically after their formative years. Nowadays the most remarked of its latter-day prophets is Ekaku Hakuin (1686–1769). From the age of 24, and after a succession of acute identity crises, he applied himself – as a monk at Shoinji temple – to the reformation of Rinzai Zen, the aim being to preclude its becoming too austere. Adopt a gentle mien. Allow one's disciples some flexibility regarding the paths they take to buddhahood. Meantime, have them examine deeply their own personalities. Continue to blend meditation with creative labour. Continually break the gridlock of routinized thought by means of *kōan*. The most famous of Hakuin's many contributions to this singular genre was "What is the sound of one hand clapping?" He was a prolific writer; and his essays still read beautifully in English translation. As he grew older, he sought to express Zen enlightenment through quality brushwork, many painted scrolls and much fine calligraphy being the result.

One of the early crises alluded to above centred round Hakuin's then believing he had lately attained *satori*. Later on he described to his pupils the frame of mind one may be in as one draws close to that consummation: "The four corners of the Earth and its ten directions, the height and depth of the universe, will be to you the great cave in which you are performing your meditation – they will be in very truth, the substance of your real self."[17]

Hakuin may not have been a great philosopher in the analytic sense. But he was certainly an inspirational teacher. Further, he will have been one of the influences that has inspired and sweetened the application of Zen to workaday pursuits. A most striking aspect of contemporary Japan is the extent to which artistic originality blends so successfully with functionalism in all forms of design from the layout of a snack bar to the crafting of a lunch box. The maxim "beauty is usability" is still very heartfelt.

Nor is it only in such peaceable pursuits that the Zen spirit is made manifest. Although to much Western understanding, Buddhism as a whole is pretty much a pacifist creed, two commentators, in particular, have famously averred that it suffuses the martial arts. In 1924, Eugene Herrigel (1884–1955) went from Heidelberg out to Tohoku Imperial University where he taught philosophy and classics for five years. He took up archery as an educative pastime. He found that shooting arrows was seen as a way to a mystic union with divinity, the attainment of buddhahood. He cited an archery master on the need "to become mindless, to wait until the bow leaves the arrow of its own accord". Yet at the same time, Herrigel himself came to see the selflessness thus required as fostered by cultural traits which pre-dated the arrival of Buddhism.[18] Maybe there is again a parallel with the terracotta army in Xi'an, one and all serene of countenance.

Through much of the last century, Daisetz Suzuki was familiar in the West as an all-round interpreter of Japanese Buddhism, especially Zen. He wrote about swordsmanship in a similar though even stronger vein.[19] The swordsman has to be above himself, above a "dualistic comprehension of the situation" (p. 97). Indeed, the swordsman must be "a perfect man in the Taoist sense: he must be above life and death as the Buddhist philosopher is above *nirvāna* and *samsāra*" (p. 142) – i.e. beyond transcendental bliss as well as cycles of incarnation.

Meanwhile, the sword itself has two functions: "The one relates to the spirit of patriotism or sometimes militarism while the other has a religious connotation of loyalty and self-sacrifice . . . " (p. 91). It was needful for the latter to control the former. Otherwise, Suzuki apprehends, military conduct in war will lapse into brutishness or cowardice or both. We all recall how the Japanese armed forces in World War Two could all too often be chillingly brutish, albeit never cowardly. Yet across the centuries, the samurai/bushido code here alluded to has been upheld better overall than are most codes of military conduct, given the exigencies and agonies of mortal conflict.

Besides, the prime input into the 1937–45 syndrome was the State Shintō officially established in 1871 as a national faith standing over and above Temple Shintō and Sect Shintō and, of course, all other obediences. At the outset, the Meiji government had moved to separate Shintō and Buddhism, despite their having been considerably syncretic since early in the Heian period (794–1185). The intention was to underpin the divine status of the Emperor, this in order to temper and modulate the inflow of Western knowledge and skills the Meiji basically sought to promote. Under the Dual Shintō (Ryōbu Shintō) system, Buddhist priests had exercised control over many Shintō shrines. But a March 1868 decree had required them to relinquish this authority. Also all Buddhist images had to be removed from the shrines.

Unfortunately, this gave rise in various localities to *Haibutsu Kishaku*, an "abolish the Buddhist statues and monks" upsurge that often involved district officials and led to the destruction of many temples, statues and movables. However, through 1871 this reaction largely died down. Meantime, State Shintō was cast in a mode more ceremonious and less authoritarian than might have been the case.

A religious settlement, if one may call it that, was effectively thrashed out by the Meiji intelligentsia and officialdom. It had implications for wider aspects of life. The gist was that the individual was entitled to personal beliefs but the regime had the prerogative of curbing public dissent in so far as this might appear needful in the interests of orderly progress. Everything was supposed to rest on due constitutionalism; and this was the safeguard shattered by xenophobic military zealots as they pitched Japan into the "dark valley" of fascism from 1931.[20]

In the wake of the 1945 surrender, Emperor Hirohito moved swiftly to renounce his divine descent from Amaterasu. At the time, the homeland destruction and desolation were such that the Japanese were widely expected to settle long-term for living standards near the Afro-Asian norm. Yet, spurred on by Douglas MacArthur, the country embarked on an economic boom which, until the oil shock of 1973, averaged close to 10 per cent a year. Such a departure had been anticipated in an essay published not long before her death by the ranking American anthropologist, Ruth Benedict.[21] She foresaw that, when circumstances allowed, the Japanese people would again seek their "place of honor among the nations", though this time not through military adventurism but via peaceful yet purposive economic growth.

The term "essay" has been used advisedly. For one criticism leveled against Ms Benedict, notably in a UNESCO study,[22] concerned her disregard of numerical techniques. Prepared for this, she had insisted that resolution of seeming contradictions in underlying Japanese attitudes would never be achieved via opinion polls geared to Occidental parameters and semantics.

Since the oil shock, Japan's economic progress has been rendered less steady by institutional shortcomings but also perhaps by cultural change. Once again, she stands in need of a reorientation of her national values and of the religious and metaphysical precepts behind them. A void opened up in 1945 but was to be effectively enveloped by a sense of economic triumph. However, this sense was much weakened by the international oil shock of 1973 and then by the bursting in the early nineties of a Japanese "economic bubble".

A major theme in any revaluation of her national ethos will perforce be

Nature. The naturalistic proclivities of Shintō were remarked above. About Zen, Suzuki had this to say: "Zen wants us to meet Nature as a friendly well-meaning agent whose inner being is thoroughly like our own, always ready to work in accord with our legitimate aspirations."[23] Conversely, it has been argued in some depth that the Japanese "praise of Nature" is too ritualized and institutionalized, too obviously a metaphor for other yearnings.[24] Maybe Suzuki unwittingly conceded as much when he spoke of Nature as the Japanese experience it as being friendly and well-intentioned, with an inner self akin to our own. So perhaps did he when, at the beginning of the same essay, he wondered if his compatriots' "love of Nature" might not owe much to the location of Fuji, quite the highest mountain, in the very middle of the main island.[25] It rather makes one think of a hierarchy of symbols.

A tolerably recent exemplar of a truly authentic naturalism coupled with a wider sensibility was Miyazawa Kenji (1866–1933). From a family both pious and prosperous, he applied a training in agricultural science to the alleviation of the plight of the rural poor. He was a writer of distinction, poetry and children's stories being his genre. He moved from Jōdo Shinshū to Nichiren but waxed rich in multi-faith linkages. He was renowned for his love of skyscapes, not least of our galactic spiral which is perceived in Japan as the "Silver River" and in the West as the "Milky Way".[26] He was conscious of the need for Religion and Natural Science to effect concordance, but did not himself contribute anything to this over and beyond a steadfast celebration of both these quests for understanding.

PART FIVE

To Here from Eternity

OVERVIEW

To Here from Eternity

As national communities are brought into closer contact by global shrinkage, they will have to become either more open to external influences or else a lot more closed to them. Much may depend on the response of the great religious obediences. Positive interaction between them should encourage openness especially if linked to a deeper engagement with Science and with Free Thought. It should also be conducive to a general revival of spiritual values as well as to a sound understanding of the historical past. Otherwise the world could be taken over by narrowly rational consumerism and thoroughly irrational cultishness. The scene could thus be set for an authoritarian backlash, political and/or religious.

Constructive inter-faith togetherness needs gain a better airing for a variety of philosophic issues. Prominent among them might be how, within the context of a world survival agenda, the exercise of armed force may relate to social and ecological goals. Underpinning everything, however, should surely be informed perspectives on cosmology and metaphysics. What does humankind count for in a cosmos comprising a hundred billion billion stars; and which is probably but one in an infinitude of universes? What part is played within ours by that phenomenon of exquisite complexity we know as Life? What part is by the elusive quality we know as Consciousness? Not to address these questions would be to disregard Occam's Razor, the neo-Aristotelian precept most basic to scientific enquiry, in our cosmos at least.

16

YEARNINGS FOR BELIEF

But already this much seems clear. Customary belief systems derive from cognitive capacities which have developed in various parts of the brain mainly in order to handle mundane information. Religion-related imagery will therefore recur as between different cultures. The souls of the dead lurking nearby would widely be recalled. So would witches and evil eyes.[1] The confirmation of group identity by means of demanding rituals is another standard trait. Religious movements which insist on them may long thrive better than those whose approach is more relaxed.[2]

Self Concern

Yearnings to belong cannot just be ignored. They reveal too much about our spiritual anguish. Yet perhaps the least sure way to accommodate them is the one which entrances the modern world, consumerism *à l'outrance* with the Self displacing the Soul. Institutional religion is swept along. Over the rainbow is Beverly Hills.

However, those of a more philosophic bent can likewise be overly concerned to find self-oriented ways through. This prospect exercised Martin Buber (1878–1965), an Austrian Jewish theologian and *inter alia* liberal Zionist keen for rapport with the Palestinians. Writing in 1932, he well apprehended the God concept becoming but a private, as opposed to a communal, construct. He cited Benedict Spinoza as grasping the "anti-anthropomorphic" imperative of understanding God to be an infinitely multifarious presence, but one still able to relate to us "finite, natural and spiritual" individuals. For anyone predisposed towards agnosticism or atheism, this convolution might compound their reservations.[3] Buber himself was uneasy.

Grounds for his apprehension can be seen in the privatization trend much remarked *vis-à-vis* contemporary Christianity as observed in the USA. Thus even "within the churches, the variant and subjective interpretations of belief need be neither challenged nor exposed: the private individual can make his claims and need not divulge his sources". A religious fellowship thereby "becomes a dispersed community of individuals".[4] So what happens next? Either decay or else reassertion through inapposite collective aggressiveness? Or might the churches turn the current situation to advantage by expressly accepting it, dialoguing with individuals in a quest for enlightenment all round?

Cultishness

Also to reckon with is the possibility that, within the vacuo of belief left by the decline of traditional faiths, more cults (usually irrational and tendentious) will arise. Sometimes one such may derive from a metaphysical progression which ends up half-baked yet which started with more going for it than the badinage currently informing the likes of the Falun Gong and Raelians. In which connection, one cannot endorse unreservedly the old Leftist nostrum (embossed by H. G. Wells) that the future will hinge on "a race between Education and Catastrophe". Education must lead on to Fulfilment. The latter may often be the harder to realize. Yet reason and reasonableness may critically depend on it.

The life and legacy of Arthur Schopenhauer (1788–1860) affords a stark demonstration of how tragically wrong things can go in a milieu crowded out by frustrated aspirations, in this instance those of the German people for full nationhood. Himself contemptuous both of religious hierarchies and of nascent democracy, this Danzigger German was impelled by war and personal loneliness into abject pessimism, a state of mind he saw as pretty much in line with the Christian outlook: "Whatever torch we may kindle and whatever space it may light, our horizon will always remain bounded by profound night."[5] The supreme truth, he insisted, was the will to survive. The ultimate hope was nirvana via Eastern mysticism.

Schopenhauer deftly pinpointed sundry philosophic truths. Thus "the intellect is not originally intended to instruct us concerning the nature of things but only to show us their relations with reference to our will".[6] Then again, and most presciently, "Physics cannot stand on its own feet but requires a metaphysic to lean upon, whatever airs it may give itself towards the latter."[7] The tragedy alluded to above is derivative. F. W. Nietzsche (1844–1900) built on Schopenhauer's preoccupation with will. Next, Nietzsche himself was eclectically exploited by Nazism.

An oblique encounter with modern cultish tendencies came out of a journalistic visit this author paid to South Vietnam in April 1966. The course of the war then hung in the balance. Among the swing factors were several social assemblages, each locally possessed of a fair measure of cohesion and even combat power. Among them was the Cao Dai ("High Tower"), a polyglot religious movement dating from 1926. It was centred on Tay Ninh, a holy city in the Mekong basin 10 to 15 miles from the jungly border with a fragilely neutral Cambodia. This border, American officials in Saigon believed, was already in use by Hanoi for armed infiltration. It almost certainly was.

The Cao Dai's emphasis on spiritism as well as anti-colonialism will have given it primal appeal. So no doubt will elaborate rituals (along with séances), mainly derived from Daoism. Its social ethic was characterized as Confucian; and its commitment to karma and rebirth, Buddhist. Among its many intermediate gods had already been Joan of Arc, Winston Churchill, Mahatma Gandhi, Vladimir Lenin and William Shakespeare – hardly syncretism of a mature and creative kind.

After the Communist victory in 1975, the Cao Dai was suppressed throughout Vietnam, its extensive land holdings and properties being confiscated and many priests arrested. Headquartered today in Canada, the sect reportedly lays claim still to six million adherents.[8] Even if this claim has some substance, the commitment of those concerned is bound to be contingent. The Cao Dai's true significance lies in its having typified the fractured cultural geography of South Vietnam as an East Asian frontier zone. At this juncture, one could ask how far the internet might encourage such splintering even in societies not rendered prone by their historical geography.

Political Religions

As the above reference to Nazism implies, the ultimate danger is that a given country will slide out of chaos into the grip of a cult which is burgeoning into a monolithic political religion – Red or Blue or hybrid. Some attention is given to the likely mechanics in Chapter 7. So perhaps it is best for now to wax anecdotal again, drawing on impressions of Stalinist Communism in practice.

Mine derive from a tortuous trip made in August 1951 to a massive

World Peace Congress being staged in East Berlin. Then about to enter University College London, I had contrived to enlist with an official observer group being sent by Britain's National Union of Students. An illicit, insistent and ultimately forlorn endeavour by the US Army of Occupation in Austria to block the progress eastwards to Berlin of ourselves and some 70 French students occasioned one or two quite vicious skirmishes, followed by a bizarre sequence of *mittel European* spy clichés interlaced with Ealing Studios/B movie farces. Immediately afterwards, this episode was written up from what effectively was our consensual point of view.[9] Lately it has been outlined in *Global Instability and Strategic Crisis* (pp. xii–xiii).

So for now let me concentrate on the ten days or so spent the other side of the Iron Curtain. The said peace congress was a manifestation of holistic intolerance many orders of magnitude grander and truer to form than American excesses in Austria. Its forceful imagery, sound and sight, resounded through one's consciousness for months if not years. The signature tune, heavily syncopated and endlessly repeated, was a celebration of menace entitled *Freedom's Song*. A short extract from its lyric says the lot:

> All who cherish the vision, make the final decision.
> Down with their lying, end useless dying.
> Live for a happy world.

Then came the refrain, "Everywhere the Youth is singing Freedom's Song . . . ".

In short, a Marxist-Leninist dictatorship will decide the Earth's future; and responsibility for the current situation lay entirely with a mendacious West. The one Berlin peace proposal which sticks in my mind was that all US forces should leave Europe forthwith. The possibility that the Soviet army might care to pull back a few kilometres received no airing. Previously I have said there were but trivial opportunities to put alternative viewpoints. On reflection, I feel even this rendering accords the event a pluralism it never possessed.

Lest anyone should find this account extravagant, allow me to recount an ancillary experience. One by-product of the confrontation in Austria referred to above was our all spending a day on the station of a Red Army infantry company. Again, the memory mix is rich. The company commander, a bemedalled veteran of the Great Patriotic War, was a benignly capacious personality, standing much in contrast with his young conscripts suffused with callow innocence. Their fare was plain. Their virtuosity on the assault course was remarkable. The dirtiness of their latrines was unbelievable.

However, the point for the moment is the walls of their barrack rooms. These were covered comprehensively by cartoons depicting the West in the most ghoulish terms. A recurring stereotype was your top-hatted Wall Street banker, his bloodied hands clutching atomic bombs. His visage would usually be anti-Semitic as per the Shylock and Fagin mode.

Looking back on those festooned walls reminds me of my one and only visit to the USSR itself, this on a Fabian Society school laid on in the *détente* summer of 1963. In Moscow we visited two large museums, each one occupying fully a palatial building one imagines once to have been the city residence of a post-boyar grand prince. My recollection is that one site encompassed the History of the Revolution and the other the History of the Red Army.

At all events, the evidence of our own eyes seemed to confirm what we were told on the side. Neither display included any depiction of Leon Trotsky, Soviet commissar of war from 1918 to 1925. Then again, the one carried no portrayal of Josif Stalin while the other confined him unrecognizably to the back row of a small group photograph. Let us recall that this devastation of the "whole truth" principle came nearly a decade into the "liberal" post-Stalin Khruschev era.

For political religionists, Truth is not ascertained through observation and experience so much as revealed by secular canons. During the 1930s, indeed, this enhanced the appeal of Communism to progressives rendered impossibly lonely, in a spiritual or cosmic sense, by the acuteness of their despair over the world scene. Such temptation was starkly discordant with the scruples most of them otherwise felt about freedom, truth and human dignity.

This fateful contradiction within a godless religiosity was reviewed in 1948–9 by the late Richard Crossman, a Labour MP and former Oxford don who was then emerging as something of an ideological lighthouse on the mainstream British Left. He discerned some hope of avoiding a recurrence of this dilemma in the supposition that "Western democracy today is not so callow or so materialist as it was in that dreary armistice between the wars".[10] Could he still have said that today?

Mainly Complaisant?

Acute doubts were entertained in Chapter 12 as to whether Evolutionary Humanism can proffer a viable alternative to revolutionary political religions. What then of the traditional religious obediences? Straightaway one has to say that their record on this score in the century past was weak overall.

They often proved unable to withstand the surges of new-found dogma, often because they never tried hard enough to understand. Nazism was never a phenomenon one could afford to ignore.

The case of Imperial Japan is likewise salutary. The path to militant nationalism can be traced back to when, soon after the Meiji restoration of 1868, the new regime announced its *saisei itchi* goal, "the unity of religious ritual and government administration". The next several decades were to witness a whole succession of measures which locked Shintō into the state and separated it clearly from Buddhism. A culmination came in 1932 with a Ministry of Education rescript to the effect that Shintō shrines were non-sectarian institutions, their prime purpose being to inculcate patriotic duty.

During World War Two, the leaders of Sōka Gakkai (the derivative of the Nichiren Shōshū school of Buddhism discussed in Chapter 15) were imprisoned for criticizing State Shintō. Indeed, the sect's founder, Makiguchi Tsuneaburō (1871–1944), died in captivity. Regardless of this, however, State Shintō continued to motivate the imperial forces. Their martial zeal was almost undiminished until the nuclear bombs were dropped.

The pliant record of Russian Orthodoxy *vis-à-vis* Stalin is also considered, in Chapter 12. What ought next to receive some attention is the on-going disputation about the other Churches' relations with fascism. May one start with the proposition that one reason why Mussolini's Italy never galvanized its people the way Hitler's Germany did was that the Vatican, along with the monarchy, acted as alternative foci of loyalty to the Italian dictator. No matter that personality-wise Mussolini was much the more solid of the two men.

Even so, many of us still find it hard to understand why the Vatican failed to condemn root-and-branch the Nazi extermination camps, once their existence and purpose had been confirmed. This satanic archipelago broke a raft of pre-existing norms, even fascist ones. It contributed nothing to any licit or rational war stratagems or aims. An appropriate declaration at Christmas 1943, say, might have weakened much further Hitler's crumbling New Order.

Moral Relativism

Beyond this, however, it is none too obvious how far ecclesiastical declamations should have extended. One quandary was bound to be even-handedness, not least as regards Hitler versus Stalin. Another would be the irreducible truth that, whenever war is waged, even the highest morality involves the choosing of lesser evils. Some very senior clerics in

Britain (among them the Archbishop of Canterbury) were persuaded that the area bombing of German cities was illegal under international law as well as being damnably cruel and vandalic. They were undeniably right on the second point and hard to gainsay on the first. Nevertheless, not a few of us would still apprehend that, without this strategic campaign through 1942–4,[11] Nazi Germany could well have survived by contriving an irreparable split between the West and Moscow. Resolving such an impasse later would have posed awesome problems, ethical as well as pragmatic, with or without nuclear bombs to hand.

Attempts by Christian denominations corporately to identify and apply ethical absolutes were proved similarly forlorn during the Cold War. Take *The Church and the Bomb*, a report delivered in 1982 to the General Synod of the Church of England by a Working Party headed by the Bishop of Salisbury.[12] Starting from an unqualified rejection of nuclear "first use", the authors understood the Soviet leaders not to "believe in flexible response" (p. 145). If attacked, they would "almost certainly retaliate with all the forces at their disposal" (p. 45). The "idea that nuclear war can be kept limited" had therefore to be dismissed as "implausible" (p. 18). We were even advised that "No one denies" that the prospects for checking escalation short of "all-out war" were "poor" (pp. 29–30).

In fact, hundreds of analysts in the West would have disavowed this last judgment. Nor may they have lacked implicit support from their Soviet opposites. On three separate occasions already, the Kremlin had threatened to launch long-range rockets in selective response to a specific challenge. In each case, the rockets could have had little material impact unless fitted with nuclear or strong chemical warheads.

Most remarked was the threat to fire rockets at London and Paris, made as the Suez campaign drew to its close in 1956. Then in 1960 down came two more gauntlets. During the crisis that May over U-2 reconnaissance flights across the USSR, Defence Minister Malinovsky spoke of instant retaliation against whichever bases any further flights came from. Two months later, Khrushchev himself threatened rocket reprisals should the Americans attack Cuba.

Correspondingly, any attempt to relate Soviet nuclear doctrine to the size and configuration of Warsaw Treaty armed forces and hence to the conduct of modern war (logistics; fire and movement; sea control . . .) would require an interpretation less naïve than *The Church and the Bomb* sought. It would run something like this. Though Moscow was scornful of American-inspired notions about the intricate modulation of deterrence, it did see nuclear firepower as separable into distinct steps, not all of which would come actively into play during a particular conflict.[13]

All of which reinforces scepticism about whether the respective Churches can helpfully define corporate positions on issues posed by scientific and technological advance, valuable though the inputs from individual Christians may often be. Nor, regarding matters military, can anybody these days forget how frequently organized religion has condoned armed aggression, its contradictory excesses included. Most of the barbarities inflicted by Imperial Japan post-1931were against the Chinese people. Japan did not share Shintō with them. But from China she had derived Buddhism, Confucianism, much of her language and many other key strands in her culture. The Japanese literati had been very aware of this.

Reformation Europe saw its full share of internecine violence with Germany its central cockpit. At what proved to be the transient Peace of Augsburg (1555), Catholics and Lutherans hammered out a compromise. Its core precept was *cuius regio, eius religio:* "to each prince, his own religion". In other words, the ruler(s) of each component statelet within the mosaic that was the Holy Roman Empire in Germany would determine its ecclesiastical allegiance. With the Peace of Westphalia in 1648, this principle was developed and consolidated.

Soon a corollary was being added, notably in England and Scotland. This was that religion must not obtrude on politics. Over the years, it has become entrenched in Britain and beyond. Tony Blair, himself a pro-American Christian, says he would not want to see issues like abortion figure in national elections the way they seem to in the USA. Not that this stance is without its difficulties. A further rider tends to be "Keep politics out of sport". Yet this may be unsafe and unsatisfactory if dealing with the Robert Mugabe's of this world. Then again, keeping religiously-derived precepts off the hustings could be read as contrary to logic. Arguably the same applies to preserving a clear divide between public services and faith-based charities. Do not these instances illuminate the shortcomings of narrow deductive logic as applied to human affairs?

No Religion?

One can expect established religion to decline further in the near future. But this is by no means to say that one looks towards its disappearance. A prison doctor-cum-confirmed atheist has encapsulated well what many broadly sense: "Religious belief is seldom accompanied by the inflamed egotism that is so marked and deeply unattractive a phenomenon in our post-religious society. Although the Copernican and Darwinian revolutions are said to have given man a more accurate appreciation of his true place in nature, in fact they have rendered him not so much anthropocentric as individually self-centred."

From which, several riders follow. The secularist personality can be the more embittered by life's dissatisfactions, the religious person the more accepting. While a secularist extends compassion to recognized victims, a religious person does to all humanity. Again, the former "de-moralizes" the world, leaving the vulnerable more exposed and therefore more reliant on inherently ineffectual professional protection.[14]

A religious framework may also afford a setting within which people can relate to their ancestral past more naturally than via museums or theme parks, not to mention Temples of Reason (see Chapter 17). Admittedly, previous generations have often been too deferential towards their past, too given over to ancestor worship or "golden age" mythology. Nowadays, however, danger comes from the other direction. We too easily lose sight of our heritage, even a couple of generations back. We thereby miss out on its concrete experience. We also lack perspective in a more holistic sense. Unfortunately, too, perspective is being lost the more due to weakening links with landscape. How many people die these days in a setting recognizably akin to the one they were born into?

What could happen, alas, is that the wheel will go full circle; and a footloose Humanity will swing back towards what then could well be a too mythological account of how things were originally. That could be grist for fascism.

Granted, twentieth-century experience (e.g. State Shintō) indicates that organized religion may fail to check this process, its having gained momentum. Instead the former may then be instrumental in promoting the latter. However, this is not to say that local religious influence cannot work against such negative tendencies ever getting under way. In Britain for over three centuries now, religious pluralism has been conducive to political moderation.

Still the most pressing consideration is that, during times of dynamic change, many people look for a new "mental map", a revisionary account of where it is all leading. There may also be a wish for such an account then to be considerably metaphysical, exploring the linkages between themes such as language, mind, logic, nature, consciousness, mentalism and God. But the pre-existing creeds will not come properly on board if they fail to bring their teaching into line with contemporary science: meaning, above all, astrophysics, astrobiology and consciousness studies.

When talking of "dynamic change", one needs look towards the next several decades: an era likely to be prone to social and ecological instability and, maybe later on, the uniquely macabre threat of insurgent biowarfare.

Yet it is not too soon to give some thought to the world that may await our descendants in the twenty-second century if our civilization manages to negotiate the nearer term. It could well be a quietist ambience, though one anxious to find "moral substitutes" for war but also for hunger, disease, illiteracy and other customary challenges. It could sink into neurotic introspection, randomized aggressiveness and cultishness unless furnished with fresh metaphysical perspectives, suitably subtle blendings of ancient and modern insights. Positive anticipation thereof could even now yield dividends. Among them could be an added incentive to grapple with more immediate concerns.

Harshly Heavenly

The likelihood that our universe is but one of an infinitude (albeit scattered in dimensions very largely or even entirely non-contactable, one to another) has huge metaphysical implications. All else apart, we may have to turn to what can be depicted as a two-tier deity. Arguably, there is a Consciousness within and across our cosmos which we may be pleased to characterize as godlike and feel we can relate to in one way or another. Yet there must also be a formula (totally impersonal and, indeed, unlifelike) which sets any and every new universe in motion. A solution broadly along these lines has been proposed by Nicolas Maxwell, Emeritus Reader in the Philosophy of Science at University College London (see Chapter 17).

One could imagine that, in divers locales throughout this totality, clones of ourselves (as we of them) will be doing everything we have done, are doing, wish we had done or wish to do. However, an alternative rendering to this Stoicism could be that endless alternatives make the likelihood of any replication infinitely low.

Yet suppose one insists instead that our cosmos stands alone after all, its scale is still daunting in relation to us humans or earthlings. There are estimated to be 100 billion stars in what may be our large but not jumbo galaxy; and there are many billions of galaxies. So we should probably allow for a hundred billion billion stars, a fair proportion of which will have planets revolving around them. Talk to lay people along these lines and one finds such calculations are frequently a crunch reason for believing in nothing except that we count for nothing.

Whichever way, the connotations are profoundly humbling. In principle, they ought to be most so to any Hitler or Mao or Saddam aspiring to establish himself as a peerless Man of Destiny or Lord of Creation. Yet a caveat to enter here is that one fell day their ilk may be able so to manipulate data as to neutralize any astral allusions within our world culture. May such a consummation still be a century or two down the line, even on worst-case assumptions.

Registration by Life

Perforce, everything hinges on who or what is "we". If it is just humans or earthlings in general, insignificance is hard to deny. If it is Life cosmically, the perspective alters. In Chapter 7 it was proposed that Life had a signal role, underpinning the Consciousness of our universe at large. When Life came into play, cosmic expansion and cooling could have been making any Consciousness possessed by energetic or organized inanimate matter progressively less adequate for the registration function which, according to the form of Idealist philosophy here adopted, is basic to the registration and therefore the existence of everything.

Not that this hypothesis is complication free, whether one assumes a single cosmos or infinitely many. What was the situation at the very instant of the Big Bang well before, if one has understood aright, any transition into matter from energy had begun or, indeed, the four great forces physicists identify had separated out? Was energy then so concentrated in certain dimensions as to be possessed of Consciousness? Is not energy more fundamental than matter in that the latter comes from the former and reverts to it? Obversely, matter is much more structural.

Was information (and therefore structure, memory and Consciousness) imparted initially from outside? Then again, what will happen as the cosmos continues its expansion for many billions of years more, something which appears to be in train whether one believes in an Open Universe programmed to expand forever or whether you find more likely its being on course for eventual contraction culminating in the next Big Crunch/Big Bang? Must it not eventually pass through a threshold loss of Consciousness as (a) Life progressively dies out and (b) the Consciousness of inanimate matter diminishes critically as it becomes more diffuse and ever less energetic? And in accordance with the hypothesis here proposed, does this not *ipso facto* mean the universe ceases to exist?

Life After Death?

Usually when we tell whether or not "we believe in God", we are really answering a different question: "Do we believe in life after death?". Whether this could be experienced just transiently is a marginal judgement best deferred a few pages. More fundamental is the subject of one's individual life putatively being eternal or essentially so, this being effected either through acceptance into Heaven (or Hell) or else by repeated reincarnations, the Transmigration of Souls.

This latter expectation figures in a considerable number of early religious cultures from Aboriginal Australia to Pythagorean and Orphic Greece, the pre-Christian Celtic world, and so on. However, it is especially associated with the Indian subcontinent. In Hinduism, an individual's soul is reborn after their death, the circumstances of rebirth being determined by *karma*. This overriding law of cause and effect ordains that one's thoughts and deeds in past incarnations will shape one's present situation and destiny. Eventually, the truly devout may achieve *moksha*, liberation from this constant cycle. Buddhism, of course, builds on this tradition.

Analytically, transmigration is not difficult to challenge in terms of what we have long known about sexual partnership, conception and inherited characteristics *vis-à-vis* ourselves and, indeed, other species. Besides, it is open to the charge that *karma* allows it to be said that those born deprived have only themselves to blame because of their previous iniquities. Even so, one cannot say that notions of reincarnation have no impetus left in them. On the contrary, in the year 2000 the results were published of a survey by Nottingham University and the BBC of religious attitudes in Britain. More than half the respondents indicated belief in a life after death, the same as half a century before. But now belief in a soul was more often linked to rebirth, less to heaven or hell than previously. Never mind that Christianity has very consistently ruled reincarnation out while the Judaic mainstream, too, has given it short shrift.

The three-storey division of the metaphysical realm into Hell, Earth and Heaven is very much a construct of the "two-armed" religions of the Arabian desert fringe. However, it is Christianity that over the centuries has highlighted most strongly this configuration (see Chapter 12). Hell figures in the Koran though less than Paradise. Sometimes the Ancient Jews envisioned a Sheol or Tophet, a gloomy realm for departed souls closely akin to the Graeco-Roman underworld. This gradually evolved into a scenario more like the Christian Hell.

A particularly disturbing aspect of this triad has to be a stark contrast in destiny between eternal damnation in hellfire and everlasting celestial bliss. We all know that, even definition-wise, human evil and good cannot be distinguished that starkly. One has only to think of the ambiguities as between egotism and altruism. The Roman Catholics eased the dichotomy somewhat by developing the concept of purgatory, an intermediate location in which souls not quite up to the mark as yet are stringently made ready. Eastern Orthodox teaching offers something similar, though Protestant tends not to.

Most important, however, is a sea change in Christian perspectives which visibly gathered pace through the middle of the last century. Hell is now mentioned little if ever, save within hard-line Protestant fundamentalism. Likewise, most modern theologians (the late Pope John Paul included) do not mention Heaven much; and when they do, it is more in terms of the beatific vision (meeting God face-to-face) rather than Life Everlasting or, indeed, the New Jerusalem of the *Book of Revelation* 21, 22 with its city walls built on jewels; city gates each hewn from a single huge pearl; streets of translucent gold; and "no more night".

It is worth recalling now, underlying the Heaven–Hell dichotomy, there has always been a quite contrary notion of death as a doorway to eternal rest for everyone. As Thomas Gray so poignantly observed in 1751, "The rude forefathers of the village sleep." By virtue of such imagery, his *Elegy Written in a Country Churchyard* rates as perhaps the most quoted poem in the English language.

Human Outreach

The instinctual human conviction that we stand apically above the rest of Creation finds clear expression in countless expectations of God or Godliness, however defined. The beatific vision is within the long Christian tradition of the "Kingdom of God", be it a state of mind in this life or blissfulness post-death. In Islam, no human visions can comprehend Allah, though a *wali* (i.e. saint) will be His protected friend. Hindus speak of the "Eternal link of the Soul to God". In Jainism and Buddhism there is the striving towards nirvana, blissful absorption into an impersonal ultimate. And so on. So can any kind of outreach transform thus our experience of God understood either as a "cosmic consciousness" or else as a more traditional entity?

The outreach mode that will come most readily to some minds is mathematics, a corpus of reasoning which Platonists stress is enduringly robust (in its exquisite precision and consistency) in a way no individual can match nor any mundane scene replicate.[15] A less deep-seated understanding of mathematical perspective was afforded by Jeans, "Physicists who are trying to understand Nature may work in many different fields and by many different methods . . . But the final harvest will always be a sheaf of mathematical formulae. These will never describe Nature itself but only our observations on Nature; we can never penetrate beyond the impressions that reality implants in our minds."[16]

Still, the big problem nowadays lies not in these differences in perspective. Nor does it in how counter-intuitive a lot of raw cosmological data feels to us mere Earthlings. It resides rather in being required to introduce extra dimensions, maybe to accommodate the equations needed to resolve some important topics. This requirement can be traced back to 1919 when Theodor Kalusa backed up by Oscar Klein, both of them Swedish, proposed an extension of the Space–Time dimensions in general relativity from four (time plus three spatial) to five in order to incorporate electromagnetism. The fifth was visualized as "curled up into a tiny loop" of which our senses would be unaware.

After a few years, this Kalusa–Klein theory fell into abeyance partly because the scientific community was ill at ease with an added dimension. However, around 1980 it was revived and extended to comprehend ten dimensions, in accordance with superstring theory. Obviously all this articulation is a human artefact. But this is not to say any human being (still less an average one) can view it synoptically en route to ascertaining how God is manifested. Its esoterics inevitably weaken the pristine import of what otherwise remains the inspirational Idealist introduction to the Johannine Gospel: "When all things began, the Word already was. The Word dwelt with God, and what God was, the Word was."[17] As things now are, we cannot but follow C. S. Lewis and admit that we can conceive but not imagine.

Mysticism

So what of the alternative approach, seeking God through mystical experience? Music comes readily to mind. Appreciation of it seems innate throughout human society, present or past. Indeed, its roots are probably set deep within our hominid but pre-human ancestry. Many composers have been profoundly religious; and the place of music in religious ceremony hardly needs dwelling on.Yet even in that context, its primary role may be social bonding. St Augustine has been among those doubtful about the net impact of music on spirituality, anxious lest the gratification it gave dulled deeper sensibilities.[18] Lately we are advised that "when a symphony's dénouement gives delicious chills, the same kinds of pleasure centres in the brain light up as when eating chocolate, having sex or taking cocaine".[19]

One stark ambivalence to be observed is that of the Protestant Reformation towards music. Take Ulrich Zwingli of Zurich, a relatively moderate leader of the chiliastic Puritanism of the Upper Rhine–cismontane region, the movement brutally suppressed (with Luther's encouragement) in the Peasants' War, 1524–6. Biblical study and political pragmatism had induced this erstwhile hymn composer to head moves to proscribe sacred music.[20] As time went by, however, community hymn singing came to feature strongly in much Protestant observance. It enhanced social solidarity and for some at least smacked of a mystical experience. A Beethoven or a Wagner would well understand. So would today's Christian charismatics.

What is "mysticism" anyhow? Certainly the term can be used too loosely for scholarly taste, being too extensible into magic or the occult. Yet the theme cannot be ignored, partly because a mild mysticism readily conflates with the holistic perspectives of counter-culture progressivism. Its essence is an individual quest for oneness with God or the Absolute, a melding sought through focused commitment. It does not correlate with quietism. On the contrary, mystic literature is suffused with energy and imaginings. The poetry and artistry of William Blake (1757–1827) is archetypal, his anti-science obsession notwithstanding.

Mysticism was a powerful tendency through the late medieval centuries, perhaps in reaction against dated religious formalism. It was so in Islam, Judaism, Eastern Orthodoxy and Catholicism. Saints Augustine, Bernard of Clairvaux, Francis of Assisi, Thomas Aquinas and Joan of Arc are readily locatable in the Roman Catholic context as, still more so, is the theologian, Meister Eckhart (*c.* 1260–1328). This tradition continues through the Counter-Reformation, especially in Spain. St Ignatius Loyola (1491–1556), founder of the Jesuits, wrote influentially on mysticism. The intensely active St Theresa of Avila (1515–82) was deeply involved as was her friend, St John of the Cross (1542–91).

For most of them, mysticism was preponderantly a matter of the soul reaching out. But George Fox (1624–91), founder of the Quakers, stressed rather a mystical quest for God within. Likewise, Eckhart had told how "In the depths of the soul God creates the entire cosmos, past, present or future". Long before, a similar mystic reorientation had been effected within Vedanta, one of the six classic Hindu systems. Its Shankara school saw the outside world as Maya, a cosmic illusion. Buddhist mysticism has evolved within the Asian meditative tradition, originally under Vedanta influence.

Reference has already been made (Chapter 11) to Sūfism within Islam. It tends to see Allah as softer and closer to us earthlings than He otherwise appears. Within Byzantium, mysticism was especially strong from the tenth to the fourteenth centuries. The theologian-poet, St Symeon, averred that "He who is far outside the whole Creation takes me unto Himself and hides me in His arms". The crowning glory for a Christian would be coming face-to-face with the "divine light" of Christ.

External Assessment

Confessing to never having had a mystical experience himself, William James sought to delineate the genre. Though a particular mystical state rarely lasted beyond a couple of hours, some recollection regularly survived. At the time, the inner experience was of ultimate reality being unknowable yet somehow known. Familiar perceptions assumed a new and weightier significance. The passage of time became transcendentally unimportant as one attained a "cosmic consciousness" (a term a Canadian psychiatrist, R. M. Burke, had coined in 1901). An affected individual came to be at one with the Absolute, a consummation "hardly altered by differences of clime or creed". The imagery of an event was rather anti-naturalist. Even so, such pantheistic nature lovers as Walt Whitman and Richard Jefferies have laid claim to the experience.[21]

James' overview is effectively endorsed by Dr Peter Fenwick of London's Institute of Psychiatry. Fenwick has also been a pioneer in the correlation of mystical awareness with sectoral brain response; and of gaining acceptance for the view that subjective evidence from a diversity of clients must be seen as a licit and vital part of further investigation.[22] Almost by definition, this evidence must include certain perceived phenomena putatively within the realm of the paranormal. Fenwick would insist on the need to record, to test as far as possible, and only then to judge.

In a study first published in 1917, Bertrand Russell evinced a surprising measure of empathy with mysticism as a concept. He recognized that mystical "absorption in inward passion" could appear to yield insights into "higher reality and hidden good". He notes how mystical states involved denial of "opposition or division anywhere", including any distinction between past and future. This leads on to an absence of indignation or protest, to the acceptance of everything with joy.

Yet in the final analysis he remained disposed to see the experience as "little more than a certain intensity and depth of feeling in regard to what is believed about the universe." While reluctant to judge the quality of mystical insights, he did insist these had always to be followed up by sound verification procedures.[23]

Is this caveat not compelling? But does it not air doubts about whether an all-knowing, all-powerful and all-good God could be drawn close to a specific religious experience? How does one distinguish between revelatory perception and *a priori* assumption? How could God, unitary or otherwise, assume the attributes needed to make Him that accessible? Is He not often understood to be beyond the spatio-temporal world? And if one allows for His only son's authentic incarnation does this not raise insuperable problems *vis-à-vis* His ongoing relationship with the rest of the cosmos? Suppose, alternatively, one accords the said mystic a sixth sense able to register God in time at least. Would this not undermine His or Her godly properties? And how could the mystic gauge these in relation to the cosmos as a whole?[24] In 1896, the remarkable Spanish-American, George Santayana (1861–1952), claimed "a real propriety in calling beauty a manifestation of God to the senses since, in the region of sense, the perception of beauty exemplifies that adequacy and perfection which in general we objectify in an idea of God".[25] But he himself was then moving towards an "animal faith" which set no great store by intellect constructs about cosmic causes. He could therefore allow himself a goodly measure of subjective imprecision.

Experience of the paranormal ought also to be scrutinized closely in terms of otherworldly inputs. It has something of a track record of (a) internal inconsistency or (b) lack of verification or (c) mundane alternative explanation. Much the same applies to the Near Death Experiences reported by people snatched back from the brink of eternity, often after cardiac arrest. In perhaps four-fifths of these cases across all cultures, the recollections are benign: a tunnel suffused with golden light, down it departed loved ones gather to greet you . . . Glimpses into the next world, albeit ones conditioned by your own religious culture. Thirty years ago, such accounts were still being reviewed for their afterlife connotations. Nowadays the positive sensations plus a remarkable absence of pain are usually attributed to endorphins and encephalins: endogenous opiates widely distributed within the body and releasable after death.

Levitation or Out-of-the-Body Experience (OBE) seems quite the most arresting of the putatively paranormal phenomena. Very many people, Susan Blackmore for one, are convinced they have literally levitated. But such perceptions can be induced by a variety of drugs or electrical stimulation[26] or else by hypnosis or controlled meditation, a state of affairs which indicates events internal to individual bodies and not detached from them. Besides which, the whole idea of a non-corporeal existence has come under incisive attack from some very able philosophers. How could one sustain mental continuity? Why do such entities only look downwards? Could, say, a bar of iron float and still retain its identity?[27] But if one is looking for indications of a perhaps transient afterlife, should one not be pursuing other lines of enquiry, ones in which these considerations might be overridden by an encompassing teleology?[28]

Metaphysics Revisited

It is necessary here to consider afresh what contribution Life makes to sustaining our universe. First of all, how does the outreach human beings so determinedly practise relate to cosmic registration? One cannot argue that it dovetails us neatly into Cosmic Godliness because (a) the latter seems too Immanent to lend itself to this and (b) it also appears possessed of more dimensions than we are aware of using.

What then is this propensity to outreach actually for? We may trace its evolutionary origins in part to the highly flexible response our pre-human forebears so successfully made to the ecological crisis signified by the virtual dryings up of the Mediterranean some five million years ago.[29] However, it has been possessed of a momentum extending over and beyond that.

So will not outreach be an attribute manifested to significant extents by many other advanced species in numerous cosmic locales? If so, does it represent the ultimate collective expression of the registration process, the way in which a unity is achieved by the Cosmic Consciousness? Does not a need for this arise out of (a) the inability of neighbouring galactic or even stellar ecosystems readily to make contact with one another and (b) the likely inability of all parties to withstand the shock, were interstellar or intergalactic visits ever to take place as between advanced but contrasting life forms?

Should outreach thus be seen as part of a process whereby the contributions individual living things make to the Cosmic Consciousness are reabsorbed into the whole? If so, could this perhaps involve individual information structures lasting awhile after physical mortality, thereby affording at least passively a transient link between the immediate past and the ongoing present? Is this a brief measure of immortality we might reasonably hope for? Personally I would be very sceptical. But whatever answers one may feel tempted to give, with our present knowledge, a crop of further questions will remain. They include where a boundary zone might be drawn between cosmic godliness and the inanimate formula which, it is proposed, controls what may likely be an infinitude of cosmi.

Afterthoughts

Meantime, my own feeling is that telepathy could exist as a weak force, confining the term just to refer to extra-sensory cognisance of what another living, perhaps far off, person is thinking or experiencing at the actual time.

Also of interest is a possible tendency for personal belief in the paranormal to correlate with declared religious beliefs but little or no religious participation.[30] What should almost certainly be ruled out, above all on account of the huge metaphysical dilemmas thereby presented, is clairvoyance.

So what are we to make of the career of Michel de Nostradamus (1503–66), eccentric astrologer and physician. In a stringent assessment, Peter Lemesurier sees him as arguably possessed of certain prophetic powers but also as a fraudster who had mastered the arts of mystification, mumbo jumbo and personal propaganda.[31] Most quoted of several allusory and undated prophecies often adjudged to be borne out is the one linked to the attempted escape via Varennes in 1792 of Louis XVI and Queen Marie-Antoinette. Lemesurier's translation of the relevant verse is thus:

> Through woods of Reim by night shall make her way
> Herne the white butterfly, by byways sent
> The elected head, Varenne's black monk in grey
> Sows storm, fire, blood and foul dismemberment.[32]

It fits pretty well the episode and the aftermath. No doubt Nostradamus could sense those lonely woods would sooner or later get involved in a nasty Paris-based crisis. But the apparent allusion to the interception of the royals at distant Varennes is astonishing.

Writing in 1999, Lemesurier provided his own list of datable Nostradamus predictions still to be tested. Every year between 1998 and 2015, there are up to ten entries. There are to be massive military operations against Christendom by Islam, an armed clash of civilizations orders of magnitude worse than what we actually know. In just one of these years tidal waves wreak havoc. They do so to southwards, in the Mediterranean to be exact. That year is 2004.[33]

17

CREATIVE CONVERGENCE

A Dual Deity

A LONGSTANDING MYSTERY CONCERNS HOW AN ALL-KNOWING, all-loving and all-powerful God can abide all the suffering and injustice this world knows. As noted above, the Emeritus Reader in the Philosophy of Science at University College London has offered a two-fold solution. There is a totally impersonal and unchanging God of Cosmic Power that determines how all phenomena unfold. But there is also a God of Value that is "the soul of humanity embedded in the physical universe, striving to protect, to care for . . . but all too often, alas, powerless to prevent human suffering."[1]

Subject to some adjustment, this model fits the analysis here unfolding. Suppose the God of Cosmic Power becomes the God of the Multiverse/the Infinitude. Next, suppose the God of Value becomes the consciousness within our own finite cosmos. It may be helpful to talk of Godliness rather than God. It may also be helpful to ask how strong are the empirical or "bottom-up" reasons for believing that a consciousness or, following Carl Jung, unconscious pervades our universe. And how does all this relate to human awareness as discussed at the start of Chapter 16?

Cosmic Unconscious

Carl Jung believed that, below the personal unconscious of repressed wishes and memories, lay a collective unconscious informed by the psychic heritage of all mankind. He was persuaded thus by having found as a child that his dreams contained imagery from somewhere beyond. As much was allegedly confirmed to him by finding that the delusions of his psychiatric patients involved symbolism recurrent in folklore the world over.

Any species passes on genetically data operationally basic to its survival. How else could the European eel know that, after several years as a larva plus a few more as a riverine eel, it must migrate several thousand miles to spawn then die in the Bermudan waters where it hatched? However, the figurative imagery Jung often recorded went beyond this, into metaphysical realms.

A configuration to which he attached an underlying importance was the mandala: a square and circle concentrically positioned and touching at four points, depending on which shape is set inside the other. The square can connote the "four corners of the Earth" and the circle the vault of heaven. In the Hindu and Buddhist worlds, the mandala is a recognized focus for the meditative transcendence of mundane thought.[2] The symbolism also became familiar in the Greece of the sixth century BC; and under the influence of Plato and Plotinus, its circle came also to represent the human soul – an interpretation later taken up by Christianity.[3] Since when, it has often influenced public architecture in the West and lately, Jung suggested (see Chapter 7) "flying saucer" reports. In prehistoric cave art, fours and threes regularly recur: the former as symbolizing the world square and the latter what was read as a three-phase lunar cycle.

The "discovery" of a collective unconscious has been presented as "potentially" of "comparable importance to the quantum theory in physics".[4] Yet to say "potentially" could be to admit that conjecture still awaits consolidation. Jung drew heavily on uncorroborated evidence about patients and self. Moreover, his writings show less interest in this whole concept than devotees have evinced. His thoughts were often shared more through conference networking, with some of his warmest endorsements coming from philosophers, not psychologists. In 1982, Daisaku Ikeda (by then the Honorary President of Sōka Gakkai in Japan) saw the collective unconscious as part of the cosmic Mystic Law the Buddha had said sustained everything.[5]

My own inclination to believe provisionally that a species at any rate can have collective awareness has lately been informed not by human behaviour but by avian. Myriads of birds gathered above Lake Geneva as dusk approached. Flocks dived, zoomed and wheeled with sublime precision. Could this connote a collective mind? The ornithologists tell us that one wader intercepting a solitary fish will mentally react in a millisecond to any change in the tactical situation. However, reacting in toto to many fellow members tightly packed within a fast-moving flock has to be orders of magnitude harder than responding one-on-one whether co-operatively or adversarially. It is not easy to believe that those involved could manoeuvre with such finesse without some extra-sensory mutuality.

Not Morphic Resonance

A superficially similar theme has been promoted by Rupert Sheldrake, a Cambridge doctoral botanist disabused of Science because, he contends, it has lost its primal sense of wonder. He made his break in 1981 with *A New Science of Life*, a book which enraged many of his peers by beaming directly to the outside world a clutch of heterodoxies about causation. Sheldrake's response was to draw the more on support from a counterculture clientele.

In 1996, he sallied forth with a text configured as a disputation with a colleague, Matthew Fox. Sheldrake aired his distinctive theme, "morphic resonance". Its essence was that "the regularities of nature are more like habits than laws . . . Nature has a kind of inherent memory rather than an eternal mathematical mind . . . *it is well known* (my italics) that the more often you make crystals, the more easily they crystallize around the world". Likewise, if "rats in Sheffield learn a new trick, rats all round the world should be able to learn it more quickly . . . " After all, "if the Universe evolves, why should not the laws of nature evolve as well?"[6]

The exasperation of the Science mainstream is no surprise. Can one actually observe real time change in Nature's basic laws and in reaction rates for crystal formation? Can these subjects and rodent learning curves be dispatched in virtually the same breath? The plain truth is that, applied to a cosmic interpretation like the one here being essayed, morphic resonance as thus enunciated would complicate more issues than it resolved. But this would never be an argument against the incorporation of this or any other germane hypothesis where and when the empirical underpinning was substantial enough.

Creation Myths

One should urgently be looking for a "bottom up" affirmation of the panpsychic thesis advanced for metaphysical "top down" reasons in Chapter 7 of this text: the thesis that all matter within our cosmos is possessed of consciousness. From it, one might infer a collective memory going back at least to the condensing out of matter 300,000 years after the Big Bang though with an added imperative somehow to reach back into that event.

Arguably, the Creation myths set disparately within human societies could represent facets of that memory. Commonly these subscribe to tenets which seem applicable enough, starting with a moment of Creation often in a brilliant flash. It is certainly to be remarked how rarely the myth-builders adopted the Aristotelian pitch which was that everything had been and would forever be much as now. Must one not agree with Victor Weisskopf, a distinguished MIT cosmologist, that it is surprising how well the Creation myth in Genesis matches the modern scientific model, particularly as regards a huge effusion of energy precursing the formation of the Sun and other stars?[7]

A further alternative was that of envisioning a universe which repeatedly expands out of a point only to contract into one aeons later. This perspective found expression not only in the Hindu legend of Lila but also in Ancient Greece. Empedocles, Anaximander, Heraclitus and Diogenes thought very much along these lines. Also, of course, it loomed prominent in the cosmological debates of this last century. None the less, folk myths worldwide do overwhelmingly depict a single Creation event though in many such accounts, an old world order is destroyed to make way for the new.

At which point one should go beyond Weisskopf and note that, in the involuted complexity of many Creation myths, can be seen colourful reflections of the history of the Cosmos hard upon that moment of Creation. It is a sequence far more intricate than straight-down-the-line Natural Science as envisaged as late as the turn of the twentieth century (i.e. 1900). What the early myth-makers did was provide "an organic timeless flow of images and narratives within which such questions were by-passed altogether. For the answers of mythology come from deep levels of consciousness, in which universal patterns or intimations are apprehended".[8] Though couched mainly in terms of the goings-on of all too human gods, the accounts well match figuratively the inflationary hypothesis discussed in Chapter 5.

Gaia and Panspermia

Still, many people unwilling simply to fall back just on metaphysics will continue to look for a hard "bottom up" confirmation of a cosmic consciousness. Further possibilities are proffered by Gaia and Panspermia. Might not the latter supplement cosmic consciousness? Might the former be another way of rendering it?

In September 1965, James Lovelock, a British scientist-cum-engineer then working with NASA, had "The intuition that the Earth controls its surface and atmosphere to keep the environment always benign for life". In 1967, he took up a suggestion by William Golding and named the concept Gaia after the Greek goddess of life. A 1979 book with *Gaia* as its main title engendered a vigorous controversy.

Lynn Margulis, a charismatic American microbiologist who has done pioneering work on the evolution of nuclei within cells, had collaborated with Lovelock on Gaia formulation since 1971. She and her science-writer son, Dorion Sagan, now saw Gaia as under attack "not only for being unscientific and untestable but as antihuman polemics, green politics, industrial apologetics and even as non-Christian ecological Satanism".[9]

Among the doubters were certain Darwinian evolutionists. Arguably, theirs was among the least reasonable of the negative stances. Collaborative Gaia and competitive natural selection might often be complementary in terms of how individuals behave within families, species and eco-systems. James Lovelock has lately depicted "the battle between Gaia and the selfish gene as part of an outdated and pointless war between holists and reductionists".[10]

Quite the boldest claim made on behalf of Gaia is its long maintaining very steadily within the atmosphere an oxygen level well above previous equilibria. Apparently, 2.2 to 2.4 billion years ago, atmospheric concentrations of methane (CH_4) were hundreds of times today's; and the associated greenhouse warming was pronounced. Biogenic methane formation took hydrogen atoms out of the water cycle. Gradually many of these, being light and energetic, were lost into Space.[11] Meanwhile, the oxygen thus liberated from disintegrating water molecules accumulated strongly in the air.

Since when, the oxygen level seems to have kept close to 20 per cent, allowing that measurements for more than several hundred million years back are not easy to take. The said fraction is high enough to sustain well life as we know it but not impossibly high apropos fire risk. Mean surface temperature has been comparatively steady despite the Sun's getting 10 per cent brighter every billion years.[12] It does all look very Gaia excepting that the initial oxygen upsurge is acknowledged to have been mortifying for pre-existing life.

Another Gaia regulatory mechanism identified early on in the debate concerned the constraint on air temperature rise exerted by the excretion into the atmosphere of dimethylsulphide (DMS), this by primitive vegetable forms within the oceans. One end product of DMS decomposition can be sulfate particles which may act as condensation nuclei for water clouds. Smaller droplets forming round more nuclei can give clouds considerably higher albedo – i.e. capacity to reflect sunlight.[13] Quite soon Lovelock was persuaded that marine algae were the chief DMS source.[14]

Now close attention is also being paid in this connection to Great Barrier Reef coral.[15]

Better testing of Gaia *per se* waits upon Global Circulation Models of the oceans and atmosphere becoming more refined and longer term. Nevertheless, enough knowledge may be to hand to choose provisionally between strong Gaia and weak Gaia or, if one insists, no Gaia. Strong Gaia commonly involves the idea of a world-soul or superorganism which inanimate parts of the Earth act as a skeleton for. Its conceptual roots are Classical, from Pythagoras onwards. Stoic philosophers even discouraged mining as best they might, lest such abrasion of Mother Earth induced natural calamities.

James Lovelock himself has subscribed to this "geophysiology" perspective in so far as it arises out of studies of more localized ecosystems. Nevertheless, he regards New Age visions of Mother Earth as a fully integrated organism broadly akin to the human body as quasi-religious notions untestable for perhaps a long time ahead. However, he does think that homeostatic relationships like that between algae and global temperature stability do require some kind of holistic explanation. Moreover, this octogenarian without belief in a personal hereafter does see Gaia as the basis for an optimistic faith.[16] Still, we would do well to discount those who might portray him as too mystical or messianic.[17]

My own expectation is that a weakish Gaia will gradually be accepted as received wisdom. Anything "strong" will pose problems with simulation and verification but also philosophically. Gaia was, after all, a mythic creature of the pristine Space Age, of "Earthrise posters blu-tacked to a million walls".[18] But why should it be confined to Mother Earth? Why not see Gaia interactiveness between earthling life forms as an accessible expression of a cosmic consciousness underpinned by a weakish Anthropic Principle?

As many have recognized, making contact with an alien civilization hard by a distant star could impact dramatically on Humankind. Yet on any definition it would represent but the minutest sliver of whatever panpsychic consciousness one may believe exists. More indicative thereof would be confirmation of panspermia via comets, asteroids or meteorites (see Chapter 7). Moreover, the status of panspermia will be underlined if it can be shown that astral and galactic patterns are conducive to seeding. Unfortunately, however, we do not understand the cosmic structure at the galactic level too well as yet. Thus it was long unclear how far "galaxies owe their sizes and shapes to special conditions at or near the beginning" or to "physical processes in the recent past".[19] But now it does seem that galaxies were proliferating within a billion years of the Big Bang.[20]

Among the questions to resolve are these. Is our Milky Way spiral galaxy of such a structure and size as to give panspermia a fair prospect? And what of the Small and Large Magellanic star clouds which slowly revolve about it? Might panspermia also operate between galaxies? How significant is the tendency of galaxies to cluster? How representative is our local cluster with its 30 or so galaxies, some adjudged "dwarf"? What significance attaches to the superclustering lately confirmed?

Transient Immortals or Cosmic Vandals

If panspermia is gaining acceptance among astronomers, it will lend credence to the view that Cosmic Consciousness is unitary, thus broadly akin to the consciousness we experience as individuals. Moreover, a unitary Cosmic Consciousness could be thought of as affording elysian pathways for a phasing out of human individuality after clinical death, a gradual dissolution of the soul into a cosmic setting.

In so far as people look towards religion for promise of Salvation, a benign survival after death, this may be all that most inwardly wish for – preserving awareness long enough to watch their immediate legacy playing out on Earth. The thought of remaining discrete eternally (and of forever relating back to those earthling years) is downright mortifying. The early Christians endorsed a heavenly Kingdom of God as but a prelude to the Second Coming of the Messiah, this within decades. With His arrival, every parameter would be recast, every reference point moved.

With our present knowledge, nobody is entitled to declaim that the transitions here envisioned are inconceivable. However, skepticism needs be our point of departure. For one thing, the ambiguities surrounding our status as human beings impact on the broad scene. Thus the vexatious notion of the soul as a unique human attribute is thereby raised anew. For another, we never interact with the lately deceased in ways most people today find credible. It could complicate things unimaginably if we did.

To which one must add that at first sight the Cosmos as a whole stands to gain little from such a provision. Our human consciousness might seem to us exceptionally intensive and articulated. Yet it remains but a sliver of the cosmic experience. Suppose in spite of everything we do eventually surmise that "our biosphere is the unique abode of intelligent and self-aware life within our galaxy".[21] The Milky Way is but one "island universe" among many billions.

Nor can it easily be argued that, in some ethical or aesthetic sense, we have special claims to recognition, presuming the universe to work that way. Wearing clothes, living in fabricated caves and having advanced communication skills prove little *per se.* Deep contradictions arise from our complex Consciousness and related aptitudes. We talk of Gaia and the like. Yet we constantly pose a far bigger threat to terrestrial ecology than does any other advanced species.

Indeed, apprehension lingers that high energy "atom smashing" experiments could accidentally lead (via one of several causal pathways) to the Earth's destruction or even to the "fabric" of Space being rended, perhaps annihilating the Milky Way and maybe much beyond. The risks are currently guestimated at one in tens of millions.[22] Yet these odds could shorten acutely if the quest for a Theory Of Everything drives "atom smashing" to much higher intensities.[23]

Conflationary Gains

The quest to resolve some of these uncertainties might usefully involve a degree of conflation between the major world obediences. Here a prerequisite must be rapprochement between the Abrahamic faiths; and a *sine qua non* for this is a working partnership in Palestine within a stable political settlement. Beyond which, one looks towards accommodation between these monotheisms and the less God-centred credos of South and East Asia. An Abrahamic trend towards perceiving Godliness in less humanoid and less localized terms should assist. There is, in any case, a long-standing disposition within Islam (Sūfism usually excepted) to conceive of Allah's mercifulness in terms austere and remote, though not entirely abstract. Contributive, too, should a diminished concern on more or less all sides with some defined eternal life.

The dividends on the world stage from convergence should be a liberalization of doctrine and expansiveness of spirit. As much comes across in Japan through the history of Sōka Gakkai, the Nichiren Buddhist derivative mentioned earlier. Post-1950, it progressed markedly in a chaotic religious climate. But this progress owed altogether too much to *shakubuku* ("breaking down"): forceful mental and physical suasion coupled with savage critiques of rival sects. Soon Sōka Gakkai entered politics via its Kōmeitō ("clean government") party, a force to reckon with awhile.

In 1958, the sect's leadership had passed to a mild-mannered moderate, Ikeda Daisaku. However, public opinion remained quite hostile, not least because Kōmeitō was a standing affront to an emergent consensus against state involvement in religion. Ikeda broke formal ties with the party in 1970. He also insisted on non-abrasive proselytism. With a collateral shift of emphasis towards personal Buddhism (i.e. less reliance on the priesthood), some were coming to see the movement "as a kind of Protestant Buddhism".[24]

In 1979, Ikeda resigned as president of Sōka Gakkai to found Sōka Gakkai International (SGI). Although or because in 1991 the Nichiren Shōshū hierarchy excommunicated all Sōka Gakkai members, SGI now flourishes exceedingly at home and, to quite an extent, abroad. It focuses heavily on world peace, social development and ecology.

One would like to feel confident that inter-faith conflation could preclude armed conflict between peoples. Once again, however, things may be less simple. Take the intra-Protestant ecumenical movement. It effectively started in England with the launch of the interdenominational Young Men's Christian Association in 1844 and then the Evangelical Alliance in 1846. It went overtly international with the World Missionary Conference in Edinburgh in 1910. The collapse of the Pax Britannica four years later condemned ecumenism to evolve against a backdrop of wars and rumours of wars between peoples of a common Christian heritage.

Still, other developments during the 1918–39 period paved the way for a World Council of Churches which held its inaugural assembly in Amsterdam in 1948. Its members were Protestant, Eastern Orthodox and independent Catholic. During the sixties, it forged close ties with Rome. Nowadays, of course, ecumenism is much constrained or otherwise influenced by grass-roots attitudes and dilemmas. Correspondingly, there has been a lot of interdenominational activity at local level.

In the years ahead, we may hear a deal about popular "peace movements" with interfaith provenance. Some of these may be well found. But others may be unsound, not to say bogus. Witness that 1951 East Berlin youth rally, an archetypal product of political religion.

In November 2005, Sun Myung Moon launched a Universal Peace Federation with an address at Connaught Hall, London. One can allow the printed version of his text some coherence and civility even though one was bound to challenge much, not least his central thesis that the nuclear family (preferably with lots of children in it) is a pertinent model for universal peace.[25] But the address actually delivered was awful: vacuous, repetitive, manichean, sometimes menacing, lavatorial, homophobic, interminable and laced with celebration of himself as perhaps "the third Adam". It constituted to my mind a stark demonstration of the case for sustaining the religious mainstreams as opposed to ill-founded cults.

One dividend of global inter-faith conflation ought to be a recharged commitment to preserving Nature. A concern with the eternal verities can always heighten awareness of the exquisite beauty of the terrestrial order. Often, indeed, this is reflected in sacred texts. Never mind that these rarely foresee the urgency now attached to preventing Nature being completely overturned by humankind.

However, the benefits of philosophic convergence may be garnered fully, not in years but decades. A further gain should be a clearer realization on the part of people further afield that what they see as a western life style imposed by globalism is, in many respects, no less novel to Westerners. Ambivalence towards it is likewise shared.

Eurasian Dialectics

Europe's interest in East Asia goes back at least to the Silk Road linking Rome to the Han. Contacts later withered but were to an extent restored under Mongol hegemony. A definitive papal interest in China found expression with the dispatch in 1289 of six Franciscan missionaries.[26] On the lay side, Marco Polo's claim to have been with Kublai Khan from 1275 to 1292 used to be accepted uncritically. Recently serious doubts have been raised about whether he even set foot in China.[27]

As has been noted, the Renaissance as customarily delimited was casual towards the global voyages of discovery it had itself generated. This was not unrelated to a disinterest in Progress as an underlying principle, save on the part of the visionary though disreputable Francis Bacon (1561–1626). The literati were more disposed to look back to a Classical golden age. Come the eighteenth century, however, enthusiasm for the Classics engendered a zest for understanding all cultures, not least to interpret Progress. For Progress was now in the air.

From 1760, the British were geopolitically dominant in India. They duly became absorbed in its cultural diversity, a proclivity which amateur landscape painting gave rich expression to. In the quiet decades after the Indian Mutiny (1857), interest burgeoned in the Indian tradition of theosophy: seeking insights into divine nature via direct experience – workaday, intellectual, mystical . . . In 1875, the Theosophical Society was founded in New York; and in 1889 was acceded to by Annie Besant (1847–1933). From an English Protestant background, she was undeniably a lady of parts – militant secularist; Fabian socialist; birth control activist; president (in 1917) of the Indian National Congress . . .

She endorsed as the messianic incarnation of Buddha, the adolescent Jiddu Krishnamurti (b. 1909). He overtly rejected this superimposed sanctity in 1929; and began to emerge as his own man, albeit with strong theosophic leanings. He saw self-knowledge as basic to all perspectives. Belief systems distract individuals and divide peoples. Sacred books add nothing because truths repeated cease to be truths. Imbibing the latest contributions is likewise futile, "for you may wander all over the Earth but you have to come back to yourself".[28] By the time I left university in 1957, Krishnamurti had displaced Gandhi as the Indian philosopher most regarded by young savants in the West. Then in the febrile sixties, overt interest was widely shown in such derivatives of Hinduism as Transcendental Meditation and the tightly disciplined Hare Krishna.

The enduring isolation of China and Japan impeded understanding, in the post-Columban West, of Mahayana ("Big Wheel") Buddhism and other regional creeds. Nevertheless, endeavours were made. Schopenhauer has been mentioned. Well before him, Leibniz had found in early Chinese writings some corroboration of his notion of a cosmos built up from monads, its infinitesimal embodiments.[29] Eventually Nietzsche was to develop an intense love–hate attitude towards Buddhism. The rejectionist aspect derived from his reading as "nihilism" the Buddhist concepts of (a) emptiness as preparatory to reawakening and (b) nirvana.[30]

In Britain at least, Buddhism was a strongish minority interest among the generation that fought World War Two, more so than any other mainstream religion outside the Judaeo-Christian ambience. This singular curiosity perhaps survives still. Many would give weight to the judgment of Keith Ward, formerly Regius Professor of Divinity at Oxford, that "Buddhists have been much more successful than theistic believers in advocating tolerance and compassion for all living things".[31] There is recognition, indeed, of the general potential of East and South Asia in this aspect of spirituality.

The roots of science culture (not least apropos the sky sciences) regularly attract attention. A pioneering study, *The Tao of Physics* was published in 1974 by Fritjof Capra of the University of Vienna and the Lawrence Berkeley laboratory. It was to be criticized on several counts by Ian Barbour, Professor of Science, Technology and Society at Carleton College, Minnesota. He found Capra too readily equated the unity which Asian tradition (particularly in its mystical aspects) accorded the cosmos with that modern physics knows. The former is essentially undifferentiated whereas the latter is "highly differentiated and structured, subject to strict constraints, symmetry principles and conservation laws".[32]

He also believed Capra was precipitate in celebrating how "bootstrap" physical theory related well to Asian philosophic perceptions of everything as reducible to ever smaller particles, thereby rendering talk of structure hierarchies invalid. Barbour pointed out that "bootstrap" theory had since fallen out of fashion. Perhaps, however, it is now being succeeded by string theory with its more coherent connotations in this context.

Yin and Yang

One of the areas in which Capra undeniably made a valuable descriptive contribution was the Taoist notion of *yin* and *yang*, Taoism essentially being a dualist philosophy emergent in China *c.* 250 BC. The *yin yang* dynamic is symbolized in the ancient *T'ai-chi T'u* diagram. The *yin* principle (dark, maternal, and receptive) is classically associated with the Earth. The *yang* (male and creative) belongs to Heaven.

Traditional Chinese medicine seeks to preserve a *yin yang* balance within us. Overall, the inside of the body is *yang* and the outside *yin*. Balance is kept by the continuous flow of vital energy, *ch'i*, along a meridional system punctuated by acupuncture points. Should imbalances cause illness, needles applied to selected points should restore normal flow.[33]

The modern history of acupuncture has been chequered.[34] Some surmise that many results are placebo effect. And how is *ch'i* defined and measured? But however weak or strong its clinical claims, the *yin yang* model can well be adapted to dialectical relations, ranging from the sex war within partnerships all the way up to Superpower showdowns, the latter scenario being one Revisionist Marxists have addressed to effect.

Modulated action–reaction is the warp and woof of social groups, parliamentary democracies, multinational alliances and, indeed, warlike crises. However, unbounded action-reaction could soon tear our fragile world asunder.

Psychology

Fritjof Capra found that "while the flavour of Hinduism is mythological and ritualistic, that of Buddhism is definitely psychological".[35] But could one as well say Eastern religions in general are more attuned to the psychological dimension? That psychological theory and procedures advanced much further in the West this last century was because, not least in the USA, the psychiatrist's couch was often seen as a straight alternative to sound metaphysics.

Sometimes this was very misguided. In 1936, in one of his more brilliant essays, Jung considered how acquaintance with yoga spread through the West after 1850. But it was a West hamstrung by "the strict line of division between science and philosophy", a negative legacy of the Renaissance Continuum. Yoga practice therefore became either a strictly religious matter or else a kind of formal discipline. Moreover, this was a dichotomy indicative of wider truths: "The Indian can forget neither the body nor the mind, while the European is always forgetting either the one or the other". Furthermore, the European "has a science of nature and knows astonishingly little of his own nature".[36]

D. T. Suzuki wrote of mysticism Zen-style relating "to the small things of Nature" as well as to the moon and sun, storms and waves . . . He noted, too, how Zen "despised learning of letters and upheld the intuitive mode of understanding".[37] Already, for East and South Asian religions in general, contemplation was "the supreme and ultimate religious value accessible to man. Contemplation proffered entrance into the profound and blissful tranquillity and immobility of the All-one".[38] One may ask how meaningful a meaning can attach to "entrance" in this context. One can further ask how acute a contrast there really is with Western mysticism.

Multiple Pathways

Evidently, much interest is now being shown by religious leaders and theologians in different faiths interacting in a revelatory quest. The Dalai Lama, Tenzin Gyatso, stresses the importance of such convergence enhancing spirituality: contemplation, love, compassion, patience, tolerance, forgiveness, humility . . . A non-religious person may, he thinks, find this harder to sustain. Which established obedience an individual may subscribe to matters less. He sees joint pilgrimages as one means of interfaith dialogue; and the goal should be "a genuine sense of religious pluralism" which should help avoid "religious bigotry on the one hand; and the urge towards unnecessary syncretism on the other". He likes the idea of a world parliament of religions.[39]

Similarly, Keith Ward looks towards "inclusivism". This he defines as a "pluralism which accepts that there can be many ways of relating to one spiritual reality, some more adequate in one respect and others more adequate in other respects". It thus becomes an attitude "which seeks to deepen its understanding of faith by attention to other traditions".[40] However, a quest for spiritual depth is not the same as one for a better understanding of divine reality or metaphysics. None the less, each will require interaction with Science. Moreover, a full manifestation of spirituality (be it religious or secular in origin) must involve concern for Wild Nature.

Britain is currently blessed with a Chief Rabbi possessed of great liberality of political outlook and a thorough understanding of the contemporary social debates. In 2002, he published *The Dignity of Difference. How to Avoid the Clash of Civilisations.* Herein Jonathan Sacks overtly pursued inclusivism, noting how every great faith has texts which, read literally, "endorse narrow particularism, suspicion of strangers and intolerance towards those who believe differently. Every great faith also has within it sources that emphasize kinship with the stranger".[41] Correspondingly, he called for an Israeli–Palestinian peace based on a two-state division of the Holy Land closely coincident with demography. It must also include agreement on Jerusalem and its holy places; joint supervision of water and other shared resources; and an international accord about refugees.[42]

Rabbi Sacks encountered orchestrated opposition, on certain counts, from hardline orthodox theologians. Uneasy, he agreed to revise certain passages in a second edition. One sentence deleted had said, "God has spoken to mankind in many languages through Judaism to Jews, Christianity to Christians, Islam to Muslims". In its place has come the more enclosed, "As Jews we believe that God has made a covenant with the singular people but that does not exclude the possibility of other peoples, cultures and faiths finding their own relationship with God within the shared frame of the Noahide laws."[43]

Acceptance of alternative pathways to (a) fuller spirituality, and (b) better understanding of the "eternal verities" requires reciprocal humility and tolerance. Yet these virtues will apply but superficially as between the mainstreams of belief unless they also obtain inside their respective base areas, geographic and thematic. With Christianity, for instance, inclusion must extend to conservative traditionalists. The dangers they could pose cannot be negated just by walking the other way. Besides, this school of thought does have a positive contribution to make. Even Creationism comes closer to true science when recast as "intelligent design" arguments. Then again, the hard Right in the USA, political and religious, was more insightful than most of us about Stalin and then Mao Tse-tung. Take, too, abortion. As someone who favours a liberal code, one must say how little one could abide the cavalier way this remedy was acclaimed by certain regimes and many individuals through the second half of the century past.

Addressing the admixture of defensive attitudes that makes up what is loosely referred to as the Religious Right could, in fact, be the most taxing part of the quest for a free-ranging dialogue about matters philosophic as between the great obediences but fully involving, too, the Sciences and other germane viewpoints. For undeniably the fundamentalism within religious traditions – Christian, Islamic, Judaic, Hindu, Buddhist – poses increased dangers yet has an added allure in a world closing in on itself. Fundamentalist emphases can vary – cultural protection, sacerdotal purity, legal rigidity, social exclusiveness, millenarianism . . . Any can lead their adherents into morbid political intolerance. But engagement to avoid this outcome must acknowledge the fair claims of more moderate conservatism, philosophic and social.

The Creationism particularly associated with Hard Right Protestantism in the USA betokens both the difficulties of, and the possibilities within, the situation. Today we cannot just laugh off the arithmetic famously deployed by James Ussher (1581–1656), sometime Archbishop of Armagh. Drawing on Genesis, he demonstrated by extrapolation that Creation took place in 4004 BC. It is not enough to negate latter-day replications by refuting point by point in learned journals the anti-evolution arguments that stem from them. The plain fact is that the weight of geological evidence against Creation six thousand and ten years ago is now so overwhelming that to deny it all is veritably to suggest that the whole cosmos works from a fraudulent ground plan. Quite an insult to a putatively personal and loving God.

Creationists should be obliged to confront this. Yet at the same time their protagonists must recognize that, even in the American Protestant context, Creationism and Fundamentalism are not synonyms, one of another. The term "fundamentalism" was coined in the USA in 1920 by Curtis Lee Laws, a member of the Northern Baptist Convention. He was forlornly hoping it would sound less extreme than "conservative". Neither he nor most of his colleagues were disposed towards millenarian action. Nor were they six-day Creationists. Their prime concern initially was to defend such time-honoured doctrines as the Virgin Birth and the Resurrection of Christ.[44]

It was the South that was starting to experience large-scale anti-Darwin crusades. It was Dayton, Tennessee that in 1925 witnessed the Scopes Monkey Trial, a schoolteacher being tried for teaching evolution. The strong and more general upsurge of Creationism from 1960 was a nativist response to the Space Age. Defend the Genesis Firmament against the Information Firmament.

Fresh Departures

Manifestly, the values spirituality embodies are too suppressed throughout modern society. One can therefore welcome the spiritual experimentation now taking place "within and across different religious traditions, within new religious movements, and outside religions altogether".[45] The most functional distinctions are as between different spiritual foci: comparative, feminist, ecological, global . . . Mainstream religion should stay posted.

Another needle question is how best to secure due recognition for Science. Admission of Creationism *pari passu* with Evolution as another hypothesis under test is veritably to return to the "Know Nothing" obtuseness abroad awhile in mid-nineteenth-century America. Obversely, when the political or religious ultra-Left goes for overkill testing of synthetic drugs while remaining careless of checking out rediscovered herbal remedies, it is resurrecting, on both counts, anti-science.

So how about when none other than Jonathan Sacks avers that we humans are here "because someone wanted us to be . . . rather as a parent giving birth to a child. The universe is neither indifferent nor hostile to our existence" despite its "more than a billion" galaxies. For the biblical "personal" God "responds to and affirms our existence as persons".[46] Does this not run contrary to Occam's Razor? A billion or more stars were never needed to underpin our existence on Earth. Nor may we have claim enough to regard ourselves as a deliberately chosen species, some kind of contra-scientific Godly implant. Nor does cosmic godliness appear to work like that in any case.

Belonging

A truism to keep firmly in mind is that the information explosion obliges all communities (religious, political or whatever) to become either more open or else more closed. Should the latter course be preferred, it will be all too easy to apprehend with Keith Ward religions degenerating into "exclusive and literalistic competitive ideologies" and the world ending in violent catastrophe.[47] If, contrariwise, a more open approach is sustained, one can anticipate conflation very gradually becoming confluence as obediences meld with one another and with revealed Science. But for many decades, the best established should retain identities sufficient for those who wish to belong at whatever level of commitment.

How strong, in times of turmoil, a yearning to belong to belief-based institutions may be is shown by the interaction with Stalinist Communism of two of the most engaged of that tormented generation of progressives born in the last decade before 1914. Jean-Paul Sartre joined "the Party" post-1945 despite his inability to accept its doctrines and a realization that his own head would soon roll should it ever assume power.[48] Likewise during the first winter of the Spanish Civil War (1936–9), Stephen Spender had fleetingly been a Party member because he saw Spain as central to "the struggle for the soul of Europe". Yet he immediately found himself unable to accept Communist constraints on self-expression.

Nevertheless, he continued afterwards to salute the decency and the assurance of the many Communists he met during a tour of Gibraltar,

Tangier and Oran early in the Spanish conflict.[49] Many of us of the Cold War generation have similar recollections. Political religions of various hues may (like their sacerdotal counterparts) infuse young people with a sense of purpose. The uglification is usually fed in higher up. Admirable though Britain's response has been in many ways to the London bombings of 2005, there has been nothing like enough concern to drive a wedge between the young suicidals and those who egg them on from locales well to rearward.

A signal attribute that a customary religious obedience can proffer is a rich core of relevant legend to be appreciated on its own quasi-mythic terms along with any historical warranty. The Noahic story marked the end of the Genesis account of early humanity viewed in the round. As regards an historical validation, a surge of water across the Turkish Straits *c.* 5550 BC created the Black Sea, a brackish extension of the Mediterranean 130 metres higher than the freshwater lake it subsumed. This event likely engendered the Genesis/Dead Sea scrolls story of Noah,[50] including his landfall by Mount Ararat in Eastern Turkey and God's promise of no repeat. Genesis also tells us Noah lived to be 950. Whoever penned that, some three millennia ago, cannot have seen this life as merely the prelude to a better one beyond the grave. But they did have a strong sense of being a part of an ennobling historical process.

Operational Conflation?

In December 2002, the month he was confirmed in St Paul's as Archbishop of Canterbury, Dr Rowan Williams delivered the year's Richard Dimbleby BBC lecture. Focusing on the dangers inherent in a market-led society preoccupied with consumer choice, he suggested one way these might be countered would be for religious communities and governments to engage "in a new way". Institutionalized religion, he freely conceded, "has a history of violence, of nurturing bitter exclusivism and claiming powers for which it answers to nobody. So the challenge for religious communities is how we are to offer our visions, not in a bid for social control but as a way of opening up some of the depth of human choices".

What then of the widespread tendency for major obediences to decline in their traditional base areas? Could this make them more ready to interact to good effect, one with another? How much might such interaction be compromised by intra-faith struggles between progressives and fundamentalists? How far could it serve to check fundamentalism?

Might extra mental space thus be created within which to deliberate about the findings of natural science? Can there be room enough for the proper ventilation of agnostic and atheistic opinions? This could be important as a matter of principle, though also in terms of being contributory at a time of conceptual flux. Just as soldiers and pacifists agree on the salience of the war question, so do believers and non-believers on the criticality of the God question.

Here one cannot essay final answers to these considerations. Perhaps it would be useful instead to put down a few markers, drawing largely on experience thus far this century. Where is Hinduism now at in regard to co-existence? What, above all, of Islam viewed thus in the round?

A Hindu Turning Point

The absence of tightly defined boundaries of belief being classically a Hindu trait ought to facilitate co-existence, not least with Islam. Yet in this instance, too, the expansiveness of a major faith is weakened by insecurity within. The very many millions of local shrines told of in the pristine India of the 1940s represented Hinduism's roots. Over the rest of that century, however, this linkage diminished for a raft of reasons, urbanization a dominant one. A reaction against the resultant anomie helps explain why the Hindi nationalists of the Bharatiya Janata Party (BJP) could weld together a coalition government after the 1998 general election. Through the turn of that year, there were numerous reports of communal clashes, not a few involving Hindus attacking Christians.[51]

Every twelve years, the thousand year old Kumbh Mela takes place in north India. This festival derives from a Hindu creation story in which Vishnu carrying a pot of sacred nectar, spills four drops upon encountering other gods. Where each drop supposedly landed, a fair or mela is held. The most important locale is Allahabad, a point of confluence of sacred rivers.

Some 70 million Hindus gathered there in January 2001. Overall, the accent was on religious co-existence. Ominously, however, Hindu extremists discussed the building of a temple at Ayodhaya, this in place of a medieval mosque on that site destroyed by Hindu zealots a decade before.

The Ayodhaya affair came to an ugly head on 27 February 2002 when a train carrying Hindus returning from there was set on fire in Gujarat killing 58 people. The next several weeks a wave of attacks on Muslims was sustained in that western state, home of Mahatma Gandhi, some 2000 people being killed. There were grounds for believing these attacks had been contingently pre-planned. Neither at this state nor at national level had BJP administrations handled the situation with unequivocal resolve.

At the time, some of us were very apprehensive lest this presaged the final descent of the subcontinent into communal violence and war. However, another big factor came into play, nuclear deterrence. Eight weeks after its formation under Prime Minister Atal Vajpayee, India's BJP-led government had conducted five military nuclear tests. A fortnight later, Pakistan responded with its own series. Neither state had ever signed the Non-Proliferation Treaty because of its discriminatory character.

In the aftermath of 9/11, there was a heightening of tension in Kashmir as well as elsewhere on the subcontinent. Over a million troops were now amassed one side or the other of the India–Pakistan border and huge numbers of landmines had been laid. But by the autumn of 2002, the two sides were scaling things down. Many commentators would agree with Pakistan's President Musharraf that reciprocal nuclear deterrence had contributed to this backing off. That year each nation state had tested nuclear-capable missiles.

In November, Sonia Gandhi, the President of the Indian National Congress, lectured with assurance in Oxford about India's vibrant traditions of multifaith inclusion and the inputs Islam had made to virtually every aspect of Indian culture.[52] Come the 2004 election, she would herself be instrumental in securing the BJP's defeat. Now it languishes in disarray with mutual deterrence its sole though important legacy.

Meanwhile India is "shining" as incipiently a "software superpower", an emergence persuasively portrayed as a legacy of a Brahmin tradition discernible from 3500 BP. In this, an ability accurately to transcribe and recall vedic texts was combined with a facility for abstract and innovative philosophic thought. The modern counterpart is joining up computerized linguistic precision with creative programming.[53]

Islam and the World

The broader question of Islam's interaction with the world at large is here examined at no great length because many have discussed it elsewhere. Nevertheless, several facets ought to be revisited. Among them is how a crisis of reciprocal mistrust particularly emanates from within the swathe of territory extending from the Horn of Africa to Central Asia. Here the impact of the West was for long low key by dint of the general impediments to access though more specifically because, between the Battle of the Nile in 1798 and the completion of the Berlin to Baghdad railway in 1913, no other power could even attempt to use the region as a fulcrum to unhinge British India. An encroaching Russia was episodically active in or near Persia and Afghanistan. But as Britain's Foreign Secretary, Lord

Salisbury, said in 1878 (with the Russians close by Merv, an ancient Central Asian city): "Mervousness does not stand the test of large-scale maps".

Therefore this zone was long left pretty much to its early historic sites, resplendent Islamic art and architecture, traditional *mores* and anti-modernizing sects. Then the advent of the internal combustion engine was to change too abruptly the scene. It progressively facilitated ingress while generating a mushrooming demand for regional oil. The result was to expose the whole area to continual culture shocks. My own experience of the new milieu was largely confined to three journalistic visits to Aden and South Arabia during the eighteen months of insurgency preceding Britain's withdrawal from that territory, on schedule, in November 1967. But even this has afforded a medley of memories expressive of the regional confusion and stress.[54]

It was in a taut Arabian milieu that Osama bin Laden grew up, his situation the unenviable one of marginalized privilege. How distracted this left him is evidenced by the vacuity of the statement he made endorsing the 9/11 strikes. All he really did was link them to the Palestine situation, no doubt to divert potential recruits to his cause and away from the allure of American lifestyles.[55]

However, it is important also to remember that far beyond the inner Islamic zone here identified, terrible communal conflicts had broken out in the sixties as Islamic people became overly anxious to catch up with a fast-changing world. Within two big countries – Nigeria and Indonesia – the upshot was extremely bloody. The Nigerian civil war, 1967–70, reportedly caused the deaths in one way or another of a million people in the country's Eastern Region; and Hausa Islamic militancy in Northern Nigeria had been a factor in conflict genesis. Meanwhile in Indonesia, maybe scores of thousands of Chinese had been massacred in 1965–6, many in spontaneous local eruptions. By then, too, in the neighbouring Federation of Malaysia, a forceful drive was under way to consolidate the national language and to build mosques. Lately Singapore had been expelled from the Federation. Then after the 1969 election, there were vicious anti-Chinese riots in Kuala Lumpur.

Clashing Civilizations?

Comparing all this carnage with experience of communal conflict of late, one can actually feel things have improved. This sense is reinforced by the degree to which established regimes in the Islamic world have backed (sometimes very overtly) President Bush's "war on terror". Obversely, Samuel Huntington has developed (*c.* 1993–6) the thesis that the world order should be viewed most fundamentally as a mosaic of discrete civilizations brittlely in contact along neatly defined fault lines. The key to any lasting accord lay in acknowledging this. What must be avoided is "the widespread and parochial conceit that the European civilization of the West is now the universal civilization of the world".[56]

He has evidently drawn upon the precedent of Arnold Toynbee (1889–1975), a British historian largely based at the Royal Institute of International Affairs. The basic unit in his systemic treatment of the broad sweep of history (mainly in a ten-volume *A Study of History, 1934–1954*) was the "full-blown civilization", a discrete entity that is born, grows and decays – its last age usually sliding into inchoate violence. This formulation long attracted fierce criticism, much of which could similarly be directed against the Huntington *weltenschaung*. Even armed conflicts take place more within than between identified civilizations. What distinct civilizations are listed varies from author to author. Also such studies are misleadingly two-dimensional. "Mental maps" are still maps. They cannot record how deeply young people in somewhere like Iran aspire for more personal freedom.[57] Likewise, Toynbee and Huntington in turn have ignored or discounted key scientific developments – e.g. in cosmology and data transmission. Arthur Koestler, publishing in the year in which the Soviets obtained the first pictures ever of the far side of the Moon, deplored the omission of Copernicus, Descartes, Galileo and Newton from *A Study of History*. He saw this as typifying "the gulf that still separates the humanities from the philosophy of nature".[58] He could have said much the same of Huntington's *The Clash of Civilisations.*

Besides which, synoptic visions are always subject to personal idiosyncrasies. One of Huntington's is to look at the world through American spectacles to an extent that is almost surreal. John Gray was not far wrong in saying of him that a "community or a culture qualifies as a civilization if it has established itself as an American minority".[59] Still, the overriding contrast is that, as from a fraught 1938, Toynbee was claiming to discern an underlying movement towards a better future guided by a thoroughly ecumenical God.[60] By not being concerned to go this extra mile in search of accommodation, Huntington may have left himself open to exploitation by those ready to combine a too aggressive conservatism with technological virtuosity. The rub is that, in our shrinking world, the choice between fanaticism and conciliation is becoming ever starker.

Acts of God?

Interest in the dialectic between Reason and Faith has undoubtedly revived of late. Already in the new millennium, it has been recharged by a succession of shock events, most notably the Twin Towers, the Boxing Day tsunami, the Caribbean hurricanes, and the South Asian earthquake. In the wake of the *tsunami*, sundry references were made to Voltaire's sardonic enquiry as to what sins Lisbon had committed to bring upon itself a death toll of 50,000 in the earthquake of 1755. But what, in principle, is the difference between that and the death of one infant from a congenital defect? If he or she had been the creation of a caring God, what would He or She have to have done to show themself to be uncaring?

One argument advanced (*vis-à-vis* the tsunami) was that it invoked upwellings of faith but still more so of compassion, which we were advised, "gives an insight into the divine". Richard Harries further quotes a rabbinic saying that, when God remembers those in misery, "He causes two tears to fall into the ocean and the sound is heard from one end of the world to the other".[61]

Granted, one can allow that, however "conspicuous" (see Chapter 7) it may be, the tsunami compassion did prove materially enduring through the early months.[62] What cannot be confirmed is how this translates into new-found harmony regionally. The areas worst affected, East Sri Lanka and Aceh, had long been torn by violent political tensions. What is evident is that the world's reserves of compassion were too depleted to address properly the South Asian earthquake.

A Creator Now Dead

Terms like "God" and "Godliness" are sure long to remain in parlance to connote the supreme entity which governs our cosmos albeit not other ones. This continuity may be reassuring. However, it will always be important to recognize that "God" is but a humanoid metaphor for something which lacks exact definition but cannot really resemble ourselves.

From what we observe at every level from individual tragedy to cosmic configuration, the Judaeo-Christian picture of a universe Creator who readily zooms into direct involvement with our societal and personal affairs is hard to sustain. What we have instead is a Being closer to the notion of Cosmic Consciousness or to a Weak Anthropic Principle. Indeed, it may not be far beyond the mainstream Islamic view of Allah as He to whom one is faithfully submissive since, in His austerely transcendental way, He is sublimely merciful and compassionate. "Islam" is translated as "submission to" or "having peace with" God.

Remarkably widespread among what we normally think of as primitive religions is the concept of a Creator who is now dead, His life's work complete. Versions were subscribed to by the Inuit and Hottentots. Ancient Egyptian and Classical Greek deities also tended to be mortal,[63] though on Assyro-Babylonian ones was bestowed the doubtful privilege of immortality.[64] A Creator "who is now dead" could be seen as a metaphor for the big single act of cosmic origination.

18

FULFILMENT REGAINED?

QUEEN OF PURE SCIENCE

Astronomy is poised to reaffirm its Copernican status as Queen of Pure Science. This is chiefly by dint of the burgeoning of Space research, American-led but with sizeable inputs from elsewhere – China, Japan, Europe, Russia, India, Canada. Much of philosophic import will be discovered. Obversely, finite limits on knowledge acquisition will loom sectorally. Academe as a whole will divide more starkly between subject areas with scope left for discovery and those which are over-exploited, worked out or plain inaccessible.

Astronomy currently knows all these states. Soon it should be possible to get a good indication of how strong an Anthropic Principle applies within our largish galaxy; and what this tells us about the part advanced Life may play in the cosmic scheme of things. However, assessing other galaxies will be far harder. Broad similarity can be presumed, given what we know about chemical and architectural correspondence cosmos-wide. Still, this is not proof positive. Then again, those concerned should prove able to map black holes better. These may or may not relate to the formation of universes beyond our own. Nor may we ever truly comprehend Consciousness, cosmic or otherwise.

To which one ought ruefully to add that astronomy may in due course witness a squeezing out of amateur observers. At present, they can still most usefully contribute – e.g. by looking directly for supernovae or scanning automated data. But further leaps forward in computing power will, before too long, undercut amateur scanning except (and it is an exception to celebrate) as an exhilarating educational tool.[1] Also, the amateur will always be well placed to contribute to the wider interpretation of the subject area.

Rejectionist Intelligentsia

A possibility ever latent is that a fair proportion of the literati will turn militantly against Space work. Several factors will bear on how far discord develops. What effect is globalization, promoted via the Information Firmament, having on peaceability? What profile is Space development assuming, especially in regard to matters military and to orbital tourism? Is cosmological research contributing to philosophic convergence worldwide? Will any protest be part of a general backlash against high technology? What else will the literati have to think about?

An intriguing exemplar of *yin yang* interaction in this domain was Theodore Roszak, the Californian *avant garde* sculptor (b. 1907), who became a prophet of the "flower power" side of the youth revolt which peaked in 1968. Then he roundly condemned our technocratic civilization, careless himself of how for very many (not least in California) it had ushered in a life style more comfortable and expansive than any previous. Among his more fanciful averrations was that "the strange youngsters who don cowbells and primitive talismans and who take to the public parks or wilderness to improvise outlandish communal ceremonies are in reality seeking to ground democracy safely beyond the culture of expertise".[2] What they were actually doing, over and above larking around, was keeping the voice of protest so callow as to be open to hijack by any extremist mountebank (Red or Blue or hybrid) purporting to embody a Rousseauist general will. Throughout this pronouncement, Roszak ignored Space research.

Come 1971, he is scorning "too much limelighted posturing by the astronauts and the research teams and the Nobel laureates . . . the boyish modesty, the understatement, the winsome embarrassment at all the applause".[3] Yet by 1979 he is persuaded that "the vast, many sided adventure we call *science* is far from being a monolithic establishment".[4] He tells of the "rights of the planet". Approvingly, he quotes those doyens of British interwar astronomy, Jeans and Eddington, as extolling mind as more fundamental than matter.[5] This revaluation evidently owed much to the USA's high technology military involvement in Indo-China finally being over.

More obstinately obtuse was the pitch Bertrand Russell adopted. In his *History of Western Philosophy,* he evinced a keen awareness of the import of cosmology historically. Nevertheless, he approached manned Spaceflight in the same spirit of blanket negation as when, in 1924, he had invoked classical mythology to warn against aeronautics. Icarus, having been enabled to fly by his father Daedalus, flew so close to the Sun that his wings melted.[6] Come the sixties, Russell's distaste for aerospace visions was supercharged by anger over the geopolitical ulterior motives.

Things came to a head with Apollo-11. Russell scorned the "silly cleverness" of Neil Armstrong's descent to the lunar surface on the grounds that Immanuel Kant "never travelled more than ten miles from Konigsberg".[7] He should have said sixty. No matter, jet-setting intellectuals ought not to deny the masses a modicum of mobility, be it first-hand or via their televisions. To do so is again to give vent to Whiggish or Renaissance hauteur.

Bertrand Russell might better have noted Kant's absorption in astronomy and in peace, together with his acceptance of an assessment of Japan made by Engelbert Kaempfer, the German physician who had travelled the country while serving with the Dutch East India Co. (1690–2). Kant had concluded that philosophic unity was basic to Japan's peace and progress. Russell could next have considered how far cosmology might help to forge a similar unity throughout the modern world.

Likewise Arnold Toynbee condemned Space programmes as "morally offensive" gambits in a "childish" competition between the superpowers to see who would be "ascendant on this planet".[8] Then shortly before his death in 1975, he dialogued with Daisaku Ikeda, the President of Sōka Gokkai. Space exploration came up with Ikeda insisting that experience showed how multinational programmes could yield "exceedingly good" practical results. However, Toynbee's response was that all Space exploration should be deferred at least until the material well-being of "the poverty-stricken three-quarters of mankind has been raised to the present level of the affluent minority".[9] Wait for the Greek Kalends?

Pragmatic Considerations

In December 2004, a Specialist Discussion on "The Scientific Case for Human Space Exploration" was held at the Royal Astronomical Society in London, NASA being well represented. The flow of the argument was to the effect that astronauts still retained tangible advantages over robots (in terms of curiosity, dexterity and adaptiveness), advantages which contingently could justify the added expense. However, concern was expressed that human beings entering a microbial ambience could be more liable than metallic artefacts to become carriers of infection. The emphases placed on this consideration varied but there was no outright dissent. For one thing, our terrestrial experience suggests that, if one did encounter ambient microbes, these could assume a myriad of forms. So it is inadequate to suggest, as some have done,[10] that the risk of any particular form proving terrestrially pathogenic was very low.

Anxiety was not being aired for the first time, not by any means. The provisional plan has been that by 2020 a NASA unmanned probe will have brought back a Martian soil sample. A panel convened by the US National Research Council warned in 1997 that such an exercise must be subject to stringent precautions.

However, nothing here would be an argument against establishing facilities manned and unmanned on the Moon, a celestial body which is almost certainly microbe free. A lunar manned observatory would have pronounced advantages over terrestrial ones as regards the absence of local electromagnetic interference; no atmospheric absorption; no debris in suspension; and no overlay of return signals from communications satellites. This last-mentioned impediment is worsening fast on Earth, considerably because of the spread of mobile phones. More particularly, low interference could help make lunar facilities critically useful for the detection, tracking and – if needs be – interception of Near Earth Objects (see below). The only caveat to enter is that facilities on the airless Moon could be exposed to bombardment even by micrometeorites.[11]

The Americans are now preparing for further manned missions to the Moon. These will initiate an extended presence using, to an extent, local raw materials. Among the several other countries with an active lunar interest is China. Apparently her targets include robotically retrieving lunar samples by 2020.

Meanwhile, the expectation remains that an American manned landing on Mars will have occurred by 2030; and during the present decade several more NASA and ESA orbiters and landers will visit the Red Planet. Already there are signs of water being, or having been, widespread there. Evidence that at least microbial life may still be extant is principally in the form of a methane presence. Isotopic tests to indicate whether this is, in fact, of biogenic origin are awaited.

Manned versus Unmanned

Microbial concerns apart, professional attitudes towards manned Spaceflight are probably more stringent now than in the heady days of the Sputniks and Apollos. Nobody any longer looks towards manufacturing in the zero-gravity ambience of orbital flight, a prospect initially enthused about in both the West and the USSR. Nor does one hear much about assembling and servicing Solar Power Satellites (SPS) intended to microwave the Sun's energy to terrestrial transformers. Yet this idea was much in vogue twenty years ago. Though Soviet previsions dating back to 1929 were claimed,[12] a paper delivered to the 1968 International Astronomical Congress by Peter Glaser, a US engineer, was what had triggered this upsurge of interest.[13]

As discussion progressed, however, it became clear that the econometric calculations were far from positive. An SPS array intended to replace a large (2500 megawatt) power station could cover 20 square kilometres of sky and have a mass of 15,000 tons. Soviet scientists were reckoning that such an investment could only become competitive if the cost of placing mass in orbit could be cut by nine-tenths.[14] Yet what did as much to foreclose this option were ecological considerations. Reining back global warming was all very well. But what about microwave overspills onto radio transmissions and, indeed, life forms? Regarding the biogenics, nobody seemed sure what thresholds, if any, applied.

But now apprehension that acute climate change is nearer in time than had been thought has stimulated anew an interest in remedial Space engineering. Among the approaches lately considered at the Lawrence Livermore National Laboratory in California is reflecting sunlight outwards by means of a diaphanous mirror parked in high atmospheric orbit. One a thousand kilometres across might suffice to compensate for a doubling in greenhouse gas levels. Already the sceptics have raised doubts, however, not least about possible aggravation of the temperature falls in the stratosphere apparently in train as temperatures in the lower atmosphere rise. Such falls could further compromise the delicate ozone layer so vital to shielding the biosphere from deadly Ultra-Violet C radiation. Nevertheless, James Lovelock is quoted as giving broad support to Space engineering solutions.[15]

All concerned would probably accept that humans will long be superior to robots and automatons for repair and construction in Space or, indeed, the free reconnoitre of other celestial bodies. Even so, talk of mining the asteroids for high value metals is surely premature.[16]

Nor is so much heard these days about manned spaceflight being peculiarly able to generate invaluable spin-off across the manufacturing base. For all the present players, China not excepted, technological progress is on a broader front regardless. All the same, manned Space programmes may do more than unmanned to underpin the rest of a nation's aerospace industry; and this, in turn, may be an important factor in technological sovereignty. Lastly, manned flight can be peculiarly expressive of international competition or collaboration or something ambivalently in between. As in the Apollo days, there is indeed apprehension of meteoritic bombardment when the sites sought are airless or nearly so. But this problem may not be too serious for short stays.

Take the International Space Station (ISS), an Earth-orbiting platform with a chequered history ever since it was proposed by President Reagan in 1984. A decade later its survival in Congress hinged on Russia being invited to join, an ulterior motive being to induce Moscow to come inboard metaphorically by more actively constraining the proliferation of nuclear warheads. Scientists in the USA and elsewhere have generally seen the ISS as an otiose diversion. Nor have matters been helped by cultural divides in orbit, especially between scientists and astronauts.[17] It has been, alas, another example of an exercise in technological virtuosity being embarked upon without a well-defined operational purpose.

Intra-Galactic Travel?

From 1957, Freeman Dyson was a lead scientist in Project Orion, the US government-funded paper study of what was conceived of as a 4000-ton nuclear-driven Spaceship (with a 1000-ton payload) intended to explore the solar system. After the expenditure of eleven million dollars, it finally died in 1965, faced with the opposition of NASA, the Pentagon hard core, the test-ban community and scientists in general.[18]

A major weakness of the concept was its being uneconomic to scale down because of the relatively constant shielding costs incurred in a vehicle impelled forward by nuclear energy. Instead, Dyson and others saw it as a pilot scheme for bigger spaceships, some of which (driven by hydrogen bombs) might visit other stars.[19] But with Sirius, the nearest being 4.7 light years away, alternative means of propulsion are mooted. One might be to exploit (in the twenty-second century?) explosive interactions between matter and anti-matter. Another could be nuclear fusion released more steadily. Fusion power, in a static terrestrial setting, has been worked on for several decades by an able and scrupulous international community. So far, however, the results have been inconclusive. Meanwhile on the fringes, there has too often been acute hype.

Still, the toughest problem might not be choosing a means of propulsion able to attain a high fraction of the speed of light; and coping, at such velocities, with what could easily be catastrophic collisions with even small particles. It could rather be managing the human stress. Round trips extending into decades would be phenomenally taxing in themselves. Moreover, the cumulative strain would be acutely compounded by landing on some extra-solar planet only to discover there life forms way outside the limits of human imagination. Suppose one did accept unreservedly the Simon Conway Morris perspective that evolutionary convergence from a variety of structural backgrounds renders almost inevitable the eventual advent wherever of animals with visual and mental capabilities akin to our own.[20] The outward appearance of these creatures could still induce dreadful culture shock. Yet the greater likelihood of landing and finding no life nor anything else extra special could be even less bearable.

However, negative reactions just might the better be contained if, come the time, much of Humanity (the cosmonauts included) was disposed to believe in the Cosmic Consciousness being proposed by some of us as a core theme in a new philosophic vision. Orbiting Sirius, say, could then have something close to spiritual connotations. Additionally or alternatively, it could reinforce the sense of togetherness on a Spaceship Earth we have all felt in some measure since the Sputniks and Apollos.

It could be instructive to quiz Renaissance historians about the impact on the European mind of the trans-oceanic voyages of discovery. Syphilis and silver apart, they have usually been dismissive. The actual explorers, Columbus equivocally excepted, are seen as having been little in touch with book learning. Nor were most of the intellectuals of the day much inclined to enquire about new-found ways of life. Therefore it behoves the advocates of deep Space exploration to make their case good when the time comes. What such a mission might comprise is a couple of spaceships plus a clutch of unmanned probes.

Astronomy in the World

Regardless of any such programme, however, astronomy/cosmology could eventually experience in very full measure the limits to knowledge liable to affect ever more subjects as we move past mid-century. Take that most ancient and abstract of related disciplines, Pure Mathematics. In 1900, twenty-three problems were famously identified as awaiting solution. In 2000, seven were. By 2050, all bar one or two may either have been identified or adjudged unsolvable. Granted, more basic aspects of subject philosophy could then be weighing in.[21] So could more sophisticated risk assessment, not least for astronomic application. Even so, knowledge horizons will not be boundless. It will therefore be important to ensure that the straightforward pursuit of knowledge does not give way too much to untestable speculation.

But the other side of this coin is the option of casting more broadly, in disciplinary terms, university teaching programmes, not least ones intended to equip young people not so much for academe as for more workaday professions. In England at least, Engineering and Economics is long established. So how about Theology and Life Science? Or Cosmology and Economics? Could Cosmology come to be seen much as the study of Classical Antiquity customarily has? In other words, it is a field sufficiently rigorous and distinct to be a good preparation for mental adaptiveness across the board.

Granted, there are academic risks in casting interdisciplinary studies wide, even on the bipolar basis here envisioned. The same goes for other modern tendencies. One which involves cosmology is the use of algorithms to process computer-acquired data and extrapolate from this. It can tie field work too closely to the theories from which the algorithms derive. Meanwhile, increased scope for fraud may be afforded by the Internet. Also, Science may draw too close to governments preoccupied with cost–benefit assessments, applied narrowly and short-term. Cosmology has the leverage and therefore the obligation to fight for appropriate norms, in this respect as in others. It will be a searching test of its status as Queen of Pure Science.

Meanwhile, it should ease its twentieth-century taboo against involvement with philosophy in whatever form. It is time to reconsider the received wisdom enunciated by the biologist, Jacques Monod: "Any mingling of knowledge with values is unlawful, forbidden."[22] Often cosmologists have defined philosophic positions – Atheist, Theist, Logical Positivist, Marxist, Judaic, Protestant, Catholic, Islamic, Buddhist . . . Nor did last century's proscription against overt linkages mean the individuals concerned never conducted a private dialectic. Still more general openness could sometimes clarify themes like the Anthropic Principle, Cosmic Consciousness, Multiverses . . . Indeed, explicitness can be an antidote against faith (i.e. religion, atheism, political religion or whatever) acting as a barrier to truth.

CULTURAL REVOLUTIONS

Science and Free Society

The importance of Science as a primary means of proffering society more material choice is too evident to dwell on. Hardly less obvious, yet in need of continual demonstration, has been its pursuit *à l'outrance* and for their own sake of hard fact and sound analysis. How important this remains may be gauged from a countercurrent of hostility "fed by a curious confluence of individual alienation, religious fundamentalism, extreme environmentalism, and even elements of postmodern scholarship".[23]

A standard post-modern critique of Science is that it is pervasively culture-bound, an allegation always liable to carry overtones to the effect that objective truths and absolute values are mythic. The proponents thereof have read Thomas Kuhn but maybe little Science. Few scientists, in my observation, have read Kuhn.

Meanwhile, tight-knit extremist minorities within the environmental movement pitch the other way. Certain panaceas are revealed to them so irrefutably that they must adopt whatever means to ensure their unimpeded application. Yet the sad reality is that, these next few decades, real world Environmentalism can only be a vast exercise in damage limitation. There will be little chance to attain absolute goals and much pressure to select the least bad of some very disagreeable options. Very like a war, in fact.

Take the identified alternative approaches towards meeting, in Africa and elsewhere, a world food demand liable to double by 2050; and to do this while preserving wildlife as best one may. You can go for making farmland wildlife-friendly, accepting less agricultural output per unit area than might have been achievable. Otherwise you can limit the amount of wilderness brought under the plough by working farmland as intensively as possible. The former approach will always be favoured by eco-militants yet the latter may sometimes damage wildlife less.[24]

Often militancy can be at its most fanatic, not least against Science, behind the banner of religious fundamentalism. A favoured battleground is education in high schools. Within the USA, Southern Baptists and their like want Creationism taught at least on par with Darwinism, never mind that professional scientific opinion worldwide is solid for the latter with good reason. Their cause ebbs and flows but is not breaking through decisively. Meanwhile, across *dār al Islām*, the incipient "Islam and Science" movement sees science teaching in Muslim religious schools – *madrasahs* – as basic to general modernization.

Happiness?

An area of academic enquiry which has mushroomed this last decade is what makes humans "happy". Lately the findings therefrom have attracted media attention. This makes all the more urgent a need to ask whether Happiness is the word which best connotes the kind of aims, societal or personal, one should be promoting.

Etymologically, the word is traceable to the Old Norse for fortuitous circumstances, those lucky breaks which make a hard life comfortable awhile. A good harvest or hunt. A mild winter. An absence of epidemics or marauders. Happiness is as it happens. Apparently, its Greek synonym, *endaimonia*, is also linked semantically to the chance factor. The late Trevor Huddleston, an Anglican bishop who campaigned against Apartheid tirelessly within South Africa and without, used ironically to tell how a good harvest in a tribal reserve could engender joy of an intensity unknown in modern western society.

Happiness is mentioned by Spenser, Shakespeare and Milton. However, it really goes centre stage during the neo-classical eighteenth-century Enlightment. It features in the American Declaration of Independence and then in the French Declaration of the Rights of Man. Soon the Utilitarians took it up. But does Happiness as a perennial political aim meld too readily into hedonism, into uninhibited self-indulgence?

The nineteenth-century Romantics thought so. As a mainstream precept, Happiness came across to them as trite and complacent. Worse, as mouthed by Jacobin extremists, it was an obscene malapropism, redolent – we would say – of Stalinist celebrations of freedom. What the Romantics themselves usually looked towards was a world transfigured "through our renovated perception of it".[25] The term "joy" regularly reappears but strictly in the transcendent and perennial sense in which it is attributed to Jesus in St John's Gospel: "These things I have spoken unto you that *my joy might remain with you*" (my italics).

A salient theme for Buddha was the transience of hedonism. Around the same time, the author of *Ecclesiastes* observed, "All human toil is for the mouth, and yet the appetite is not satisfied." The Bhagavad Gita says that escaping the "jungle of delusion" required abandoning the "desire for joy", here meaning hedonism. Then in the last century, Sigmund Freud's emphasis on highly conservative instincts precluded untrammelled happiness because giving vent to those instincts was not compatible with civilization survival. One fine day the human intellect might take charge but only, he feared, "in a distant, distant future".[26]

Holistic Environment

Yet as one reviews the current literature about Happiness, one finds that many who study it are not really concerned with some blissful absence of negatives. Rather they are looking for engagement and fulfilment, a measure of spirituality likely included.[27] Realization thereof may be judged by each individual within themselves but also in relation to how society around them reacts. Significant, too, will be the ecological background, man-made and otherwise. Likewise important will be any received metaphysical account; and how swayed thereby individuals may be.

For billions of people, the contemporary ambience is dysfunctional in all these respects. Arguably, the battle against the excesses of modernity has barely been joined thus far. Much scope remains for political *chevauchée* – continual skirmishes in pursuit of ultimate objectives. Not least does this apply to ecology. Take the careless suppression of Dark Skies with their starfields referred to in Chapter 8. Correcting this could go hand-in-hand with the judicious limitation of lights indoors. In 1933, an eminent Japanese novelist, Junichirō Tanizaki, wrote an evocative essay on how profoundly the traditional décor of Japanese homes, eating places and temples had depended on the interplay of soft light and deep shadow. Feeling the cause was lost already, he told how he "would call back at least for literature this world of shadows we are losing. In the mansion called literature I would have the eaves deep and the walls dark, I would push back into the shadows the things that come forward too clearly . . . ".[28]

In due moderation, shadows, silence and slowness are at the very heart of spirituality. Nevertheless something was absent. It is concern about the disappearance of the starfields which, by 1933, must have been all too well in train across urban Japan. But was Tanizaki just being archetypically Japanese in this respect? A British observer of Meiji Japan at the turn of the century had noted an awareness of the beauty and spirituality of the Moon. But the stars seemed to him "much less admired and written about in Japan than in Europe. No Japanese bard has ever apostrophised them as the poetry of heaven".[29] The pronounced seasonal weather drew attention instead. However, in the previous period of *Sakoku* or "seclusion" (1636–1867), a good deal of astronomical observation took place in Japan, using Chinese or – increasingly – Western frameworks of reference.[30] During the Heian period (794–1185), the rising influence of Buddhism from China had considerably found expression in the celebration of the Pole Star and Great Bear as sources of earthly authority.[31]

China's Mighty Experiment

Undeniably, China poses a huge question for the medium term – say, 10 to 15 years ahead. Spelt out item by item, her modernization agenda would seem ambitious even if applied to countries orders of magnitude smaller demographically.[32] On the other hand, she is currently in the hands of a purposive elite which at the top "consists almost entirely of science-based engineers".[33] It is by no means without venality and nepotism but is, in some cardinal respects, most impressive.

The approach to Ballistic Missile Defence cannot always be a measure of wider comprehension. But how it has lately been addressed in China does afford insights into the timbre of the regime. My own experience of dialogues in 1987, 2000 and 2002 with ranking Chinese officials and academics has been encouraging. Certainly the colloquium in Beijing in March 2000 involving a senior Anglo-Franco-German delegation was far more fruitful than any such event one recalls with the Soviets.[34] It was more relaxed, pragmatic and candid.

Moreover, this was in line with advice received about Chinese trends in general philosophy. They display more "epistemological optimism" than currently obtains in the West about how far systematic analysis can take one in terms of resolving philosophic dilemmas – e.g. how rapidly should China democratize? This sanguineness also extends to the interrelating of dissimilar themes.[35]

Not that any of this means China's rulers are bound to judge well when and how to move towards a free pluralistic society. Throughout modern history, privileged strata everywhere have prevaricated too long over this looked-for derivative of material progress. Neither the Tiananmen Square showdown nor the Falun Gong suppression leaves one too confident this Beijing elite will be different.

Nor should anyone simply see liberalization as guaranteed because essential to continued economic advance. This argument we know of old. After Stalin's death, we were told (by a revisionist Marxist) that "aircraft designers must be let out of the prisons, literally and metaphorically", if the USSR wished to compete in aerospace.[36] Alas, aeronautical science is never that regime sensitive. Witness Werner von Braun.

Nevertheless, aspects of the Chinese situation do warrant some optimism. Thanks to the syncretism of the formative religious cultures, much of the country is culturally homogeneous, dialect variations apart. Today the popular understanding is that "we Chinese" are revitalizing an ancient culture, not least in relation to sky science. May one add that a little remarked aspect of this legacy post-Han is the large-scale settlement of what we know as central China, a deployment into unfamiliar terrain undertaken to secure strategic depth in the face of famine-prone steppic barbarians. A pioneer historical geography of East Asia averred that "No other people have achieved a comparable transformation nor shown themselves so adaptable."[37] Her traditions of technocratic adaptation contemporary China draws on mightily.

It is incumbent on the outside world fully to engage with the manifold expressions of Chinese soft power now impacting globally. Also important is that North Korea's fanaticism be not left to boil over against a background of grudging Chinese indulgence and hesitant American stringency. Meanwhile Washington's policy towards Beijing must not come across as but a Dullesian one of comprehensive containment. This would shore up *immobilisme* on all sides.

At present, relations between Beijing and other capitals in the Confucian zone (Pyongyang, Tokyo, Taipei) are fraught. Nevertheless, bilateral dialogues within this compass probably hold the key to lasting peace, regionally and beyond. Even in its currently low profile, Confucianism still caps a syncretic tradition. It stands itself for Earth-bound pragmatism, being committed simply to the rather Aristotelian notion that, provided family ties are vibrant, virtually any form of government can and therefore must be edifying.

Max Weber, the founding father of German political sociology quoted in Chapter 15, saw the Confucianism he knew of as, like Catholicism, inimical to enterprise and liberty. But a doctoral student at Colombia has lately emphasized that, in the developing world today, Catholicism can be quite a force for democratic transformation. Furthermore, his own researches on twentieth-century China revealed how often people informed by Confucianism spearheaded movements for democratic change.[38] So maybe its principled pragmatism can further democracy throughout East Asia; and hence, too, the full liberation of its philosophic/metaphysical potential.

Eventual Limits to Growth

In the sixties, "limits to economic growth" got a bad name. For one thing, the case made was overly reliant on dogmatic but arbitrary econometrics.[39] For another, despite some scrupulous advocacy,[40] it never came across overall as judiciously weighing gains against losses. The former were usually presented as large, unqualified, and readily realizable.

Therefore an alternative concept of "sustainable development" emerged, signally at the UN's 1992 Rio conference on Environment and Development. Yet although mentioned in 11 of the 27 principles in the Earth Summit Declaration there formulated, the concept was inadequately delineated. Environmental protection was subsumed (the 4th principle). Poverty eradication was basic (the 5th). Something beyond economic growth was sought (the 12th). Otherwise, render according to taste.

There had been a lack of firm guidance from academe or officialdom.

Witness two Muscovite inputs. A 1997 Presidential *ukase* saw sustainable development as a national security objective. Yet a *ukase* the previous year had linked sustainability to mystical notions the Russian scientist, V. I. Vernadsky (1863–1965) had entertained about the "sphere of wisdom" or "noosphere". This he defined as a milieu in which the measure of wealth will be "the spiritual values and knowledge of humankind, existing in harmony with the environment".[41]

Very distractive currently is the spread of projections for world economic output the next turn of century, assuming no overall strategy for growth limitation. In 2001, the Intergovernmental Panel on Climate Change (IPCC) cited estimates fanning out by more than an order of magnitude from the present aggregate of Gross Domestic Products (GDP) (see Chapter 8). Yet not even this divergence covered every plausible possibility. How accurately can one predict long-term changes in the efficiency of energy use? Take, too, demography. Soon figures were adduced showing birth rates falling faster than previously realized. Obversely, other reports were predicting quantum extensions of mean life span via medical breakthroughs.

Extrapolations of energy demand are therefore liable to be so broadly cast, yet medianally so steep, that the precautionary principle calls for a modicum of austerity or what the religious term "asceticism". Undeniably climate change is a major complication. So is a strong possibility that oil and gas output will peak through mid-century in any case.

The Poverty of Strategy

All of which obliges one to plea yet again for a radical realignment. The subject within which my working life has largely been spent – namely, "strategic studies" – desperately needs a basic revamp. Philosophy has been described as basically a discourse about death. This characterization can be challenged as incomplete. But nobody is going to deny that strategic studies is ultimately about armed strife. Nor that for some years after its first appearance (in 1958) this effectively meant the "nuclear balance of terror". So may this erstwhile young staffer (1962–4) at the Institute for Strategic Studies (now the decidedly nodal International Institute for Strategic Studies, IISS) laud the deal of seminal thought then generated.

By 1970, however, this era was over. Indeed, some were saying that strategic studies had now lost its *raison d'être*. In fact, the subject did survive but conspicuously failed to move on sufficiently. It has failed to ever since, even though the collapse of Soviet Communism *c.* 1990 freed it to explore new horizons. Verily, its golden dawn had proved to be its high noon. Quality work is still being done but within too narrow confines.

An IISS directing staff perspective on future strategic directions was published in 1992. With no little verve and insightfulness, it argued the need to break out of the nuclear confines. Even so, it mentioned climate change not at all, not even in connection with the water factor in Middle Eastern conflict.[42] Yet in 1988, opinion worldwide had come, visibly and decisively, to recognize climate change as menacing stability. In 2004, Sir David King, chief scientific adviser to the British government, was to describe it as a bigger security threat than "terrorism".

The background to the existing gridlock was examined in my 2004 study, *Global Instability and Strategic Crisis* (Chapter 3). Suffice for now to identify interlocking themes which ought to be embraced. The Information Firmament. Tensions within Advanced Industrial Societies. Globalization and the Developing World. Climate Change and Instability. Other Environmental Stresses. Crime and Drugs. Natural Disasters. Near Space and Ocean Management. The Action–Reaction Phenomenon. Philosophies of Internal Security. Alternative World Futures.

Quite a bit of germane literature is being generated and merits incorporation.[43] Episodically this occurs, a good example being an assessment in the IISS house journal of the Boxing Day *tsunami* aftermath.[44] What is urgently required, however, is a basic recasting of priorities. Arguably, too, any such regime change necessitates name change too. My own feeling has long been that due account should have been taken of a comment in 1972 by Zbigniew Brzezinski, later to be National Security Adviser to President Carter. Contrasting the time-honoured "international" approach to global affairs with the more "planetary" one espoused by radical youth, he called for concepts to blend the two.[45] Since when, "planetary" has remained too counter-culture for many. Still, the term "world" may be gaining ground on "international". So "world survival studies" instead of "international security affairs" or "strategic studies"?

The Just War

The "Just War" concept might usefully spearhead a new departure. After all, it has figured in Chinese thought since Sun Tzu. So has it in Islamic thought since the Prophet's injunction to "Fight in the name of Allah and in the Paths of Allah" with adversaries being given a prior option of compliance. The concept of *Jihad* or Holy War against injustice eschews violence against the innocent including the elderly, women and children. Avoidance of collateral damage to Nature is also enjoined.[46] There has lately been recognition within Islam as in the West that advanced technology is making selective targeting more feasible.[47] However, those who dispatch suicide bombers are bound to say that technology gains so asymmetrical oblige the weaker side to be less discriminating.

A consummation to hope or pray for has to be that, as and when an inter-faith doctrine of the "just war" does evolve, it will embrace obligations to rebuild any shattered communities as well as protect throughout the natural and cultural heritage. Recent interventionary strategies have lapsed badly, either through omission or commission. Thus Afghanistan has suffered terrible despoliation, ecological and cultural, during the continual strife since 1979. Moreover, after the breaking of the Taliban in 2001–2, opium production increased dramatically.[48] This situation is now being addressed in a style synoptic but desperately unsure.[49]

Archaeological attrition was a most negative aspect of the long confrontation with Saddam, things coming to a head with his overthrow. Though the sequence is not entirely clear, there is no denying the terrible damage inflicted by various parties well before, but also during and after this *démarche.* No matter the urgent warnings from the art world.[50] Now one stands aghast at how little this aspect has figured in the subsequent debate about the confrontation as a whole. We are, after all, talking about Ancient Mesopotamia – Chaldeans, Sumerians, Babylonians, Assyrians, Persians, Hammurabi, Harūn ar-Rashīd . . . So much of our common background.

What is Terror?

Next, let us consider the lopsided application, especially since 9/11, of the word "terrorist". This subject, too, was explored at length in the 2004 text (Chapter 3). May one now simply reiterate a plea against a manichean disposition to make "terrorist" a corral word which takes in all insurgents everywhere but security forces nowhere. Half the legends of American history would be cast as terrorists on this crude reckoning. More immediately to the point is that such linguistic spin encourages harshness and, indeed, abuse by security personnel.[51] Christ stopped at Abu Ghraib as well as Guantanamo.

Much of the stereotyping derives from people who have never had real-life contact with a guerrilla movement. Had they done so, they would have noted a contrast between youngsters naïvely in search of fulfilment and their managers, most of whom either have become, or else always were, consumed with self-advantaging. They might also have found many within a disaffected population whose support for actual insurgency was highly contingent on the current situation, the *yin* and the *yang*.

No particularity more clearly shows strategic studies to stand in need of broad relaunch than does its failure to test the modish semantics of "terrorism". The bottom line ought to be that we should all be free to judge for ourselves when armed violence passes that unacceptable threshold, regardless of where this violence emanates from. To my mind, the kidnapping since 1999 of several thousand Chechens (by Russian security forces or pro-Moscow factions) should count as terrorism if anything is to.[52]

Other Agendae

Multilateral arms control was never an easy objective to pursue. Qualitative differences are hard to take account of. Regional geography, too, is difficult to factor in. The last 15 years or so, both complications have been accentuated.

Accordingly, the 2004 study proposed that what it termed Internal Arms Control be applied more comprehensively than hitherto (Chapter 12). This would involve naming various kinds of weapons which armed forces could use only with express government or alliance permission, much as obtained in NATO through the Cold War with theatre nuclear warheads and also poison gas.

Sometimes such constraint would be required because of how certain weapons genre inflict death or injury. But it needs also be applied to most or many of the exotic non-lethals now under development. These should be covered because their capacity to frustrate and humiliate could, if freely exploited, have decidedly totalitarian connotations. Then again, laser beams used not just for ranging, signalling or surveillance but as weaponry *per se* may pose a specific threat to eyesight. It is extremely disturbing that no free discussion appears to have taken place in NATO or elsewhere of the connotations of the AL-1A Airborne Laser the United States Air Force is now expected to receive into service from 2010 in modest numbers (2004 study, Chapter 9). St. Augustine would have been horrified.

Clandestine Biowarfare

What has just been covered clearly lends itself to cross-cultural dialogue in a variety of modes, formal and otherwise. Important in their own right, such exchanges may also serve as useful precedents for application to the realm of bio-ethics. A couple of decades down the line, the two domains may have to come together to tackle the gravest challenge humankind has ever inflicted on itself: that of biowarfare, this conducted clandestinely by insurgents or intelligence services.

How a world survival strategy may be hammered out apropos this threat remains to be seen. But it ought to comprise the following strands (2004 study, Chapter 6):

A common approach to prophylaxis. Doctrinal agreement on intervention strategy against regimes understood to be evolving offensive capabilities. Readiness to allow (as a licit alternative) some proliferation of nuclear weapons within the context of multirole alliances. Restoring a broad subjectivity to the meaning of "terrorism", avoiding *inter alia* popular backlashes against the unfairness inherent in using the term one-sidedly in situations where there is systemic injustice.

Crucial, too, will be the non-weaponization of Space. For the United States or any other major power comprehensively to deploy Ballistic Missile Defence in orbit could be to invite being cast as supremacist. Some would then insist that a capability for spreading germs at surface level would be justified as a means of undercutting this screen.

A particular danger is that of alienating China gratuitously. Granted, Beijing has not been diplomatically active on this score of late. Nevertheless, the Chinese community of defence analysts remains disposed to see their country as the one Space-based BMD will eventually be aimed at.[53]

A Critical Departure

However, the concept of Global Missile Defence (GMD) may steadily be approaching a parting of the ways with profound implications philosophically. One has hopefully in mind its being redirected away from "rogue" regimes and towards the threat posed by natural missiles from afar – meteorites, asteroids and comets. As C. S. Lewis would have put it, GMD will no longer look inwards beneath a "towering" yet enclosing firmament but outwards across a cosmic "sea which fades away". This much will surely be bad news for Creationism and, by extension, all other fundamentalisms, news that no backlash of intolerance could gainsay. On these grounds alone, one must trust this realignment will not be undermined by those who see themselves as "progressives" – the Bertrand Russells of the next generation.

The world data base on natural missile/Near Earth Object (NEO) threats is growing. A multinational Spaceguard Foundation was established in Rome in 1996. By 2004 the NASA-led Spaceguard record of NEO asteroids at least a kilometre across had risen to 700 as against 150 in 1998. Toutatis, an asteroid several kilometres wide first observed in 1989, passed within 1,500,000 km of the Earth in September 2004.

A British National Space Centre report assessed the risk thus in September 2000. An NEO 75 metres in diameter will hit the Earth on average once a millennium; and this with a force of some tens of megatons, enough to flatten a metropolis of several million. One a few hundred metres across will impact at several thousand megatons averagely every 15,000 years; and cause ocean-wide tsunamis or, on land, devastate somewhere the size of Estonia. An impact like the one which (65 million years ago) acutely accelerated the dinosaur decline typically occurs once in a million centuries, the NEO being perhaps 15 km across. Its impact could exceed a 100 million megatons.[54]

A megaton yield equals a million tons of TNT, fifty times that of the Hiroshima bomb. It is quite a representative strength for a hydrogen fusion (alias thermonuclear) warhead. In 1961, the Soviets exploded one of sixty megatons above Novaya Zemlya. By the end of the Cold War, a full-scale thermonuclear exchange would likely have expended well over 10,000 megatons. A caveat to enter is that, when comparing individual detonations, a cube root law applies. A thousand-fold increase in energy release will yield just a ten-fold increase in equivalent destruction radius.

The crux of anti-NEO defence is how to intercept objects of uncertain shapes, sizes and compositions; travelling at perhaps 25 kilometres a second; and detectable, with current technology, maybe 150 million kilometres away which could imply about ten weeks' warning time. Were interception to effect but partial disintegration, this could mean the Earth's then being struck by a salvo. Should the flight path be deflected, it would need to clear the Earth completely.

The chief modes envisaged are (a) disintegrative melting or deflection by thermonuclear explosion and (b) repeated nudges, probably involving a manned Spacecraft. Nudging could be otiose against, say, an asteroid comprised of rubble stretching back perhaps 30 km. Comets, which represent a fifth of the threat, pose similar problems. Obversely, a thermonuclear burst might not be energetic enough against an abnormally large NEO. Work at the University of Washington in Seattle indicates that a megaton impact should suffice to divert a solid body a kilometre across, no doubt disintegrating it considerably meantime. Engagement of one 10 kilometres wide could therefore require a gigaton (i.e. 1000 megatons). [55]

A neglected hazard is meteor storms, something neither of the above solutions is adapted to. Apparently such an event over Britain is what is recorded in a tract a scholarly and patriotic Romano-British monk called

Gildas wrote *c.* 540. Never properly understood by general historians, this episode was well picked up by Victor Clube and Bill Napier in their historical review of natural missilry.[56]

The sequence was likely thus. Through 441, a tense stand-off continued between the Romano-British legatees and the south-eastern enclave of English invitees. Then in 442 a meteor storm struck Romano-Britain, apparently effecting a hit on its federal council in session. Enough dust reached the stratosphere to disturb climates around the northern hemisphere. Swiftly the English moved against their erstwhile hosts, desperate to extend their own domain.[57]

Whatever strategy may be adopted, defence versus NEOs will be onerous. Manned spaceflight will be required for Earth orbiting; battle stations on Mars and/or the Moon; and maybe deep probes to contact . . . Moreover, passive defence will be needed to cope with actual impacts. Indeed, this side of things should receive high priority since it could also help contain disasters originating terrestrially: earthquakes and *tsunamis*; supervolcanoes; magnetic flips; acute climate change . . . Over time a dispersed urban order could complement specific disaster containment as well as serve to uphold community values (2004 study, pp. 263–4).

Building transnational trust will be imperative. Confidence must prevail that priorities will not be skewed nor NEOs turned towards Earthly protagonists.[58] Recourse to nuclear warheads must never be taken as a precedent for using them against other countries' Space assets, a stratagem which still arouses apprehension. [59] Rather, anti-NEO defence should serve to delineate our Earthly life force within the pantheon of cosmic consciousness. To define planetary togetherness in terms of a measure of cosmic exclusiveness is surely sound psychology and therefore good strategy.

Holistic Naturalism

In December 1983, Gro Harlem Brundtland (one-time Swedish Prime Minister) was invited by the UN Secretary-General to organize and chair a World Commission on Environment and Development. Its report identified "Outer Space" as one of the three global commons, the others being the Oceans and Antarctica.[60] This designation was part of an incremental process whereby the Space environment moved towards due recognition (2004 study, pp. 261–2). In 1967, the UN had adopted a Treaty on Principles Governing the Activities of States in the Exploration and Use of Outer Space. It stipulates that the Moon and other celestial bodies were to be explored for peaceful purposes only. It prohibits the orbiting of Weapons of Mass

Destruction. It precludes the assertion in this ambience of territorial "national sovereignties". Within ten years, 80 nations had signed up.

Various other multinational bodies have a Space involvement under the UN or otherwise. Most prominent among various regional ones is the European Space Agency, a body which in 2003 signed a keynote collaboration agreement with the EU. In 1992 and 2002, World Space Congresses were held in Washington and Houston respectively, under the auspices of the American Institute of Aeronautics and Astronautics. The Chinese were conspicuous among the 13,000 managers and researchers who attended Houston.

The Brundtland approach to more synoptic Space regulation was pragmatic: "A system of Space traffic control in which some activities were forbidden and others harmonized cuts a middle path between the extremes of a sole Space authority and the present near anarchy".[61] Two issues to be addressed are the long-standing one of orbital debris which could worsen by orders of magnitude should armed conflict ever erupt in Near Space (2004 study, pp. 174–5); and the emergent development of what would likely long be very plutocratic ($100,000 a short trip?) Space tourism. Near Space management could herald globalization in general being properly addressed, not simply left to the "invisible hand" of consumerism with all this could portend – culturally and ecologically.

This bespeaks a diametrically different approach to the use of Near Space in order to preserve a diversity fast diminishing on Earth to what has been proposed by Freeman Dyson. He has favoured our "greening the galaxy": that is, so proliferating across it that earthlings may "learn to grow gardens in stellar winds and in supernova remnants" and perhaps eventually "fill every nook and cranny of the universe".[62] Alas, it seems to me that husbanding a stellar wind or whatever would require esoteric skills and social discipline considerably at variance with "cultural diversity" as a concept to be celebrated. Nor does intergalactic travel look remotely feasible.

Important, too, will be the erosion of a cultural dichotomy whereby professional historians of the period have had no idea what likely hit Romano-Britain in 442. From whence, one may eliminate such policy blind spots as the one which precluded regimes around or near the Bay of Bengal from realizing prior to 2004 that, since the Northern Sumatra area is exceptionally prone to severe vulcanism, it must be liable to big earthquakes too, ones under the sea included.

This geophysical ignorance looks even less excusable in the light of research published five years previously. From AD 535–6, coolness and aridity beneath a thick stratospheric dust veil affected much of the northern hemisphere. Soon economic crises broke in Ireland and Britain, the Mediterranean, Arabia Felix and China. Bubonic plague went rampant in Europe and the Near East.[63] The general presumption had been that the cause was an NEO impact.[64] However, in 1999 this event, too, was definitively attributed to Krakatoa's exploding with more force than any volcano the last 72,000 years.[65] It lies between Sumatra and Java.

Lateral Thought

Part of the argument for the cultural conflation here espoused could be that outstanding minds in a variety of fields can display a valuable flair for lateral thinking. Inspirational in this regard is the etymology of "holistic", the word so much in vogue in the New Left and counter-culture days of the 1960s to connote a quasi-mystical overview of the natural order. Reportedly, it was actually coined in 1926 by Field Marshal Jan Christian Smuts of South Africa.[66] He used it in his book, *Holism and Evolution*, to connote the "whole making" balance of nature. In two world wars, this former Boer commando leader was to serve in the highest councils of the British Empire. But clearly he comprehended that world survival involved something more than military strategy.

Similarly, in the long years when the greenhouse effect was still being ignored or discounted by nearly all the meteorologists and climatologists, it was warned against by the two men identified as the respective fathers of the American and Soviet hydrogen bombs. In 1958, Edward Teller noted with Albert Latter that, if hydrocarbon combustion drove up levels of atmospheric carbon dioxide, melting of the ice sheets would eventually result.[67] A decade later, Andrei Sakharov warned that carbon dioxide from "burning coal is altering the heat reflecting qualities of the atmosphere. Sooner or later, this will reach a dangerous level".[68]

Unfortunately, the forces of irrational optimism proved too strong for either to shift. So what we must now hope is that History does not repeat itself *vis-à-vis* the multiple threats to world survival building up. We must further hope we shall not be too slow to recognize that, behind all of them, a philosophic crisis looms – one of identity, meaning and fulfilment. It is hard to escape the conclusion that, come 2106, our blue green planet will be either far more hellish or a lot more benign than now. Such divergence Winston Churchill and H. G. Wells portrayed to effect in their respective ways. Today it applies with a vengeance. The *yin* and *yang* principle may help to sustain, in various spheres, the reciprocal moderation a situation so novel requires.

Astrophysics Revisited

Throughout, the underlying question has to be whether one can make a statement about metaphysical realities which helps people customarily subscribing to a variety of different belief systems to come the better to terms with one another and all else. So the point of departure remains the classic enigma. Why is there anything at all? Why not nothing whatsoever? The answer sought could be described as classically simple, "Nothing" has to be observed in order to be nothing. Yet through that registration process, it ceases to be a nothingness. Quite Existentialist and very Zen!

A corollary must be that the "something" which is the cosmos we know has also to be registered as such. It therefore requires a built-in Consciousness. From whence, it is no enormous leap to the conclusion that Life can have a salient role in the provision of this, a role it needs in order to account for its own presence and its staggering complexity. One can hardly claim to work from the Aristotelian precept that "Nature does nothing in vain" without adducing some reason for Life's consciousness and other attributes. Clearly one cannot just say that Life has emerged merely in order to enjoy itself. Over and above all else, our cosmos is hardly that Life-friendly or, indeed, Life-oriented.

Nevertheless, the above could be read as something of a rejoinder to the gloomy postscript penned by Steven Weinberg of Harvard for the provisional yet still valuable overview he proffered in 1977 of the Standard Model for the early history of the universe. This was being formulated following the detection in 1964 by Penzias and Wilson of the predicted microwave background, the discovery which clinched the triumph of Big Bang cosmology over Steady State.

Weinberg famously allowed how hard it was to accept our Earth as "just a tiny part of an overwhelmingly hostile universe" which "faces a future extinction of endless cold or intolerable heat. The more the universe seems comprehensible, the more it also seems pointless". Accordingly, the "effort to understand the universe is one of the very few things that lifts human life a little above the level of farce; and gives it some of the grace of tragedy".[69]

To which one may respond that any system as fully enclosed as a cosmos must be pointless in the ultimate, existing for the sake of its own existence. Therefore the more pertinent question to ask is how useful a part does Life play within it. Is it not a significant factor in conscious registration? But if so, does this not logically drive one back to invoking the Panpsychic tradition whereby even inanimate matter can be accorded some Consciousness, this likely being a function of its thermal energy or level of organization at the given time. Otherwise far too much will go unregistered within the Cosmos simply because the total purview achieved by all the life forms is inherently so limited.

Life can be seen as gradually emerging to help offset awhile the relentless cooling of a universe expanding. But then the question arises of how long Life might survive in whatever mode within this ambience. Thinking of Life in the compact and individuated forms we know, one's answer could be not so very long, cosmically speaking. However, in 1979 Freeman Dyson wrote an acclaimed essay in the "thought experiment" genre. It envisaged the physical and biological connotations of the universe being regarded as "open": that is to say, subject to indefinite expansion. He deemed it, in this situation, not inconceivable that over, say, ten billion years Life would "evolve away from flesh and blood and become embodied in an interstellar black cloud or a sentient computer".[70] But could the requisite adjustments, qualitative and quantitative, conceivably be effected?

Life as we thus far observe it displays a peculiarly insistent tendency, much discouragement notwithstanding, to evolve towards more complex forms which should tend *inter alia* to achieve higher Consciousness levels. The outreach these forms are liable to essay may or may not be successful in drawing near to Godliness in any very tangible sense. Nevertheless, they may play a signal role binding collective consciousness more tightly within the kingdom of Life.

All in all, one feels drawn to the view of David Chalmers that any tolerably complete Theory of Everything (TOE) should "have two components: physical laws, telling us about the behaviour of physical systems from the infinitesimal to the cosmological; and what we might call psychochemical laws, letting us know how some of those systems are associated with Conscious experience".[71] What must be admitted, however, is that the justification for Life here propounded is very much a product of deductive reasoning, the top down approach. To date we lack the materials to test it against bottom up empiricism.

Riders and Corollaries

A major snag is that a general definition of Consciousness still eludes us, whether as applicable to particular Life-systems or to Creation at large. Much else, too, is unfinished. Gaia and Panspermia are concepts which seem *a priori* to proffer backing to the quasi-intuitive thought that it may be valid to talk of a Cosmic Consciousness reaching back through time. Additionally, we have a general sense of critical thresholds within Consciousness, both apropos individual living systems and across Creation at large. Yet we remain quite unable to attach units of measurement to any of this or to formulate laws concerning its operation across Space and Time. How fast and far we may now progress in these directions is problematic.

Undeniably, a striking aspect of the universe as we know it is how finely honed, in relation to various structural requirements, the constants of Nature are. Not infrequently, indeed, one could have postulated several orders of magnitude less refinement and still been struck by how low a random probability of the outcome observed would have been. The strength of the Penzias–Wilson microwave background varies around the sky by one part in 100,000. Were there no variation, coherent aggregations of matter could not have formed up in the early universe. Slightly more variation and the result might have been too lumpy – or too bestrewn with Black Holes – to give planetary Life much chance to get going.

The upshot of this, coupled with the singular values attaching to other constants, has been something close to the Aristotle/Aquinas precept of an "unmoved mover", a determining influence on all Creation under it. Correspondingly, this Godliness is utterly Immanent, suffusing everything. How else can one explain, for instance, every electron bearing precisely the same "negative" electrical charge: "precisely" to the utmost limits of modern observation and measurement?

Furthermore, our being virtually 100 per cent sure our universe was created in an instant of time opens up the likelihood that other universes are being generated continually. This may allow of a representative spread of various constants. It will preserve as well the continuity that a single ephemeral creation would *ipso facto* compromise. One must bear in mind, too, that Black Holes could conceivably act as conduits for the seeding of information into new universes, albeit disaggregated in flows that might be quite smooth in relation to the enormous forces involved yet highly turbulent by any other standards. Surely we should discount speculation, even by ranking astronomers, that someone might pass intact through a black hole to emerge in another space–time dimension.[72]

If the registration through Consciousness hypothesis be valid for our cosmos, it very likely applies to any other cosmi. Nor is it very easy to imagine the mass versus energy distinction not assuming prominence elsewhere. But there the rub is that, in the Einsteinian world we still inhabit, the dichotomy between energy and matter is less absolute than was classically assumed. That, after all, is the main import of the leap forward we may know as quantum theory.

Moreover, in pristine circumstances this axiom can be decidedly non-trivial. Take the "false vacuo", concepts hypothesized to facilitate analysis of the very early history of the universe as per the Standard Model. A "false vacuum" would simply be an extremely high energy state within what is otherwise a nothingness, a state so high that one cubic centimetre thereof could have a mass equivalent of 10^{67} tons. The matter in the entire observable universe today has a mass of 10^{50} tons.[73] The neglect of this aspect thus far does not *ipso facto* invalidate the argument here advanced. But "false vacuo" are among various features that would have to figure in the astrophysical side of a full interpretation.

The seeding of basic information via black holes or however could connote the emergence of multiverse families, each possessed of a degree of structural mutuality. In the light of what has just been said, however, one cannot presume any such progression will be smooth and narrowly logical. In particular, one cannot assume that the constants of Nature would stand entirely unaltered during a seeding process. Lately some tentative evidence has been forthcoming that certain of our terrestrial constants have subtly varied across spans of hundreds of millions of years. Confirmation thereof would increase expectations that even cosmi within the same family could exhibit major structural differences.

This layman's perspective tends to be that we stand little chance of ever learning much definitive about universes "parallel" to our own. One has to acknowledge, none the less, that scope for somewhat informed speculation remains. Even the solidly circumspect Martin Rees has visualized a quantum computer which could, in effect, share "the computational burden among a near infinity of parallel universes".[74] A basic assumption appears here to be that computation could be the sticking point, not information. But is this likely? Again, one hits big questions about thresholds of transition.

A keynote proposal just restated is that the historic concept of the unmoved mover looks applicable to the governance of Creation in terms of modern knowledge, an admixture of speculation included. Even so, within a multiverse or infinitude setting, any such Godliness might best be treated as binary. At the deepest level, the focus is on the everlasting principle or formula which covers absolutely everything. At the other, one's prime concern is with the spirit of Life (*élan vital*, as Bergson would have said) within our own cosmos. The former will be quite inaccessible to our outreach. The latter could be approachable by ourselves and by many other forms of complex life extant within this cosmos.

Not that approaches seem likely to culminate in the mystic melding between Humanity and the Divine so often envisaged by religious seers. Even the cosmic Godliness is too austere, immanent and mysterious; and can hardly, in any case, be at all humanoid. Still less plausible is a personal God as per the Abrahamic tradition. Albert Einstein sometimes talked of

God but hotly denied belief in a nexus so intimate.[75] Nor did he ever show much interest in having Life or Consciousness figure in his proposed Theory of Everything.

On the other hand, if there is a prospect of sentient survival awhile at least after death, it probably relates to such outreach. However sceptical some of us may remain, nobody is entitled at this stage in our understanding to be peremptorily dismissive. For instance, some will be encouraged in this thinking by the current disposition to treat material substance as an aspect of information. So may they be by the school of thought which says that, in due course, advanced extra-terrestrials may help to extend our outreach.[76]

The Privilege of Human-ness

Bearing in mind the cosmic norms in so far as we may know them, it is certainly a considerable privilege to be a human being; and, indeed, to be ensconced on so salubrious a planet. The more's the pity then that we Lords of our Local Creation currently desecrate and plunder our Earth so badly; and that we are ourselves so prone to prejudicial irrationality and overreactive violence.

Mercifully, there are some shaky but tangible grounds for believing we could be poised to enter an era in which relations much improve within and between the great religious obediences; between them and philosophic agnosticism; and between all of these and the natural sciences. Any contrary trend towards separatism will not prevail. If this comes to pass, the emphasis can be more on dialogue, less on dogma. Also, value and belief systems will become more individual, more or less regardless of any person's formal allegiance. Already, too, various priesthoods are becoming more part time and sometimes more in touch in consequence. If all this is indeed in train, surely it can be turned to good account. After all, individuals yearning for meaning and fulfilment is what it all adds up to. It is on individuals that the stars look down.

NOTES

Part One The Heavens in History

1 AGES OF STONE

1 Robert Graves (introducing), *New Larousse Encyclopedia of Mythology* (London: Hamlyn, 1968), pp. 244–5.
2 John Adair, *The Pilgrim's Way* (London: Book Club Associates, 1978), p. 95
3 Arden Reed, *Romantic Weather. The Climates of Coleridge and Baudelaire* (Hanover: University Press of New England, 1983), pp. 274–6.
4 Evan Hadringham, *Early Man and the Cosmos* (London: William Heinemann, 1983), pp. 14–15.
5 Jules Cashford, *The Moon, Myth and Image* (London: Cassell, 2003), pp. 324–5.
6 Gregg De Young in Helaine Stein (ed.), *Astronomy across Cultures* (Dordrecht: Kluwer Academic Publishers, 2000), pp. 467–8.
7 L. V. Maksimov and N. P. Smirnov, "A Contribution to the Study of the Causes of Long-Period Variations in the Activity of the Gulf Stream", *Oceanology* 5, 2, (1965): 15–24.
8 Robert G. Currie, "Lunar Tides and the Wealth of Nations", *New Scientist* 120,1627 (5 November 1988): 52–5.
9 Robert G. Currie, "Luni-Solar 18.6 and Solar 10–11 year signals in USA temperature records", *International Journal of Climatology* 13,1 (January 1993): 31–50.
10 Cashford, *The Moon, Myth and Image*, pp. 15–18.
11 *Ibid.*, p. 282.
12 Graves, *New Larousse Encyclopedia of Mythology*, pp. 499 and 496.
13 Claude Lévi-Strauss (Monique Layton trans.), *Structural Anthropology*, 2 Vols (London: Allen Lane, 1977), Vol. II, p. 215.
14 John Jirikowic and P. E. Dawson, "The Medieval Solar Activity Maximum", *Climate Change* 26 (July 1994): 309–16.
15 James P. Kennett and Robert C. Thunell, "Global Increase in Quaternary Explosive Vulcanism", *Science* 187, 4776 (14 February 1975): 497–503.
16 Kam-bin Liu and P. A. Colvinaux, "Forest changes in the Amazon Basin during the last glacial maximum", *Nature* 318, 4776 (12 December 1985): 556–7.
17 Stephen Oppenheimer, *East of Eden* (London: Constable, 2003), Fig. 0.1.
18 *Ibid.*, pp. 8–9.

19 Rebecca L. Cann *et al.*, "Mitochondrial DNA and Human Evolution", *Nature* 325, 6099 (1 January 1987): 31–6.
20 "Neanderthal art alters the face of archaeology", *New Scientist* 280, 2425 (6 December 2003): 11.
21 Richard G. Klein in Takeru Akazawa *et al.* (ed.), *Neanderthals and Modern Humans in Western Asia* (New York: Plenum Press, 1998, chapter 33.
22 *New Scientist*, 152, 2054 (2 November 1996): 6.
23 Lawrence H. Robbins in Helaine Selin and Sun Xiaochun (eds.), *Astronomy across Cultures* (Dordrecht: Kluwer, 2000), p. 35.
24 Roger Lewin, *The Origins of Modern Humans* (New York: Scientific American Library, 1993), pp. 150–4.
25 Dirk J. Struik, *A Concise History of Mathematics* (London: G. Bell, 1965), p. 7.
26 Glyn Daniel, "Megalithic Monuments", *Scientific American* 243, 1 (July 1980): 64–76
27 John A. Eddy, "The Astronomical Alignment of the Big Horn Medicine Wheel", *Science* 184, 4141 (7 June 1974): 1035–43.
28 Gerald S. Hawkins, "Stonehenge Decoded", *Nature* 200, 4904 (26 October 1963): 306–8
29 Alexander Thom and Archibald Thom in E. C. Krupp (ed.), *In Search of Ancient Astronomies* (London: Chatto and Windus (1979), p. 68
30 Elizabeth Chesley Batty, "Archaeoastronomy and Ethnoastronomy so far", *Current Anthropology* 14, 4 (October 1973): 389–41.
31 Timothy Taylor, "The Gundestrup Cauldron", *Scientific American* 266, 3 (March 1992): 66–71.
32 Subhash C. Kak, "The Astronomy of the Age of Geometric Altars", *Quarterly Journal of the Royal Astronomical Society* 36 (1995): 377–84.
33 M. Senda, "Perceived Space in Ancient Japan" in Alan R. H. Baker and Mark Billinge (ed.), *Period and Place* (Cambridge: Cambridge University Press, 1982), pp. 212–19.
34 Daisetz T. Suzuki, *Zen and Japanese Culture* (Princeton: Princeton University Press, 1959), p. 337.
35 T. E. Lawrence, *Seven Pillars of Wisdom* (London: Jonathan Cape, 1935), p. 357.
36 George Adam Smith, *The Historical Geography of the Holy Land* (London: Hodder and Stoughton, 1894), pp. 29–30.
37 Werner Keller (William Neil trans.), *The Bible as History* (London: Hodder and Stoughton, 1961), pp. 158–61.
38 John Garstang and J. B. E. Garstang, *The Story of Jericho* (London: Hodder and Stoughton, 1940), pp. 136–8 and 165.
39 J. Z. Boer, J. R. Hale and J. Chanton, "New Evidence for the Geological Origins of the Ancient Delphic Oracle", *Geology* 29, 8 (2001): 707–11.
40 Donald W. Olson, "When the Sky Ran Red", *Sky and Telescope* 107, 2 (February 2004): 28–35.

2 HAMMURABI TO PTOLEMY

1 Subhash C. Kak, "Knowledge of the Planets in the Third Millennium BC", *Quarterly Journal of the Royal Astronomical Society* 37 (1966): 709–15.
2 R. L. van de Waerden in A. C. Crombie (ed.), *Scientific Change* (New York: Basic Books, 1963), chapter 3.
3 A. A. Aaboe, "Scientific Astronomy in Antiquity", *Philosophical Transactions of the Royal Society* A 276 (1974): 21–42.
4 G. Hellman, "The Dawn of Meteorology", *Quarterly Journal of the Royal Meteorological Society* XXXIV, 148 (October 1908): 221–32.
5 J. M. Steele and F. R. Stephenson, "Lunar Eclipse Times predicted by the Babylonians", *Journal of the History of Astronomy* 91 (May 1997): 119–31.
6 Peter Whitfield, *Astrology, A History* (London: British Library, 2001), p. 33.
7 E. W. Webster (trans.), *The Works of Aristotle*, 11 vols. (Oxford: The Clarendon Press, 1908), Vol.III *Meteorologica*, Book I, Section 14.
8 *Ibid.*, Section 10.
9 Joseph Needham, *Science and Civilisation in China* (Cambridge: Cambridge University Press, 1959), Vol. III, p. 462.
10 Bertrand Russell, *History of Western Philosophy* (London: Unwin, undated), p. 206.
11 *Ibid.*, p. 25.
12 J. L. Bintliff in Anthony Harding (ed.), *Climatic Change in Later Pre-History* (Edinburgh: Edinburgh University Press, 1982), pp. 149–50.
13 Cameron Shelley, "The Influence of Folk Meteorology in the Anaximander Fragment", *The Journal of the History of Ideas* 61:1 (January 2000): 1–17.
14 Thomas Heath, *Aristarchus of Samos, The Ancient Copernicus* (Oxford: Clarendon Press, 1913), chapter IV.
15 Arthur Koestler, *The Sleepwalkers* (London: Hutchinson, 1959), p. 26.
16 *Ibid.*, pp. 35–7.
17 Bertrand Russell, *History of Western Philosophy*, p. 151.
18 Friedrich Solmsen, "Plato and the Concept of the Soul (Psyche): Some Historical Perspectives", *Journal of the History of Ideas* 44, 3 (July 1983): 355–67.
19 Thomas Heath, *Aristarchus of Samos*, p. 141.
20 Plato, *The Republic*, Book VII, 529A–530B.
21 Plato in *Epinomis*, 982C, D, quoted in Heath, p. 185.
22 R. G. Bury (trans.), *Plato Book VII Timaeus* (London: William Heinemann, 1929), 34B.
23 Sir Ernest Barker (trans.), *The Politics of Aristotle* (Oxford: Clarendon Press, 1948), p. xii.
24 Koestler, *The Sleepwalkers*, p. 62.
25 *Ibid.*
26 Barker, *The Politics of Aristotle*, p. 132.
27 *Ibid.*, Book V.
28 *Ibid.*, p. 287.
29 John Maxwell O'Brien, *Alexander the Great: The Invisible Enemy* (London: Routledge, 1992), pp. 172–7.

30 Heath, *Aristarchus of Samos*, p. 228.
31 Ian Barbour, *Religion in an Age of Science* (San Francisco: Harper, 1991), pp. 218–19 and 230.
32 Koestler, *The Sleepwalkers*, p. 61.
33 Mike Edmunds and Philip Morgan, "The Antikythera Mechanism", *Astronomy and Geophysics* 41, 6 (December 2000): 6.10–6.17.
34 Owen Gingerich, "The Trouble with Ptolemy", *Isis* 93, 1 (March 2002): 70–4.

3 LATE ANTIQUITY TO COPERNICUS

1 Sean McClachlan and Almudena Alonso-Herrero, "The Century the Sky Disappeared . . . ", a presentation in the Proceedings of the Conference on *The Inspiration of Astronomical Phenomena.* Edited by Nick Campion (Bath, University of Bath, 2003), pp. lviii–lxix.
2 Peter Brown, *The World of Late Antiquity* (London: Thames and Hudson, 1971), pp. 41–4.
3 Gerald Bonner, "Augustine and Millenarianism", in Rowan Williams (ed.), *The Making of Orthodoxy* (Cambridge: Cambridge University Press, 1989), pp. 235–54.
4 Theodore E. Mommsen, "St Augustine and the Christian Idea of Progress: The Background to the City of God", *Journal of the History of Ideas* XII, 3 (June 1951): 346–76.
5 Trevor Rowe, *St. Augustine: Pastoral Theologian* (London: Epworth Press, 1974), p. 111.
6 Henry Chadwick (trans.), *St Augustine. Confessions* (Oxford, Oxford University Press, 1991), p. 54.
7 Johan Huizinga (James S. Holmes and Hans van Marle, trans.), *Men and Ideas* (London: Eyre and Spottiswoode, 1960), p. 255.
8 Arthur Koestler, *The Sleepwalkers* (London: Hutchinson, 1959), Part II.
9 Henri Pirenne, "Medieval Cities" in Alfred F. Havigshurst (ed.), *The Pirenne Thesis* (Lexington: D.C. Heath, 1976), pp. 1–26.
10 Eliyahu Ashtor, *A Social and Economic History of the Near East in the Middle Ages* (London: Collins, 1976), p. 102.
11 G. E. Fussell, "The Classical Tradition in Western European Farming: the 14th and 15th centuries", *Agricultural History Review* 17, 1 (1969): 1–8.
12 Bradley Schaefer, "The Crab Supernova in Europe: Byzantine Coins and Macbeth", *Quarterly Journal of the Royal Astrolonomical Society* 36, 4 (December 1995): 369–76.
13 Michael Frasseto (ed.), *The Year 1000* (New York: Palgrave MacMillan, 2002), Introduction, p. 2.
14 Norman Cohn, *The Pursuit of the Millennium* (London: Paladin, 1972), p. 282.
15 A. C. Krey, "Urban's Crusade – Success or Failure?", *American Historical Review* LIII, 2 (January 1948): 235–50.
16 Steven Runciman, *A History of the Crusades*, 3 vols. (Cambridge: Cambridge University Press, 1957), Vol. 1, p. 51.

17 G. G. Coulton, *The Medieval Scene* (Cambridge: Cambridge University Press, 1965), p. 103.
18 Lynn White, "Natural Science and Naturalistic Art in the Middle Ages", *American Historical Review*, LII, 3 (April 1947): 421–35.
19 Julien Green (Peter Heinegg, trans.), *God's Fool* (London: Hodder and Stoughton, 1986), pp. 151–4.
20 Philotheus Boehner (ed. and trans.), *Ockham: Philosophical Writings* (London: Nelson, 1957), p. 2.
21 Evelyn Edson, *Mapping Time and Space* (London: The British Library, 1999), p. 121.
22 William Egginton, "On Dante, Hyperspheres and the Curvature of the Medieval Cosmos", *Journal of the History of Ideas* 60, 2 (April 1999): 195–216.
23 Francis Lee Utley (ed.), *The Forward Movement of the 14th Century* (Columbus: Ohio State University Press, 1961), p. 82.
24 *Ibid.*, p. 108.
25 H. R. Trevor-Roper, *The European Witch Craze of the 16th and 17th Centuries* (Harmondsworth: Penguin, 1969), p. 112.
26 Peter Whitfield, *Astrology. A History* (London: The British Library, 2001), p. 141.
27 Ted Hughes, *Shakespeare and the Goddess of Complete Being* (London: Faber and Faber, 1992), Appendix II.
28 Eric Fromm, *Escape from Freedom* (London: Kegan Paul, 1942), pp. 40 *et seq.*
29 Simone Weil, *The Need for Roots* (New York: Harper and Row, 1971), p. 45.
30 Gavin Menzies, *1421* (London: Bantam Press, 2002), pp. 323–30.
31 Nayan Chanda, "Sailing into Oblivion", *Far Eastern Economic Review* 162, 36 (9 September 1999): 44–6.
32 For a fuller discussion of the themes in this section, see the author's *History and Climate Change*, pp. 294–5.
33 Marc U. Edwards, *Printing, Propaganda and Martin Luther* (Berkeley: University of California Press, 1994), p. 1.
34 David E. Sopher, *The Geography of Religions* (Englewood Cliffs: Prentice-Hall, 1967), p. 41.
35 John Herman Randall, "Scientific Method in the School of Padua", *Journal of the History of Ideas* 1 (1940): 177–206.
36 Cited in Hermann Kesten (E. H. Ashton and Norbet Guterman, trans.), *Copernicus and his World* (London: Martin Secker and Warburg, 1945), p. 6.

4 A RENAISSANCE CONTINUUM

1 C. S. Lewis, *The Discarded Image* (Cambridge: Cambridge University Press, 1964), pp. 98–9.
2 *Ibid.*
3 Thomas S. Kuhn, *The Structure of Scientific Revolutions* (Chicago: University of Chicago Press, 1970).
4 Thomas S. Kuhn, "Second Thoughts on Paradigms" in Frederick Suppe (ed.),

The Structure of Scientific Theories (Urbana: University of Illinois, 1977), pp. 459–62.

5 Thomas S. Kuhn, *The Copernican Revolution* (Cambridge, MA: Harvard University Press, 1957), p. 185.

6 Arthur Koestler, *The Sleepwalkers* (London: Hutchinson, 1959), p. 191.

7 Peter Barker, "Copernicus and the Critics of Ptolemy", *Journal of the History of Astronomy* 30, 101 Part 4 (November 1999): 343–58.

8 Robert Westman in chapter 2 "Competing Disciplines" in Marcus Hellyer (ed.), *The Scientific Revolution* (Oxford: Blackwell, 2003).

9 Hermann Kesten, *Copernicus and his World* (London: Martin Secker and Warburg, 1945), p. 181.

10 Peter Whitfield, *Astrology. A History* (London: The British Library, 2001), pp. 165–9.

11 Keith Thomas, *Religion and the Decline of Magic* (London: Penguin, 1991), p. 416.

12 John D. Barrow, *The Book of Nothing* (London: Jonathan Cape, 2000), pp. 87–91.

13 Lawrence S. Lerner and Edward A. Gosselin, "Galileo and the Spectre of Bruno", *Scientific American* 255, 5 (November 1986): 116–23.

14 Alexander Koyré, "Galileo and Plato", *Journal of the History of Ideas* 4, 1943: 400–28.

15 Margaret C. Jacob, *The Newtonians and the English Revolution, 1689–1720* (Hassocks: Harvester Press, 1976), pp. 102–3.

16 Dava Sobel, *Longitude* (London: Fourth Estate, 1991), p. 41.

17 S. F. Mason, "Science and Religion in 17th-century England", *Past and Present* 3 (February 1953): 28–44.

18 Phyllis Allen, "Scientific Studies in the English Universities during the 17th century", *Journal of the History of Ideas* X (1949): 219–53.

19 Christopher Hill, *The World Turned Upside Down* (London: Maurice Temple Smith, 1972), pp. 238–9.

20 Benjamin Wardhaugh, "Mr Birchensha's Ear", *The Owl* 7 (Trinity 2005): 5–8.

21 Brian J. Ford, *Images of Science: A History of Scientific Illustration* (London: The British Library, 1982), p. 170 *et seq.*

22 J. U. Nef, *Cultural Foundations of Industrial Civilisation* (Hampden: Archon Books, 1974), pp. 16–17.

23 James Rodger Fleming, *Historical Perspectives on Climate Change* (New York: Oxford University Press, 1998), pp. 34–5.

24 A. R. Hall, "Sir Isaac Newton's Note-Book, 1661–1665", *Cambridge Historical Journal* 2 (1948): 239–50.

25 Walter H. Bucher in *The Planet Earth* (New York: Simon and Schuster, 1957) for *Scientific American*, p. 72.

26 Bernard Eccles, "Astrological physiognomy from Ptolemy to the present day", *Culture and Cosmos* 7, 2 (Autumn/Winter 2003): 15–36.

27 S. J. Tester, *A History of Western Astronomy* (Woodbridge: Boydell, 1987), p. 240.

28 Thomas, *Religion and the Decline of Magic*, p. 418.
29 John Redwood, *Reason, Ridicule and Religion* (London: Thames and Hudson, 1976), p. 18.
30 Roland Mousnier (Brian Pearce trans.), *Peasant Uprisings in 17th-century France, Russia and China* (New York: Harper and Row, 1970), pp. xvii–xx, 312 and 331–2.
31 J. B. Bury, *The Idea of Progress* (London: Macmillan, 1920), p. 64.
32 Stephen Gaukroger, *Descartes: An Intellectual Biography* (Oxford: The Clarendon Press, 1995), p. 4.
33 Thomas D. Lennon, "The Cartesian dialectic of Creation, Part II" in A. C. Crombie (ed.), *Scientific Change* (New York: Basic Books, 1963), pp. 333–9.
34 George MacDonald Ross and Richard Franks in chapter 17, Nicholas Bumin and E. P. Tsui-James, *The Blackwell Companion to Philosophy* (Oxford: Blackwell, 1966).
35 Roger Saltau, *Pascal: The Man and the Message* (London: Blackie, 1927), p. 1997.
36 Maurice Dumas in A. C. Crombie (ed.), *Scientific Change* (New York: Basic Books, 1963), chapter 13
37 For a discussion of calculations made, see the author's *History and Climate Change* (New York: Routledge, 2001), pp. 289–90.
38 David Goodman and Colin A. Russell (ed.), *The Rise of Scientific Europe, 1500–1800* (London: Hodder and Stoughton, 1991), p. 242.
39 *Ibid.*, p. 267.
40 G.B. Sansom, *The Western World and Japan* (London: The Cresset Press, 1950), p. 234.
41 Boleslaw Szczesniak, "The penetration of Copernican theory into feudal Japan", *Journal of the Royal Asiatic Society* Parts 1 and 2 (1944): 52–61.
42 J. E. Thornes, "Luke Howard's influence on Art and Literature in the early 19th century", *Weather* 39, 8 (August 1984): 252–5.
43 D. E. Pedgeley, "Luke Howard and his Clouds", *Weather* 58, 2 (February 2003): 51–5.
44 F. Richard Stephenson, "Historical eclipses and Earth's rotation", *Astronomy and Geophysics* 44, 2 (April 2003): 22–7.
45 *Shorter Oxford English Dictionary* (Oxford: Clarendon, 1950), 2 Vols. Vol. 2, p. 2002.
46 Roy Jenkins, *Churchill* (London: Pan, 2002), pp. 33–4.
47 Marlana Portolano, "John Quincy Adams' Rhetorical Crusade for Astronomy", *Isis* 91, 3 (September 2000): 480–503.
48 A Pannekoek, *A History of Astronomy* (New York: Dover, 1961), p. 371.
49 Michael J. Crowe, *The Extra-Terrestrial Life Debate, 1750–1900* (Cambridge: Cambridge University Press, 1986), pp. 206–7.
50 Percival Lowell, *Mars and its Canals* (London: Macmillan, 1906), p. 377.
51 *Ibid.*, p. 545.
52 Carl Sagan and Paul Fox, "The Canals of Mars: An Assessment after Mariner 9", *Icarus* 25 (1975): 602–12.

53 Thomas A. Dobbins and William Sheehan, "The Canals of Mars Revisited", *Sky and Telescope* 107, 3 (March 2004): 114–17.
54 René Albrecht-Carrie, *The Meaning of the First World War* (Englewood Cliffs: Prentice-Hall, 1965), pp. 42–3.
55 In April 2005, an Anglo-Canadian team reached the North Pole in 36 days, vindicating Peary's claim to be the first person to reach the North Pole.
56 Rosalynn D. Haynes, *H. G. Wells: Discoverer of the Future* (London: Macmillan, 1980), p. ix.
57 Neville Brown, *The Future Global Challenge* (New York: Crane Russak, 1977), p. 50.
58 H. G. Wells, *The First Men on the Moon* (London: Macmillan, 1920), p. 118.
59 John Archibald Wheeler (with Kenneth Ford), *Geons, Black Holes, and Quantum Foam* (New York, W.W. Norton, 1998), pp. 89–90.

5 THE COSMOLOGICAL REVOLUTION

1 Ian Barbour, *Religion in an Age of Science*, Gifford Lectures 2 vols. (New York: Harper San Francisco, 1990), Vol. 1, p. 194.
2 Tony Hey and Patrick Walters, *Einstein's Mirror* (Cambridge: Cambridge University Press, 1997), Fig. 5.9.
3 Albert Einstein (Robert W. Larson, trans.), *Relativity* (London: Routledge, 2004), Appendix 4. First German edition was in 1916.
4 Larry Abbott, "The Mystery of the Cosmological Constant", *Scientific American* 258, 5 (May 1988): 82–8.
5 Lawrence M. Krauss and Michael S. Turner, "A Cosmic Conundrum", *Scientific American* 291, 3 (September 2004): 70–7.
6 William Eggington, "On Dante, Hyperspheres and the Curvature of the Medieval Cosmos", *Journal of the History of Ideas* 60, 2 (April 1999): 195–216.
7 Denys Wilkinson, *Our Universe* (New York: Columbia University Press, 1991), pp. 122–3.
8 Yu-Ying Brown, *Japanese Book Illustration* (London: British Library, 1988), p. 46.
9 Robert Graves (introduced), *New Larousse Encyclopedia of Mythology* (London: Hamlyn, 1974), pp. 88–9.
10 Sartre actually declined the prize in advance for "personal reasons".
11 Jean Paul Sartre (Hazel Barnes, trans.), *Being and Nothingness* (London: Methuen, 1957), p.16.
12 Henri Bergson (Arthur Mitchell, trans.), *Creative Evolution* (London: Macmillan, 1922), pp. 298–9.
13 Edward P. Tryon, "Is the Universe a Vacuum Fluctuation?", *Nature* 246, 5433 (14 December 1973): 396–7.
14 Alan Kostelecký, "The Search for Relativity Violations", *Scientific American* 291, 3 (September 2004): 92–101.
15 *Astronomy Now*, vol. 19, May 2005, Supplement on the Universe: 14.
16 Frank Close, *Particle Physics* (Oxford: Oxford University Press, 2004), pp. 102–5.

17 John D. Barrow, "Enigma Variations", *New Scientist* 175, 2359 (7 September 2002): 30–33.
18 John Archibald Wheeler and John Ford, *Geons, Black Holes and Quantum Foam* (New York: W. W. Norton, 1998), chapter 13.
19 Leonard Susskind, "A Universe Like No Other", *New Scientist* 180, 2419 (1 November 2003): 34–41.
20 Brian Clegg, *A Brief History of Infinity* (London: Constable and Robinson, 2003), p. 54.
21 Michio Kaku, *Hyperspace* (Oxford: Oxford University Press, 1994), p. 262.
22 Andrei Linde, "The Self-Reproducing Inflationary Universe", *Scientific American* 271, 5 (November 1994): 32–9.
23 Kaku, *Hyperspace*, p.263.
24 Barbour, *Religion in an Age of Science*, p. 137.
25 Editorial, "The Certainty Principle", *New Scientist* 183, 2457 (27 July 2004): 3.
26 Milton K. Munitz, "One Universe or Many", *Journal of the History of Ideas* XI, 2 (April 1951): 231–55.

Part Two The Life Dimension

6 ASTROBIOLOGY

1 Freeman Dyson, *Disturbing the Universe* (New York: Harper and Row, 1979), p. 250.
2 John D. Barrow and Frank J. Tipler, *The Anthropic Cosmological Principle* (Oxford, Clarendon Press, 1986), p. 181.
3 See the author's *The Future Global Challenge* (London: Royal United Services Institute, 1976), p. 181.
4 Barrow and Tipler, *The Anthropic Cosmological Principle*, p. 18.
5 John D. Barrow and Joseph Silk, *The Left Hand of Creation* (London: Heinemann, 1984), p. 206.
6 Wendy F. Freedman and Michael S. Turner, "Cosmology in the new Millennium", *Sky and Telescope* 107, 3 (March 2004): 42–7.
7 Stephen J. Hawking, *A Brief History of Time* (London: Bantam Books, 1988), p. 167.
8 Barrow and Tipler, *The Anthropic Cosmological Principle*, p. 675.
9 Martin Rees, *Before the Beginning* (London: Simon and Schuster, 1997), p. 206.
10 John D. Barrow, *The Book of Nothing* (London: Jonathan Cape, 2000), pp. 308–11.
11 Amanda Gefter, "The Riddle of Time", *New Scientist* 188, 2521 (15 October 2005): 30–33.
12 Helen J. Fraser *et al.*, "The Molecular Universe", *Astronomy and Geophysics* 43 (April 2002): 2.10–2.18.
13 "Life without water is almost unthinkable, isn't it?", *New Scientist* 180, 2473 (29 November 2003): 8.
14 David Koerner and Simon Levy, *Here be Dragons* (New York: Oxford University Press, 2000), chapter 9.

15 P. C. Sylvester-Bradley in I. G. Gass *et al.*, *Understanding the Earth* (Sussex, Artemis, 1972), chapter 9.
16 The Archaean era is deemed to have ended 2500 million years ago. The succeeding Pre-Cambrian ended 600 million years ago.
17 Manfred Schidlovski in Gerda Horneck and Christa Baumstark-Khan (eds.), *Astrobiology* (Berlin: Springer, 2002), p. 284.
18 J. William Schoff and Bonnie M. Packer, "Early Archean (3.3 to 3.5 billion years old) microfossils from Warawoona group, Australia", *Science* 237, 4710 (3 July 1987): 70–73.
19 David C. Catling *et al.*, "Biogenic methane, hydrogen escape and the irreversible oxidation of early Earth", *Science* 293, 5531 (3 August 2001): 849–53.
20 Dirk Wagner *et al.* in Horneck and Baumstark-Kahn (eds.), *Astrobiology*, p. 148.
21 Dana Desonie, *Cosmic Collisions* (New York: Henry Holt for *Scientific American*, 1996), p. 28.
22 "The Earth's Hidden Life", *The Economist*, 341, 7997 (21 December 1996): 133–8.
23 Editorial, *International Journal of Astrobiology* 1, 1 (2002): 1–2.
24 Andra J. Wolfe, "Germs in Space", *Isis* 93, 2 (June 2002): 183–205.
25 F. Hoyle and Chandra Wickramasinghe, "Panspermia 2000" in their *Astronomical Origins of Life* (Dordrecht: Kluwer, 2000), pp. 1–17.
26 Stuart Ross Taylor, *Destiny or Chance* (Cambridge: Cambridge University Press, 1998), pp. 96–7.
27 Baruch S. Blumberg in Horneck and Baumstark-Kahn (eds.), *Astrobiology*, p. 2.
28 Christopher Chyba and Carl Sagan, "Endogenous production, exogenous delivery and impact-shock synthesis of organic molecules: an inventory for the origins of life", *Nature* 355, 6355 (9 January 2002): 125–32.
29 Michael H. Engel and Bartholomew Nagy, "Distribution and enantiomeric composition of amino acids in the Murchison meteorite", *Nature* 296, 5860 (29 April 1982): 837–40.
30 Claude Perron in Paul Murdin (ed.), *Encyclopedia of Astronomy and Astrophysics*, 4 vols. (London: Nature for the Institute of Physics, 2001), Vol. 2, p. 1717.
31 Chandra Wickramasinghe, *A Journey with Fred Hoyle* (Singapore: World Scientific Publishing, 2005), p. 140.
32 Ian Gilmour and Mark A. Sephton (ed.), *Astrobiology* (Cambridge: Cambridge University Press and Open University, 2004), Table 1.5.
33 Wickramasinghe, *A Journey with Fred Hoyle*, p. 208.
34 *Ibid.*, pp. 204–5.
35 Stephen K. Donovan (ed.), *Mass Extinctions: Processes and Evidence* (London: Belhaven, 1989), p. 50.
36 Omer Blaes, "A Universe of Disks", *Scientific American* 291, 4 (October 2004): 22–9.
37 David Pilbeam, "The Descent of Hominoids and Hominids", *Scientific American* 250, 3 (March 1984): 60–9.

38 Richard Dawkins, *The Ancestor's Tale* (London: Weidenfeld and Nicolson, 2004), p. 55.
39 Vladimir Lytkin *et al.*, "Tsiolkovsky, Russian Cosmism and Extraterrestrial Intelligence", *Quarterly Journal of the Royal Astronomical Society* 36, 4 (December 1995): 369–76.
40 Koerner and Levy, *Here be Dragons*, p. 176.
41 Charles H. Lineweaver *et al.*, "The Galactic Habitable Zone and the Age Distribution of Complex Life in the Milky Way", *Science* 303, 5654 (2 January 2004): 59–62.
42 Stephen Jay Gould, *The Structure of Evolutionary Theory* (Harvard, MA: Belknap Press, 2002), p. 898.
43 Richard Dawkins, *The Blind Watchmaker* (Harlow: Longmans, 1986), chapter 9.
44 Richard Dawkins, *A Devil's Chaplain* (London: Phoenix, 2004), p. 233.
45 Barrow and Tipler, *The Anthropic Cosmological Principle*, p. 16.
46 *Ibid.*, p. 21.
47 Hawking, *A Brief History of Time*, p. 137.

7 CONSCIOUSNESS

1 Sir James Jeans, *The Mysterious Universe* (London: Penguin, 1937), p. 16.
2 Michael J. Crowe, *The Extraterrestrial Life Debate* (Cambridge: Cambridge University Press, 1986), p. 540.
3 *Ibid.*, pp. 47–56.
4 Bernard J. Baars, "The Double Life of B. F. Skinner", *Journal of Consciousness Studies* 10, 1 (2003): 5–25.
5 David J. Chalmers, "The Puzzle of Conscious Experience", *Scientific American* 273, 6 (December 1995): 62–8.
6 Francis Crick, *The Astonishing Hypothesis, The Scientific Search for the Soul* (London: Simon and Schuster, 1994), p. 3.
7 Roger Penrose, "Mechanisms, microtubules and the mind", *Journal of Consciousness Studies* 1, 2 (1994): 241–9.
8 Roger Penrose interviewed by Jane Clark, *Journal of Consciousness Studies* 1, 1 (1994): 22–3.
9 Rick Grush and Patricia Smith Churchland, "Gaps in Penrose's Toilings", *Journal of Consciousness Studies* 2, 1, 1995: 10–29.
10 Roger Penrose, *The Road to Reality* (London: Jonathan Cape, 2004).
11 Susan A. Greenfield, *Journey to the Centers of the Mind* (New York: W. H. Freeman, 1995), pp. viii–ix.
12 John Smythies, "Space, Time and Consciousness", *Journal of Consciousness Studies* 10. 3 (2003): 47–56.
13 Steve Fuller, *Kuhn vs. Popper* (Duxford: Icon, 2003), p. 74.
14 Hermann Bondi, *Relativity and Common Sense* (London: Heinemann, 1964), p. 58. footnote 2.
15 Gerald Halton, "Mach, Einstein and the Search for Reality" in Colin Chant and

John Fauvel (ed.), *Darwin to Einstein, Historical Studies on Science and Belief* (Harlow: Longman/Open University Press, 1980), chapter 5.2.

16 Sir Arthur Eddington, *The Nature of the Physical World* (London: J.M. Dent, 1935), p. 319.

17 Sir James Jeans, *The Mysterious Universe* (London: Pelican, 1937), pp. 171–84.

18 Roger Penrose, *The Emperor's New Mind* (Oxford: Oxford University Press, 1989), pp. 97 and 112.

19 Greek term for "master constructor". Introduced into *Timæus* by Plato to mean "Cosmic Creator".

20 R. G. Bury, Plato VII *Timæus etc.* (London: William Heinemann, 1949), p. 5.

21 Tim Folger, "Does the Universe exist if we're not looking?", *Discovery* 23, 6 (June 2002): 44–8.

22 Seth Lloyd and Y. Jack Ng, "Black Hole Computers", *Scientific American* 291, 5 (November 2004): 31–57.

23 Alexander Campbell Fraser (ed.), *The Works of George Berkeley D.D.* 4 vols. (Oxford: The Clarendon Press, 1871), Vol. 1.

24 Erwin Schrödinger, *What is Life?* (Cambridge: Cambridge University Press, 1951), chapter VI.

25 Kate Douglas, "Death Defying", *New Scientist* 83, 2462 (28 August 2004): 40–3.

26 Susan Blackmore, *Consciousness* (London: Hodder and Stoughton, 2003), pp. 178–9.

27 Eran Pichersky, "Plant Scents", *American Scientist* 92, 6 (November–December 2004): 514–21.

28 Vladimir Shulaev *et al.*, "Airborne signalling by methyl salicylate in plant pathogen resistance", *Nature* 385, 6618 (20 February 1997): 718–21.

29 Luis P. Villarreal, "Are Viruses Alive?", *Scientific American* 291, 6 (December 2004): 77–81.

30 Bertrand Russell, *History of Western Philosophy* (London: Unwin, n. d.).

31 *The Ethics of Benedict de Spinoza* (D. D. S. trans.), (New York: D. Van Nostrand, 1888), p. 4.

32 Roger Scruton, *Spinoza* (Oxford: Oxford University Press, 1986), p. 42.

33 David Skrbina, "Panpsychism as an underlying theme in Western Philosophy", *Journal of Consciousness Studies* 10, 1 (2003): 4–46.

34 *Ibid.*

35 Ralph Barton Peary, *The Thought and Character of William James* 2 Vols. (Boston: Little Brown, 1935), Vol. 2, pp. 78 and 295.

36 Steven Pinker, *The Blank Slate, the Modern Denial of Human Nature* (London: Allen Lane, 2002).

37 Keith Ward, "Meeting of Minds", *New Scientist* 184, 2475 (27 November 2004): 9.

38 Greenfield, *Journey to the Centers of the Mind*, p. 137.

Part Three UTOPIA LOST?

8 AN END TO NATURE?

1 David Howarth, *Tahiti. A Paradise Lost* (New York: Viking Press, 1983), chapter 3.
2 Margaret Mead, *Coming of Age in Samoa* (London: Penguin, 1943), p. 21.
3 Derek Freeman, *Margaret Mead and Samoa. The Making and Unmaking of an Anthropological Legend* (Cambridge, MA: Harvard University Press, 1983), pp. 206–9.
4 Phyllis Grosskurth, *Margaret Mead* (London: Penguin, 1988), pp. 90–3.
5 Anthony Storr, *Human Aggression* (London: Allen Lane, 1968), Introduction.
6 Claire and W. M. S. Russell, *Violence, Monkeys and Men* (London: Macmillan, 1969).
7 Robert Ardrey, *The Territorial Imperative* (London: Collins, 1967), chapter 9.
8 Konrad Lorenz (Marjorie Latzke, trans.), *On Aggression* (London: Methuen, 1967), pp. 206–7.
9 Arthur Koestler, *The Ghost in the Machine* (London: Hutchinson, 1967), p. 307.
10 David Cesarini, *Arthur Koestler: the homeless mind* (London: William Heinemann, 1998), p. 494.
11 Koestler, *The Ghost in the Machine*, pp. 61–3.
12 *Ibid.*, p. 301.
13 William Beebe (ed.), *The Book of Naturalists* (Princeton: Princeton University Press, 1988), pp. 395–416.
14 Koestler, *The Ghost in the Machine*, p. 297.
15 Mary Lutyens, *The Life and Death of Krishnamurti* (London: John Murray, 1970), pp. 92–3.
16 Bill McKibben, *The End of Nature* (New York: Viking Penguin, 1990), chapter 2.
17 Stephen H. Schneider, "The Greenhouse Effect and the US Summer of 1988: Cause and Effect as a Media Event?", *Climate Change* 13,2 (October 1988): 113–5.
18 Fred Pearce, "Harbingers of Doom", *New Scientist* 183, 2457 (26 July 2004): 45–7.
19 Robert T. Watson (ed.), *Climate Change 2001: A Synthesis Report* (Cambridge: Cambridge University Press for IPCC, 2001), pp. 62–3.
20 H. H. Lamb, *Climate, History and the Modern World* (London: Routledge, 1995), p. 179.
21 Neville Brown, *History and Climate Change* (London: Taylor and Francis, 2001), chapter 9.
22 C. Pfister *et al.*, "Winter Severity in Europe : the 14th Century", *Climate Change* 34, 1 (September 1996) : 91–104.
23 Richard A. Kerr, "Sea Change in the Atlantic", *Science* 303, 5654 (2 January 2004): p. 35.
24 George Kimble, *The Weather* (Harmondsworth: Penguin, 1951), pp. 216–17.

25 John G. Harvey, *Atmosphere and Oceans. Our Fluid Environment* (London: Artemis, 1985), p. 45.
26 Matthew Sturm *et al.*, "Meltdown in the North", *Scientific American* 289, 4 (October 2003): 43–9.
27 Francis Spufford, *I May Be Some Time. Ice and the English Imagination* (London: Faber and Faber, 1996), p. 187.
28 Ulrike Kronfeld-Goharini, "The On-Going Nuclear Contamination of the Arctic Region", *Inesap Briefing Paper* 11 (October 2003), 8 pp.
29 Ransom A. Myers and Boris Worm, "Rapid Worldwide Depletion of Predatory Fish Communities", *Nature* 623, 6937 (15 May 2003): 280–3.
30 "Eels slide towards extinction", *New Scientist* 180, 2415 (4 October 2003): 14.
31 Rob Edwards, "Sonar kills whales", *New Scientist* 180, 2416 (11 October 2003): 10.
32 Johan Huizinga (James S. Holmes and Hans van Merle trans.), *Men and Ideas* (London: Eyre and Spottiswoode, 1960), p. 255.
33 "Is Ptolemy's view of the Universe resurgent?", *Asahi Shimbun* (22 September 2004) online.
34 P. Cinzano, F. Falchi and C. D. Eldridge, "The First World Atlas of the artificial night sky brightness", *Monthly Notices of the Royal Astronomical Society* 328 (2001): 689–707.
35 David J. Travis *et al.*, "Contrails reduce daily temperature range", *Nature* 418, 6898 (8 August 2002): 601.
36 Michael Bakich, "Reclaim the Night Sky", *Astronomy* 32, 6 (June 2004): 38–43.
37 Christopher Dewdney, *Acquainted with the Night: An Introduction to the World after Dark* (London: Bloomsbury, 2004).

9 RETREAT FROM REASON

1 "Radical thoughts on our 160th birthday", *Economist Survey*, 28 June 2003, p. 5.
2 Teodor Shanin, "How the other half lives", *New Scientist* 175, 2354 (3 August 2002): 45–7.
3 Michael Bond, "The pursuit of happiness", *New Scientist* 180, 2415 (4 October 2003): 40–3.
4 Rodger Doyle, "Calculus of Happiness", *Scientific American* 287, 5 (November 2002): 17.
5 Adam Potkay, "Whatever happened to Happiness?", *Philosophy Now* 27, (June/July 2000): 24–7.
6 Thomas Schwartz and Kiron K. Skinner, "The Myth of the Democratic Peace", *Orbis* 46, 1 (Winter 2004): 159–72.
7 London, Chapman and Hall, 1902.
8 George Dangerfield, *The Strange Death of Liberal England* (London: Constable, 1936), pp. 355–6 and 409.
9 E. P. Thompson quoted in Dorothy Rowe, *Living with the Bomb. Can we live without Enemies?* (London: Routledge and Kegan Paul, 1985), p. 146.

10 George Birdwood, *The Willing Victim* (London: Secker and Warburg, 1969), p. 162.
11 Neville Brown, *History and Climate Change* (London: Taylor and Francis, 2001), p. 79.
12 Lewis Mumford, *The City in History* (London: Pelican, 1961), p. 622.
13 Patrick West, *Conspicuous Compassion* (London: Civitas, February 2004), 80 pp.
14 Neville Brown, *Global Instability and Strategic Crisis* (London: Taylor and Francis, 2004), p. 77.
15 "Pushing a different sort of button", *The Economist* 373, 8399 (20 October 2004): 106.
16 George Orwell, *The Road to Wigan Pier* (Harmondsworth: Penguin, 1976), pp. 169–71.
17 Brown, *Global Instability and Strategic Crisis*, pp. 103–11.
18 Michael Shermer, "Smart People Believe Weird Things", *Scientific American* 287, 3 (September 2002): 19.
19 Erich Fromm, *The Fear of Freedom* (London: Kegan Paul, Trency and Trubner, 1942), p. 40.
20 Eric Fromm in Ralph K. White (ed.), *Psychology and the Prevention of Nuclear War* (London: University Press, 1986), p. 238.
21 See e.g. Richard F. Hamilton, *Who Voted for Hitler* (Princeton: Princeton University Press, 1982).
22 Jacques le Goff (Arthur Goldhammer, trans.), *Time, Work and Culture in the Middle Ages* (Chicago: University of Chicago Press, 1980, pp. 189–90.
23 George Watson, "Were the Intellectuals duped?", *Encounter* LXI, 6 (December 1973): 20–30.
24 Arthur Haupt, "Hollywood launches into Space", *Dialogue* 11, 4 (1978): 58–74.
25 Gerard K. O'Neill, *The High Frontier: Human Colonies in Space* (New York: William Morrow, 1977), chapter XII.
26 Frank Brady, *Citizen Welles* (London: Hodder and Stoughton, 1990), pp. 162–79.
27 E. C. Krupp in E. C. Krupp (ed.), *In Search of Ancient Astronomies* (London: Chatto and Windus, 1974), p. 222.
28 Immanuel Velikovsky, *Worlds in Collision* (London: Victor Gollancz, 1950), pp. 7–8.
29 E. C. Krupp in E. C. Krupp (ed.), *In Search of Ancient Astronomies*, p. 256.
30 John Gribbin and John Plagemann, *The Jupiter Effect* (London: MacDonald, 1974), p. 116.
31 John Gribbin and John Plagemann, *Beyond the Jupiter Effect* (London: MacDonald, 1983), p. 51.
32 Quoted in Richard Dawkins, *A Devil's Chaplain* (London: Phoenix, 2004), pp. 198–9.
33 Peter Whitfield, *Astrology. A History* (London: British Library, 2001), chapter 5.
34 *Ibid.*, p. 199.

35 Robert Sheaffer in Ben Zuckerman and Michael H. Hart (ed.), *Extraterrestrials. Where are they?* (Cambridge, Cambridge University Press, 1995), chapter 3.
36 Martin Walker, *The Guardian*, 4 February 1987.
37 Sheaffer, *Extraterrestrials. Where are they?*, p. 20.
38 C. G. Jung (R. F. C. Hull, trans.), *Flying Saucers* (London: Routledge and Kegan Paul, 1959), pp 18–19.
39 Information supplied by Yu-Ying Brown at the British Library.
40 Shozui Togando, "Mandala" in *Kodansha Encyclopedia of Japan* 9 vols. (Tokyo: Kodansha, 1983) Vol. 5, pp. 99–100.
41 Paper by Tsuda Tetsueí at the workshop on *The Worship of Stars in Japanese Religious Practice* (School of Oriental and African Studies: University of London, September 2004).
42 Maria Hsia Chang, *Falun Gong. The End of Days* (New Haven: Yale University Press, 2004), chapter 3.
43 Edward Irons in Christopher Partridge (ed.), *Encyclopedia of New Religions* (Oxford: Lion, 2004), pp. 265–6.
44 Vladimir Lytkin *et al.*, "Tsiolkovsky, Russian Cosmism and Extraterrestrial Intelligence", *Quarterly Journal of the Royal Astronomical Society* 36, 4 (December 1995): 369–76.
45 Leon Trotsky, *Literature and Revolution* (Ann Arbor: University of Michigan, 1960), pp. 210–11.
46 J. G. Crowther, *Soviet Science* (Harmondsworth: Penguin, 1942), p. 16.
47 *Ibid.*, p. 58.

10 CONFLICT IN HEAVEN?

1 Mike Moore, "Unintended Consequences", *Bulletin of Atomic Scientists* 56, 1 (January/February 2000): 58–60.
2 James A. Abrahamson, "SDI could trigger a Space Renaissance", *The Futurist* XIX, 5 (October 1985): 16.
3 Leonard E. Schwartz, "Manned Orbiting Laboratory", *International Affairs* 43, 1 (January 1967): 51–64.
4 George Dyson, *Project Orion* (London: Allen Lane, 2002), chapter 19.
5 E.g., Yevgeni Velikov *et al.*, *Weaponry in Space. The Dilemma of Security* (Moscow: Mir Publishing House, 1986).
6 *Scientific American* 261, 6 (December 1986): 64.
7 Margaret Thatcher, *The Downing Street Years* (London: HarperCollins, 1993), p. 467.
8 The Tanner Lectures, Brasenose College, March 1992.
9 Lt. Col. Cynthia A. S. McKinley, "The Guardians of Space", *Aerospace Power Journal* XIV, 1 (Spring 2000): 37–45.
10 Maj. Gen. John L. Barry and Col. Darrell L. Herriges, "Aerospace Integration, Not Separation", *Aerospace Power Journal*, XIV, 2 (Summer 2000): 42–7.
11 Maj. Howard D. Belote, "The Weaponization of Space", *Aerospace Power Journal* XIV, 1 (Spring 2000): 46–51.

12 Richard L. Garwin and Hans A. Bethe, "Anti-Ballistic Missile Systems", *Scientific American* 218, 3 (March 1968): 21–31.
13 Richard L. Garwin, "Holes in the missile shield", *Scientific American* 291, 5 (November 2004): 48–57.
14 Henry F. Cooper, "To build an affordable shield", *Orbis* 39, 1 (Winter 1995): 85–99.
15 Dave Dooling, "Space Sentries", *IEEE Spectrum* 34, 9 (September 1997): 50–9.
16 Geoffrey Forden, "Laser Defenses: What if they work?", *Bulletin of the Atomic Scientists* 58, 5 (September/October 2002): 49–53.
17 Sudip Mazumdar and Melinda Liu, "Seeds of Invention", *Newsweek* CXLIV, 5 (22 October 2004): 55–60.
18 Patricia M. Mische, *Star Wars and the State of our Souls* (Minneapolis: Winston Press, 1985), p. 130.
19 Joshua Teitelbaum, "Duelling for Da'wa: State vs Society on the Saudi Internet", *Middle East Journal* 56, 2 (Spring 2002): 223–39.
20 G. D. Quartly *et al.*, "Measuring Rainfall at Sea. Part 2 – Spaceborne Sensors", *Weather* 57, 2 (October 2002): 363–70.
21 See the author's *The Future of Air Power* (London: Croom Helm, 1986), pp. 142–9.
22 B. H. Liddell Hart (ed.), *The Soviet Army* (London: Weidenfeld and Nicolson, 1956), chapter 21.
23 Tara Patel, "Nuclear treaty flounders as Asia steps up demands", *New Scientist* 149, 2020 (9 March 1996): p. 6.
24 Gro Harlem Brundtland, Chair, *Our Common Future* (Oxford, Oxford University Press for World Commission on Environment and Development, 1987), pp. 18–19.

Part Four A Dissolving Herigage?

11 JUDAISM FULCRAL

1 Karen Armstrong, *In the Beginning* (London: HarperCollins, 1997), p. 14.
2 Hans Küng (John Bowden trans.), *Judaism* (London: SCM Press, 1992), pp. 6–12.
3 Werner Keller (William Neil, trans.), *The Bible as History* (London: Hodder and Stoughton, 1956), pp. 125–7.
4 Rhys Carpenter, *Discontinuity in Greek Civilization* (Cambridge: Cambridge University Press, 1966), pp. 3 and 5–18.
5 Colin Burgess, "Volcanoes, Catastrophe and the Global Crisis of the Late Second Millennium BC", *Current Archaeology* X, 10 (November 1989): 325–9.
6 Albertine Gaur, *A History of Writing* (London: The British Library, 1984), p. 17.
7 *Ibid.*, Fig. 55.
8 Küng, *Judaism*, p. 60.
9 Friedrich Heiler, *Prayer* (New York: Oxford University Press, 1932), pp. 142 ff.
10 Robert Graves (intro.), *New Larousse Encyclopedia of Mythology* (London: Hamlyn, 1968), pp. 80–2.

11 Küng, *Judaism*, p. 29.
12 Peter Stanford, *Heaven* (London: HarperCollins, 2002), pp. 30–8.
13 George A. Barton, *The Religion of Israel* (New York: MacMillan, 1918), chapter XV.
14 Anna Sapir Abulafia, *Christians and Jews in the Twelfth Century Renaissance* (London: Routledge, 1995), pp. 128–33 and 90–1.
15 Philip Ziegler, *The Black Death* (London, Collins, 1969), chapter 5.
16 Amy Chua, "Vengeful majorities", *Prospect* 93 (December 2003): 26–32.
17 Norman Solomon, *Judaism* (Oxford: Oxford University Press, 1996), Appendix A.
18 Lecture by Warren Kenton at "Sky and Psyche" conference, University of Bath, July 2005.
19 Hilary L. Rubinstein *et al.*, *The Jews in the Modern World* (London: Arnold, 2002), p. 59.
20 Küng, *Judaism*, p. 172.
21 Martin Gilbert, *Winston S. Churchill* (London: Heinemann, 1975), chapters 31 to 33.
22 Robert Rhodes-James, *Churchill. A Study in Failure, 1900–1939* (London: Weidenfeld and Nicolson, 1970), p. 133.
23 Albert Einstein (Alan Harris, trans.), *The World As I See It* (London: Watts, *c.* 1940), pp. 96 and 105.
24 Abraham Pais, *Einstein Lived Here* (Oxford: Clarendon Press, 1994), chapter 19.
25 Camilio Dresner (ed.), *The Letters and Papers of Chaim Weizmann* (Jerusalem: Jerusalem University Press, 1978), XV–Series A, pp. 404–8.
26 Einstein, *The World As I See It*, pp. 104–5.
27 Hannah Arendt, *The Origins of Totalitarianism* three parts (New York: Harcourt Brace, 1958), Part 1, p. 6.
28 Hannah Arendt in Michael Selzer (ed.), *Zionism Reconsidered* (London: Macmillan, 1979), pp. 213–49.
29 Rubinstein *et al.* (ed.), *The Jews in the Modern World*, p. 143.
30 *Ibid.* p. 429.
31 Henry Near, *The Kibbutz Movement, A History,* 2 Vols. (Portland: Vallentine Mitchell, 1997), Vol. II, chapter 12.
32 Isaiah Berlin, *Karl Marx* (London: Oxford University Press, 1963), p. 267.
33 Neville Brown, *Strategic Mobility* (London: Chatto and Windus for Institute for Strategic Studies, 1963), chapter 4 (1).
34 Ninian Smart, *The World's Religions* (Cambridge: Cambridge University Press, 1998), p. 425.
35 For example, Arie Issar, "To calm troubled waters", *New Scientist* 183, 2454 (3 July 2004): 45–7.
36 Neville Brown, *History and Climate Change* (London: Taylor and Francis, 2001), chapter 7.1

12 CHRIST AND THE HUMANISTS

1 "Special Report, Holy Writ", *The Economist* 374, 8407 (1 January 2005): 28–30.
2 St Matthew, chapter 5, verses 40–45 in *The New English Bible. New Testament* (Oxford and Cambridge: The University Presses, 1961).
3 E. Royston Pike, *Ethics of the Great Religions* (London: Watts, 1948), p. 172.
4 Edward Gibbon, *The Decline and Fall of the Roman Empire,* 6 vols. (London: J. M. Dent, 1987, Vol. 1, Chapter III.
5 Karen Armstrong, *The Gospel According to Woman* (London: Elm Tree, 1986), pp. 12–13.
6 *Ibid.*, p. 15.
7 Maurice Wiles in John McManners (ed.), *The Oxford Illustrated History of Christianity* (Oxford: Oxford University Press, 1990), pp. 560–1.
8 Steven Runciman, *A History of the Crusades,* 3 vols. (Cambridge: Cambridge University Press, 1955), Vol. III, p. 469.
9 Charles Homer Haskins, *The Renaissance of the 12th Century* (Cambridge, MA: Harvard University Press, 1927), p. 11.
10 R. W. Southern, *Medieval Humanism and Other Studies* (Oxford: Basil Blackwell, 1970), p. 31.
11 Neville Brown, *History and Climate Change* (London: Routledge, 2001), chapters 8 and 9.
12 John Hedley Brooke, *Of Scientists and their Gods,* Inaugural Lecture, 21 November 2000 (Oxford: Oxford University Press, 2001), p. 18.
13 James A. Leith, *Space and Revolution* (Montreal and Kingston: McGill and Queens, 1991), pp. 120 and 138.
14 Richard Harries quoted in *Oxford Today* 12, 1 (Michaelmas 1999): p. 29.
15 Phil Zuckerman, "Secularisation: Europe – Yes, United States – No", *Skeptical Enquirer* 28, 2 (March–April 2004): 49–52.
16 *The Daily Telegraph,* 31 July 2002.
17 "The odd couple", *The Economist* 374, 8412 (6 February 2005): 46.
18 Fareed Zakaria, "Americans Eat Cheese, Too", *Newsweek* CXLIV, 15 (11 October 2004): 25.
19 John Rennie, "15 Answers to Creationist Nonsense", *Scientific American* 287, 1 (July 2002): 62–9.
20 Alan Wolfe, "Dieting for Jesus", *Prospect,* 94 (January 2004): 52–7.
21 Peter Hinchcliff in McManners (ed.), *The Oxford Illustrated History of Christianity,* p. 486.
22 "Crossing the Communists", *The Economist* 375, 8423 (23 April 2005): 68.
23 Rex Ambler, *Global Theology* (London: SCM Press, 1990), chapter 5.

13 COUNTERVALENT ISLAM

1 For example, A. C. Bouquet, *Comparative Religion* (Harmondsworth: Penguin, 1971), p. 277.
2 David Keys, *Catastrophe* (London: Century Books, 1999), pp. 61–4.
3 Abdullah Yusuf Ali (ed.), *The Holy Qur'an* (Elmhurst: Tahrike, Tarsile Qur'an,

1987), sūra XXXIV, v. 16. The sacred text is divided into sūra or thematic chapters.

4 Neville Brown, *History and Climate Change* (London: Taylor and Francis, 2001), p. 111.
5 Ali (ed.), *The Holy Qur'an*, sūra II, note 125.
6 Alfred Guillaume, *Islam* (Harmondsworth: Penguin, 1962), p. 25.
7 Andrew Rippin, *Muslims* (London: Routledge, 2001), p. 29.
8 Fred McGraw Donner, "Mecca's Food Supplies and Muhammad's Boycott", *Journal of the Economic and Social History of the Orient* XX, III (1997): 249–66.
9 Josiah C. Russell, *Medieval Demography* (New York: AMS, 1987), p. 80.
10 Sir John Glubb, *A Short History of the Arab Peoples* (London: Hodder and Stoughton, 1969), pp. 104–5.
11 Richard Hodges and David Whitehouse, *Mohammed, Charlemagne and the Origins of Europe* (London: Duckworth, 1989), pp. 169–70.
12 Cyril Mango, *Byzantium, the Emperor of New Rome* (New York: Charles Scribner's, 1980), p. 69.
13 Brown, *History and Climate Change*, pp. 113–18.
14 Xavier de Planhol, *The World of Islam* (Ithaca: Cornell University Press, 1959), pp. 2–8.
15 R. M. Holt *et al.* (eds.), *The Cambridge History of Islam*, 2 vols (Cambridge: Cambridge University Press, 1970), Vol. 2, p. 445.
16 Anne Marie Schimmel, *Islam* (Albany: The State University of New York, 1992), pp. 121–6.
17 Ali (ed.), *The Holy Qur'an*, note 204.
18 Ronald L. Nettler, *Sūfi Metaphysics and Qur'anic Prophets* (Cambridge: Islamic Texts Society, 2003), p. 17.
19 Seyyed Hossein Nasr, *Ideals and Realities of Islam* (London: George Allen and Unwin, 1966), p. 16.
20 Jaime Vicens Vives (Frances M. López-Morillas trans.), *An Economic History of Spain* (Princeton: Princeton University Press, 1969), pp. 108–10.
21 Roger Collins, *Early Medieval Spain, Unity in Diversity, 400–1000* (London: Macmillan, 1983), pp. 164–5.
22 David Martin Varisco, "Islamic Folk Astronomy" in Helaine Selin (ed.), *Astronomy Across Cultures* (Dordrecht: Kluwer, 2000), pp. 615–50.
23 Clinton Bailey, "Bedouin Star-lore in Sinai and the Negev", *The Bulletin of the School of Oriental and African Studies* XXXVII, 3 (1974): 580–96.
24 AH, literally *Anno Hegirae.* Islamic calendar dated from the Prophet's *Hegira* (i.e. flight) from Mecca in AD 622.
25 Nasr, *Ideals and Realities of Islam*, pp. 6–7.
26 Seyyed Hossein Nasr, *Islamic Science* (Westerham: World of Islam Festival, 1976), pp. 17–20.
27 Fazlur Rahman, *Islam* (London: Weidenfeld and Nicolson, 1966), p. 186.
28 Seyyed Hossein Nasr, *An Introduction to Islamic Cosmological Doctrines* (London: Thames and Hudson, 1978), pp. 232–3.

29 S. Pines, "What was original in Arabic Science?" in A. C. Crombie (ed.), *Scientific Change* (New York: Basic Books, 1963), chapter 6.
30 Nasr, *An Introduction to Islamic Cosmological Doctrines*, p. 165.
31 Michael Marmura in R. M Savory (ed.), *Islamic Civilisation* (Cambridge: Cambridge University Press, 1976), chapter 4.
32 Oliver Leaman, *Averroës and his Philosophy* (Oxford: Clarendon Press, 1968), p. 5.
33 Bernard R. Goldstein, "The Origin of Copernicus's Heliocentric System", *Journal of the History of Astronomy* Part 3, 33, 112 (August 2002): 291–36.
34 Roger Arnoldez, *Averroës. A Rationalist in Islam* (Notre Dame Indiana: University of Notre Dame Press, 2000), p. 14.
35 Brown, *History and Climate Change*, p. 33.
36 David A. King in Selin, *Astronomy Across Cultures*, p. 589.
37 Nasr, *Islamic Science*, pp. 81 and 112.
38 Pines in Crombie, *Scientific Change*, p. 205.
39 Akbar S. Ahmed, *Living Islam* (London: BBC Books, 1993), p. 77.

14 INDIAN PLURALISM

1 Dirk J. Struik, *A Concise History of Mathematics* (London: G. Bell, 1965), p. 26.
2 Michio Yano, "Science and Religion in Ancient India", *Zen Buddhism Today* 5 (November 1987): 50–9.
3 Subash Kak in Helaine Selin (ed.), *Astronomy Across Cultures* (Dordrecht: Kluwer, 2000), p. 336.
4 Chita Takenaka, "The Relation between Religion, Philosophy and Science in Ancient India", *Zen Buddhism Today* 5 (November 1987): 36–49.
5 Ninian Smart, *The Religious Experience of Mankind* (New York: Charles Scribner, 1969), p. 128.
6 Fritjof Capra, *The Tao of Physics* (London: Fontana, 1987), p. 322.
7 David Pingree, "The Recovery of Early Greek Astronomy from India", *Journal of the History of Astronomy*, 7, 19 Pt. 2 (June 1976): 109–23.
8 Andre Gunder Frank and Barry K. Gills, "World System Economic Cycles and Hegemonial Shift to Europe, 500 BC to 1500 AD", *The Journal of European Economic History* 22, 1 (Spring 1993): 155–83.
9 Ninian Smart, *Doctrine and Argument in Indian Philosophy* (London: George Allen and Unwin, 1964), p. 49.
10 Rajesh Kochhar and Jayant Narlikar, *Astronomy in India, A Perspective* (New Delhi: Indian National Science Academy, 1995), pp. 1–9.
11 W. J. Van Liere, "Traditional Water Management in the Lower Basin", *World Archaeology*, 11, 3 (February 1980), pp. 265–80.
12 Robert Stencel *et al.*, "Astronomy and Cosmology at Angkor Wat", *Science* 193, 4250 (23 July 1976): 281–4.
13 "Gautama" is the family name. Buddha simply means "the enlightened one".
14 A. C. Bouquet, *Comparative Religion* (Harmondsworth: Penguin, 1971), p. 159.

15 Ninian Smart, *Doctrine and Argument in Indian Philosophy* (London: George Allen and Unwin, 1964), p. 28.
16 Ian Mabbott, "The Beginnings of Buddhism", *History Today* 52,1 (January 2002), pp. 24–9.
17 Apparently "Mahatma" is Sanskrit for "great souled". In other words, a person of unusual holiness. Gandhi was accorded this accolade by Sir Rabindranath Tagore (1861–1941), the Bengali Nobel Laureate in Literature.
18 Abraham Pais, *Einstein Lived Here* (Oxford: Clarendon Press, 1994), chapter 9.3.
19 Arthur Koestler, *The Heel of Achilles* (London: Hutchinson, 1974), pp. 221–4.
20 V. A. Naipaul, *India: A Wounded Civilisation* (Harmondsworth: Penguin, 1979), p. 156.
21 *Ibid.*, pp. 153–4.
22 Sonia Orwell and Iain Angus (eds.), *The Collected Essays, Journalism and Letters of George Orwell*, 4 vols (London: Secker and Warburg, date?), Vol. 4, Chapter 133.
23 Daisaku Ikeda, "Towards a World without War – Buddhism and the Modern World", *Soka Gakkai News* (17, 276, May 1992), pp. 13–20.
24 Orwell and Angus, *The Collected Essays, Journalism and Letters of George Orwell* p. 465.
25 Ronald Duncan (ed.), *Selected Writings of Mahatma Gandhi* (London: Faber and Faber, 1951).
26 Raghavan Iyer (ed.), *The Moral and Political Writings of Mahatma Gandhi*, 3 vols (Oxford: Clarendon Press, 1986), Vol.1, p. 294.
27 *Ibid.*, p. 578.
28 Naipaul, *India: A Wounded Civilisation*, p. 168.
29 "The Swamis", *The Economist* 369, 8355 (20 December 2003): 95–7.
30 "India's Shining Hopes", *The Economist* 370, 8363 (21 February 2004), Special Survey.

15 JAPANESE TOGETHERNESS

1 Joseph Needham, *Science and Civilisation in China*, 23 vols. to date (Cambridge: Cambridge University Press, 1959 - 2004), Vol. 3, pp. 459–61.
2 Robert K. G. Temple, *China: Land of Discovery* (Wellingborough: Stephens, 1986), pp. 33–4.
3 Richard Stephenson and Kevin Yau, "Oriental tales of Halley's comet", *The New Scientist* 103, 1423 (24 September 1984): 30–2.
4 Ito Shuntaro, "The introduction of Western cosmology in seventeenth-century Japan – 1", *The Japan Foundation Newsletter* XIV, 1 (May 1986): 1–9.
5 Helen J. Baroni, *Obaku Zen. The Emergence of the Third Sect of Zen in Tokugawa Japan* (Honolulu: University of Hawaii Press, 2000), chapter 11.
6 Boleshlaw Szczesniak, "The penetration of the Copernican theory into feudal Japan", *Journal of the Royal Asiatic Society* Parts I and II (1944): 52–61.
7 Basil Hall Chamberlin, *Things Japanese* (London: John Matthew, 1905), p. 441.

8 Paul Varley, *Japanese Culture* (Honolulu: University of Hawaii Press, 2000), p. 11.
9 Victor Harris (ed.), *Shintō: The Sacred Art of Modern Japan* (London: British Museum Press, 2001), p. 15.
10 Tao Hanzhang, *The Art of War* (Ware: Wordsworth, 1993), p. 105.
11 H. H. Gerth and C. Wright Mills (eds.), *From Max Weber* (London: Routledge and Kegan Paul, 1970), Chapter XVII, Sections 5 to 7.
12 Masahiko Miyake, *Kodansha Encyclopedia of Japan*, 9 vols. (Tokyo: Kodansha, 1983), Vol. 7, p. 190.
13 Rystaro Shiba (Takemoto Akiko trans.), *Kūkai.The Universal: Scenes from his Life* (New York: ICG Muse, 2003), p. 128.
14 *Ibid.*, p. 267.
15 Neville Brown, *History and Climate Change* (London: Taylor and Francis, 2001), pp. 210–11.
16 John F. Fairbank *et al.*, *East Asia. Tradition and Transformation* (London: George Allen and Unwin, 1975), p. 48.
17 Anne Bancroft, *Zen. Direct Pointing to Reality* (London: Thames and Hudson, 1979), p. 25.
18 Eugene Herrigel, "Japan's Art of Archery" in Shigeyoshin Matsumae (ed.), *Towards an Understanding of Budo Thought* (Tokyo: Tokai University Press, 1987).
19 Daisetz T. Suzuki, *Zen and Japanese Culture* (Princeton: Princeton University Press, 1959).
20 Douglas Howland, "Translating Liberty in 19th-century Japan", *Journal of the History of Ideas* 62, 1 (January 2001): 161–81.
21 Ruth Benedict, *The Chrysanthemum and the Sword* (London: Secker and Warburg, 1947), chapter 1.
22 Jean Stoetzel, *Without the Chrysanthemum and the Sword* (London: William Heinemann, 1955), pp. 10–17.
23 Suzuki, *Zen and Japanese Culture*, p. 351.
24 Arne Kalland, "Culture in Japanese Nature" in Ole Brunn and Arne Kalland (ed.), *Asian Perceptions of Nature: A Critical Approach* (London: Curzon, 1995), pp. 243–57.
25 Suzuki, *Zen and Japanese Culture*, p. 351.
26 J. Sigrist and D. M. Stroud (trans.), *Kenji Miyazawa's The Milky Way Railroad* (Tokyo: Stone Bridge Pass, 1995).

Part Five To Here from Eternity

16 YEARNINGS FOR BELIEF

1 Pascal Boyer, "Why is Religion Natural?", *Skeptical Inquirer* 28, 2 (March/April 2004): 25–31.
2 Richard Sosis, "The Adaptive Value of Religious Ritual", *American Scientist* 92, 2 (March–April 2004): 166–72.
3 Martin Buber, *The Eclipse of God* (London: Victor Gollancz, 1953), chapter 1.

4 Bryan Wilson in John McManners (ed.), *The Oxford Illustrated History of Christianity* (Oxford: Oxford University Press, 1990), pp. 592–4.
5 Arthur Schopenhauer (R. B. Haldane and John Kemp trans.), *The World as Will and Idea*, 4 vols (London: Trübner, 1886), Vol. II, p. 392.
6 *Ibid.*, p. 380.
7 *Ibid.*, p. 375.
8 David Burnett, "Cao Dai" in Christopher Partridge, *Encyclopedia of New Religions* (Oxford: Lion, 2004), pp. 234–6.
9 British Youth Festival Preparatory Committee, *The Innsbruck Story* (London: National Council of Civil Liberties, 1951).
10 Richard Crossman (ed.), *The God that Failed* (London: Hamish Hamilton, 1950), p. 11.
11 For a summation of the argument, see the author's *The Future of Air Power* (Beckenham: Croom Helm, 1986), pp. 8 - 9.
12 John Baker (Chair), *The Church and the Bomb* (London: Hodder and Stoughton, 1982).
13 Neville Brown and Anthony Farrar-Hockley, *Nuclear First Use* (London: Buchan and Enright, 1985), pp. 43–4.
14 Theodore Dalrymple, "Why Religion is Good for Us", *The New Statesman,* 21 April 2003.
15 Roger Penrose, *The Road to Reality* (London: Jonathan Cape, 2004), p. 14.
16 Sir James Jeans, *Physics and Philosophy* (Cambridge: Cambridge University Press, 1943), p. 15.
17 *The New English Bible. New Testament* (Oxford: Oxford University Press and Cambridge University Press, 1961), p. 143
18 Anthony Storr, *Music and The Mind* (London: HarperCollins, 1992), p. 22.
19 Norman H. Weinberger, "Music and the Brain", *Scientific American* 291, 5 (November 2004): 66–73.
20 G. R. Elton (ed.), *The New Cambridge Modern History, The Reformation,* vol. 11 (Cambridge: Cambridge University Press), 1962, pp. 98–101.
21 William James, *The Varieties of Religious Experience* (London: Longmans Green 1902), Lectures XVI and XVII.
22 Dr Peter Fenwick, *The Neurophysiology of the Brain Its Relationship to Altered States of Consciousness (With emphasis on mystical experience),* Wrekin Lecture, 1980.
23 Bertrand Russell, *Mysticism and Logic* (London: Routledge, 1994), chapter 2.
24 Nick Zangwill, "The myth of religious experience", *Religious Studies* 40, 1 (March 2004): 1–22.
25 George Santayana, *The Sense of Beauty* (London: Adam and Charles Black, 1896), p. 10.
26 Olaf Blanke *et al.*, "Stimulating illusory own-body perceptions", *Nature* 419, 6904 (19 September 2002): 269.
27 Joe Fearn, "Astral Projection and Out of Body Experiences", *Philosophy Now* 42 (July/August 2003): 10–13.

28 Robert Matthews. "Out of Body Experiences", *BBC Focus* 155 (October 2005): 60–5.
29 Richard Dawkins, *The Ancestor's Tale* (London: Weidenfeld and Nicolson, 2004), p. 55.
30 James E. Kennedy, "The Roles of Religion, Spirituality and Genetics in Paranormal Beliefs", *Skeptical Inquirer* 28, 2 (March/April 2004): 39–42.
31 Peter Lemesurier, *The Essential Nostradamus* (London: Judy Piatkus, 1999), pp. 12–13.
32 *Ibid.*, p. 42.
33 *Ibid.*, pp. 52–6.

17 CREATIVE CONVERGENCE

1 Nicolas Maxwell "Cutting God in Half", *Philosophy Now* 35 (March/April 2002): 22–5.
2 Fritjof Capra, *The Tao of Physics* (London: Fontana, 1968), p. 46.
3 Marie-Louise von Franz (William H. Kennedy trans.), *C. G. Jung: His Myth in Our Time* (London: Hodder and Stoughton, 1975), pp. 141–3.
4 Anthony Stevens, *Jung* (Oxford: Oxford University Press, 1994), p. 47.
5 Daiseku Ikeda (Charles S. Perry trans.), *Life. An Enigma: A Precious Jewel* (Tokyo, Kodansha International, 1982), p. 25.
6 Rupert Sheldrake and Matthew Fox, *Natural Grace. Dialogues on Science and Spirituality* (London: Bloomsbury, 1996), pp. 142–3.
7 Victor E. Weisskopf, "The Origin of the Universe", *American Scientist* 76, 5 (September–October 1988): 472–7.
8 R. J. Stewart, *The Elements of Creation Myth* (Shaftesbury: Element Books, 1989), p. 6.
9 Quoted in Jon Turney, *Lovelock and Gaia. Signs of Life* (Cambridge: Icon Books, 2003), p. 4.
10 James Lovelock, *Homage to Gaia* (Oxford: Oxford University Press, 2000), p. 3.
11 F. Tian *et al.*, "A Hydrogen-rich Early Earth Atmosphere", *Science* 308, 5719 (8 April 2005): 155.
12 James Lovelock, *The Ages of Gaia* (Oxford: Oxford University Press, 1979), pp. 116–22.
13 Robert J. Charlson in Stephen H. Schneider and Penelope J. Boston (ed.), *Scientists on Gaia* (Cambridge, MIT Press, 1991, chapter 12.
14 J. E. Lovelock, *Gaia. A New Look at Life on Earth* (Oxford: Oxford University Press, 1983, pp. 101–4.
15 "Coral reefs create clouds to control the climate", *New Scientist* 185, 2485 (5 February 2005): 2.
16 Lovelock, *Homage to Gaia*, p. 392.
17 For example, Lawrence E. Joseph, *Gaia. The Growth of an Idea* (London: Arkana, 1990), chapter IX.
18 Turney, *Lovelock and Gaia*, p. 41.

19 John D. Barrow and Frank J. Tipler, *The Anthropic Cosmological Principle* (Oxford: Clarendon Press, 1986), pp. 387–8.
20 Richard B. Larson and Volker Bromm, "The First Stars in the Universe", in *The Once and Future Cosmos*, a Special Edition of *Scientific American*, September 2004: 4–11.
21 Martin Rees, *Our Final Century* (London: William Heinemann, 2003), pp. 187–8.
22 *Ibid.*, pp. 119–24.
23 Mark Alport, "Apocalypse Deferred", *Scientific American* 281, 6 (December 1999): 17.
24 David Miller in Christopher Partridge (ed.), *Encyclopedia of New Religions* (Oxford: Lion, 2004), pp. 238 and 245.
25 *God's Ideal Family* (London: Universal Peace Foundation, 2005), p. 11.
26 J. R. S. Phillips, *The Medieval Expansion of Europe* (Oxford: Oxford University Press, 1988), p. 87.
27 Frances Wood, *Did Marco Polo go to China?* (London: Secker and Warburg, 1995).
28 J. Krishnamurti, *The First and Last Freedom* (London: Victor Gollancz, 1954), p. 38.
29 Walter W. Davis, "China, the Confucian Ideal, and the European Age of Enlightenment", *Journal of the History of Ideas* 44, 4 (October 1983): 523–48.
30 Benjamin A Elman, "Nietzsche and Buddhism", *Journal of the History of Ideas* 44, 4 (October 1983): 671–86.
31 Keith Ward, *The Case for Religion* (Oxford: Oneworld, 2004), p. 133.
32 Ian Barbour, *Religion in an Age of Science*, 2 vols. (New York: Harper San Francisco, 1991), Vol. 1, p. 119.
33 Capra, *The Tao of Physics*, pp. 119–21.
34 Adam Brookes, "Sticking Point", *Far Eastern Economic Review* 162, 23 (10 June 1999): 70–1.
35 Capra, *The Tao of Physics*, p. 105.
36 C. G. Jung, "Yoga and the West" in Sir Herbert Read *et al.*, (ed.), (R. F. C. Hull, trans.), *The Collected Works of C. G. Jung*, 12 Vols. (London: Routledge and Kegan Paul, 1958), Vol. 11, pp. 529–37.
37 Daisetz T. Suzuki, *Zen and Japanese Culture* (Princeton: Princeton University Press, 1970), pp. 238 and 50.
38 H. H. Gerth and C. Wright Mills (ed.), *From Max Weber* (London: Routledge and Kegan Paul, 1970), chapter XI.
39 Tenzin Gyatso, *Ancient Wisdom. Modern World* (London: Little Brown, 1999), chapter 15.
40 Ward, *The Case for Religion*, pp. 227–31.
41 Jonathan Sacks, *The Dignity of Difference* (London: Continuum, 2002), p. 207.
42 *Ibid.*, p. 189.
43 For a general summation of this editing, see Bishop Richard Harries in *The Scottish Journal of Theology* 57, 1 (2004): 110–11.

44 Harriet A. Harris, "Fundamentalisms" in Partridge (ed.), *Encyclopedia of New Religions*, pp. 409–15.
45 Ursula King in Partridge (ed.), *Encyclopedia of New* Religions, p. 379.
46 Sacks, *The Dignity of Difference*, p. 180.
47 Ward, *The Case for Religion*, p. 235.
48 Walter Kaufmann in Walter Kaufmann (ed.), *Existentialism from Dostoevsky to Sartre* (New York: Meridian, 1956), p. 48.
49 Stephen Spender in R. H. S. Crossman (ed.), *The God that Failed* (London: Hamish Hamilton, 1950), pp. 231–72.
50 Richard A. Kerr, "Black Sea deluge may have helped to spread farming", *Science* 279, 5354 (20 February 1998): 1132.
51 Peter Lyon in Alan J. Day (ed.), *The Annual Register 1998*, Vol. 240 (Rockville: Keesing's Worldwide, 1999), p. 329.
52 Sonia Gandhi, *Conflict and Co-existence in Our Age* (Oxford: Oxford Centre for Islamic Studies, 2002), 23 pp.
53 Cheryll Barron, "The Indian Genius", *Prospect 96* (April 2001): 14–15.
54 Neville Brown, *Global Instability and Strategic Crisis* (London: Taylor and Francis, 2004) pp. 19–20.
55 Adam Robinson, *Bin Laden* (Edinburgh: Mainstream, 2001), p. 287.
56 Samuel Huntington, *The Clash of Civilisations and the Remaking of the World Order* (London: Touchstone, 1998), p. 55.
57 Ronald Inglebert and Pippa Borris, "The True Clash of Civilizations", *Foreign Policy* (March/April 2003): 63–75.
58 Arthur Koestler, *The Sleepwalkers* (London: Hutchinson, 1959), p. 13.
59 John Gray, "Global utopias and clashing civilizations: misunderstanding the present", *International Affairs* 74, 1 (January 1998): 149–63
60 William H. McNeill, "Toynbee, Arnold J." in David L. Sills (ed.), *Encyclopedia of the Social Sciences*, 18 vols. (New York: Free Press, 1979), Vol. 18, pp. 775–9.
61 Bishop Richard Harries, University Interfaith Observance in Response to the Tsunami, *Oxford Magazine*, 4th week, Hilary Term, 2005, pp. 14–15.
62 Gareth Price and Graham Brown, *The World Today* 61, 2 (February 2005): 4–7.
63 Sir James Frazer, *The Golden Bough*, Abridged Edition (London: Macmillan, 1929), chapter XXIV.
64 F. Guirand in Robert Graves (introduces), *New Larousse Encyclopedia of Mythology* (London: Hamlyn, 1968), p. 54.

18 FULFILMENT REGAINED?

1 Nik Szymanek, "A Night on the Faulkes", *Astronomy Now* 19, 5 (May 2005): 53–61.
2 Theodore Roszak, *The Making of a Counter-Culture* (London: Faber and Faber, 1968), p. 265.
3 Theodore Roszak, "Autopsy on Science", *New Scientist* 57, 830 (11 March 1971): 536–8.
4 Theodore Roszak, *Person/Planet* (London: Victor Gollancz, 1979), p. 50.

5 *Ibid.*, p. 61.
6 Chad Trainer, "Earth to Russell", *Philosophy Now* 40 (March/April 2003): 20–2.
7 Bertrand Russell, "Why men should keep away from the Moon", *The Times*, 15 July 1969.
8 Arnold Toynbee, *Surviving the Future* (London: Oxford University Press, 1971), p. 138.
9 Richard L. Gage (ed.), *Choose Life* (London: Oxford University Press, 1976), p. 167.
10 Arie J. Suckerman, "Germ from Mars?", *Nature* 276, 5686 (23 November 1978): 326.
11 M. Beech *et al.*, "The Potential Danger to Space Platforms from Meteor Storm Activity", *Quarterly Journal of the Royal Astronomical Society* 36, 2 (June 1995): 127–52.
12 Leonid Leskov, "Power-industry orbital complexes of the 21st century", *Space Policy* 1, 1 (February 1985): 84–5.
13 Peter E. Glaser, "Evolution of the Solar Power Satellite concept: the utilisation of energy from Space", *Space Solar Power Review* 4 (1983): 11–21.
14 "Russians look out to Space for solar power plants", *The Times*, 9 December 1985.
15 Fred Pearce, "A mirror to cool the world", *New Scientist* 181, 2440 (27 March 2004): 26–9.
16 Jeffrey S. Kargel, "Metalliferous Asteroids as Potential Sources of Precious Metals", *Journal of Geophysical Research* 99, E10 (25 October 1994): 129–41.
17 "When Astronauts refuse to volunteer", *Nature* 391, 6669 (19 February 1998): 737.
18 George Dyson, *Project Orion* (London: Allen Lane, 2000), pp. 254–5.
19 *Ibid.*, pp. 102–6.
20 Simon Conway Morris, "Aliens Like Us?", *Astronomy and Geophysics* 46, 4 (August 2005): 4.24–4.26.
21 Ian Stewart, "The Mathematics of 2050" in John Brockman (ed.), *The Next Fifty Years* (London: Phoenix, 2003), pp. 29–56.
22 Quoted in Freeman Dyson, *Disturbing the Universe* (New York: Harper Row, 1979), p. 246.
23 Robert Lawrence Kuhn, "Science as Democratizer", *American Scientist* 91, 5 (September–October 2003): 388–90.
24 Rhys E. Green *et al.*, "Farming and the Fate of Nature", *Science* 307, 5709 (28 January 2005): 550–5.
25 Adam Potkay, "Whatever happened to happiness?", *Philosophy Now* 27 (June/July 2000): 24–7.
26 Martin G. Kalin, *The Utopian Flight from Unhappiness* (Totowa: Littlefield and Adams, 1975), pp. 168–9.
27 See for example Claudia Wallis, "The New Science of Happiness", *Time* 165, 6 (7 February 2005): 44–9.
28 Junichiro Tanizaki, *In Praise of Shadows* (London: Vintage, 2001), p. 63.

29 Basil Hall Chamberlin, *Things Japanese* (London: John Murray, 1905), p. 441.
30 Steven L. Renshaw and Sadri Ihara, "Amateur Astronomy in Japan", *Sky and Telescope* 93, 3 (March 1997): 104–8.
31 See the paper by Meri Arichi delivered at the September 2004 conference on the Stars in Japanese Culture, School of Oriental and African Studies, University of London.
32 Brown, *Global Instability and Strategic Crisis*, pp. 96–8.
33 Kuhn, *American Scientist* 91, 5 (September/October 2003): 389.
34 Neville Brown (ed.), *American Missile Defence. Views from China and Europe* (Oxford: Oxford Research Group, May 2000), *Current Decisions Report* No. 25.
35 Thomas Metzer, "Western Philosophy on the Defensive", *Philosophy Now* 26 (April/May 2000): 30–2.
36 Isaac Deutsher, *Heretics and Renegades* (London: Hamish Hamilton, 1955), p. 215.
37 Albert Kolb (C. A. M. Sym, trans.), *East Asia. The Geography of a Cultural Region* (London: Methuen, 1971), p. 41.
38 Wang Juntao in Daniel A. Bell and Hahm Chaibong (eds.), *Confucianism for the Modern World* (Cambridge: Cambridge University Press, 2003), Chapter 3.
39 Donella H. Meadows *et al.*, *The Limits to Growth* (London: Earth Island, 1972).
40 For example, E. J. Mishan, *The Costs of Economic Growth* (London: Staples Press, 1967), especially chapters 4 and 7 which address the nub of external diseconomies in urban growth.
41 Jonathan D. Oldfield, "Russia, Systemic Transformation and the Concept of Sustainable Development", *Environmental Politics* 10, 3 (Autumn 2001): 96–110.
42 John Chipman, "The future of strategic studies: beyond even grand strategy", *Survival* 34, 1 (Spring 1992): 109–31.
43 For example, Ann M. Florini and P. J. Simmons, *The New Security Thinking: A Review of the North American Literature* (New York: Rockefeller Brothers Fund, 1998).
44 Tim Huxley, "The Tsunami and Security: Asia's 9/11?", *Survival* 47, 1 (Spring 2005): 123–32.
45 Zbigniew Brzezinski, "The International and the Planetary", *Encounter* XXXIX, 2 (August 1972): 49–55.
46 "Interview with M. J. Akbar", *Philosophy Now* 37 (August–September 2002): 18–19.
47 John Kelsay, "Al-Shaybani and the Islamic Law of War", *Journal of Military Ethics* 2, 1 (2003): 63–75.
48 D. S. Lewis (ed.), *The Annual Register, 2003*, Vol. 245 (Bethesda: Keesing's Worldwide, 2004), p. 4.
49 "After victory, defeat", *The Economist* 376, 8435 (16 July 1005): 55.
50 For example, "Iraq's history is our history too", *The Art Newspaper, International Edition* XIII, 130 (November 2002): 1 and 4.
51 Susan T. Fiske *et al.*, "Why Ordinary People Torture Enemy Prisoners", *Science* 306, 5701 (26 November 2004): 1482–3.

52 "The War after the War", *The Economist* 371, 3419 (26 March 2003): 44.
53 Kori Urayama, "China Debates Missile Defence", *Survival* 46, 2 (Summer 2004): 129–42.
54 *Report of the Task Force on Potentially Hazardous Near Earth Objects* (Leicester: British National Space Centre, September 2000), p. 16.
55 Jeff Hecht, "Killing it softly", *New Scientist* 178, 2391 (19 April 2003): 37–9.
56 Victor Clube and Bill Napier, *The Cosmic Winter* (Oxford: Basil Blackwell, 1990), pp. 106–10.
57 Neville Brown, *History and Climate Change. A Eurocentric Perspective* (London: Taylor and Francis, 2001), pp. 90–2.
58 Project Air Force, *Space Weapons, Earth Wars* (Santa Monica: Rand Corporation, 1999), Appendix C.
59 Daniel G. Dupont, "Nuclear Explosions in Orbit", *Scientific American* 290, 6 (June 2004): 68–75.
60 World Commission on Environment and Development, *Our Common Future* (Oxford: Oxford University Press, 1987), chapter 10.
61 *Ibid.*, p. 279.
62 Dyson, *Disturbing the Universe.* Chapter 21.
63 Brown, *History and Climate Change*, pp. 92–3.
64 Emma Rigby *et al.*, "A comet impact in AD 536?", *Astronomy and Geophysics* 45 (February 2004): 1.23–1.26.
65 David Keyes, *Catastrophe* (London: Century Books, 1999).
66 *Supplement to Oxford English Dictionary*, 4 vols. (Oxford: Clarendon Press, 1976), Vol. 2, p. 120.
67 Edward Teller and Albert L. Latter, *Our Nuclear Future. Facts, Dangers and Opportunities* (London: Secker and Warburg, 1958), p. 167.
68 Andrei B. Sakharov (*New York Times*, trans.), *Progress, Co-existence, and Intellectual Freedom* (New York: New York Times Book Service, 1968), p. 49.
69 Steven Weinberg, *The First Three Minutes* (London: André Deutsch, 1977), pp. 154–5.
70 Freeman J. Dyson, "Time Without End: Physics and Biology in an Open Universe", *Review of Modern Physics* 51, 3 (July 1979): 447–60.
71 David J. Chalmers, "The Puzzle of Conscious Experience", *Scientific American* 273, 6 (December 1995): 62–8.
72 Arthur Gibson, *God and the Universe* (London: Routledge, 2000), pp. 178–9.
73 Paul Davies, *The Last Three Minutes* (London: Weidenfeld and Nicolson, 1994), pp. 33–4.
74 Martin Rees, *Our Final Century* (London: William Heinemann, 2003), p. 148.
75 Richard Dawkins, *A Devil's Chaplain* (London: Phoenix, 2004), p. 174.
76 Paul Davies, *Are We Alone?* (London: Penguin, 1995), pp. 89–90.

INDEX